Born to Explore

Outward Odyssey
A People's History of Spaceflight

Series editor
Colin Burgess

Born to Explore

John Casani's Grand Tour of the Solar System

Jay Gallentine

UNIVERSITY OF NEBRASKA PRESS • LINCOLN

The University of Nebraska Press is part
of a land-grant institution with campuses
and programs on the past, present, and
future homelands of the Pawnee, Ponca,
Otoe-Missouria, Omaha, Dakota, Lakota,
Kaw, Cheyenne, and Arapaho Peoples, as
well as those of the relocated Ho-Chunk,
Sac and Fox, and Iowa Peoples.

Publication of this volume was assisted by
the Virginia Faulkner Fund, established
in memory of Virginia Faulkner, editor in
chief of the University of Nebraska Press.

For customers in the EU with safety/
GPSR concerns, contact:
gpsr@mare-nostrum.co.uk
Mare Nostrum Group BV
Mauritskade 21D
1091 GC Amsterdam
The Netherlands

Library of Congress Control Number:
2025005763

Set in Garamond Premier Pro by A. Shahan.

Dedicated to the memory of Mr. Andrew Fidler.

A teacher of history. Perhaps my greatest teacher.

And truly one of a kind.

Let us create vessels and sails adjusted to the heavenly ether, and there will be plenty of people unafraid of the empty wastes.

—Open letter from Kepler to Galileo, published in *Conversation with the Starry Messenger*, 1610

Contents

Illustrations

Acknowledgments

Throughout my writing projects, I have been so very blessed to have the unending support of my family. Here we are, my dearest Anne, having made it through a third. And my sons, it means everything to see you off into the world doing such amazing things yet still having you be so supportive of this parsing of memos and memories that I seem compelled to do.

Huge thanks go once again to Rob Taylor and the University of Nebraska Press for believing in my research and for welcoming another installment in the Outward Odyssey series. Colin Burgess again fearlessly served as editor, adviser, general sounding board, and guy who talked me off the ledge— repeatedly. I couldn't have done it without you, sir! At the press, Taylor Martin and Sara Springsteen and Tish Fobben handled various aspects of the project to march it forward. Thank you!

Forever will I be in debt to my friend and colleague Francis French. Despite his own pressing to-do list of writing, talking, educating, and volunteering, he took the time to carefully peer-review each chapter and provide detailed comments. That he read part of chapter 25 in the shadow of a total solar eclipse honestly moved me. In my experience, few people can be your friend yet pull no punches when critically reviewing a book manuscript. I knew I could count on his unique and honest perspectives, and the end result is much better because of his efforts.

Many, *many* thanks to my interview subjects, who provided the cornerstone of this work. Some of them have been putting up with questions from me for well over a decade—long before *Born to Explore* coalesced in my head. Along the way, this effort turned into a mini biography of a man who'd been born to explore but didn't even know it, and that's John Casani. He never seemed to tire of new rounds of questioning about everything, from his westward migration to his mustache to his suspenders phase, and opened up to me in a way that only happens after some seventeen years of dialogue.

Every interview is memorable for its own reasons. Thanks to Don Gurnett, who endured his own interrogation about everything from the day he learned he'd ride on the *Voyagers* to the methods of transporting a sailplane cross-country. And then with Roger Diehl I could only laugh when, in the middle of our discussion, he commented, "Boy, you're really moving the cobwebs." I was touched when Mary Reaves told me, "Thank you for the memories. I seem to put them out of my mind until someone asks me a question. And then all the people come smiling back to me." Every interview subject of mine received the opportunity to review and comment on chapter drafts.

I am thankful to John and Lynn Casani's sons, who reviewed the manuscript, offered anecdotes about everything from the burning bus to the skunks to the New Year's parties, and supplied enough family photos to fill an album.

I send thanks for the serendipity that once put me in an airplane seat next to Caitlin J. Albrecht. A professional court reporter, she expertly transcribed the majority of raw interview recordings I made for this book. This means Caitlin is the only other person besides myself to have listened to entire conversations with the people you're about to meet. Caitlin, you allowed me to focus on the writing process itself. Much appreciated!

Major thanks also to John Whisenhunt for reviewing and commenting on draft chapters and for the day-to-day encouragement and friendship. Galilean Bob Gounley reviewed the complete manuscript as well, calling out valuable corrections and clarifications. Thanks, Bob!

I send my appreciation to Kathy Kurth at the University of Iowa Department of Physics and Astronomy. It's weird that my time as a Hawkeye should come full circle like this, and Kathy has been invaluable in scrounging up that exact document or research paper needed to complete the telling of a particular anecdote. Anne Moore and Anne Kent-Miller, both at the University of Iowa, were of great assistance in finalizing a number of image permissions.

Brothers Seaton King and James King were so very kind to tell me stories about their dad, Clyde, and to dig through their photos looking for pictures of him. Thank you so much! A grand bow of appreciation also goes out to Denise Fitzpatrick and Michael Mahoney, both at the University of Pennsylvania Athletics, for their kind assistance in determining exactly which football game John Casani had attended. Aaron Janofsky of the Committee on Space

Research provided details of his organization's 1986 conference and unearthed an important document. David Hitt helped me chase down a few details of early shuttle flights; the two of us ended up in a serious philosophical discussion about the relevance of the space shuttle program. Emily Carney's knowledge of Shuttle/Centaur and space station history was of great assistance.

I am also indebted to those on my beloved Minnetonka Masters swim team for their supportive interest in my writing projects and for their camaraderie during this process. Meeting for workouts at 5:30 a.m. on most days of the week has created a bond with these exceptional individuals whom I very likely wouldn't even know otherwise. Every time they asked how things were going with the book, it helped me remain committed and take another step toward completion. All those sets of 5x200s, or whatever we had to do, were as much mentally therapeutic as they were physical!

Every project has its own kinds of hurdles, and this one was no exception. It survived the loss of loved ones, surgery, moving, a flooded rental house, a global pandemic, a burning washing machine, cat bites, "engine-out service," and an owl in the kitchen. However, I simply would not have completed *Born to Explore* without the neurologists who have helped me bring under control a yearslong beatdown of migraine headaches. Their escalation had become a showstopper for nearly everything in life. To my neurologists, I send thanks for helping me reclaim my sanity and for your enduring compassion. I owe my day-to-day function to your efforts.

And then I must take a final moment to recognize one of the greatest teachers I ever encountered—Mr. Andrew Fidler of Millard North High School in Omaha, Nebraska. He taught a few different subjects, but I had the guy for Advanced Placement European History. Without question he was the first adult I encountered with the ability to convey an excitement for history and who communicated that to us during pretty much every classroom session. The root of his brilliance lay in *how* he presented the material. He kept things light yet informative and never dawdly. Competing popes who excommunicated each other were "popes in stereo." On the anniversary of Napoleon's death, we held a pizza party in his honor, complete with wearing black armbands. At times Mr. Fidler employed different voices in his lectures, practically acting out the history in front of us. But he never let us off easy. One of Mr. Fidler's unit tests consisted of nothing more than a face-to-face conversation about the material. It terrified me because there was no place to hide.

Another time, Mr. Fidler had just started into a new section about the American Revolution. Suddenly the door opened as the Advanced Placement *American* History teacher snuck in to grab something. Without missing a beat, Mr. Fidler proclaimed, "And now here comes Mr. Eicher to debate this from the American perspective!" After leaving for college, I actually revisited Mr. Fidler's classroom at least twice to merely sit there and let the history wash over me. His presentations were *that great.*

Mr. Fidler, thank you for everything. With you as inspiration have I endeavored to contribute to the historical record and am so very sorry that you aren't here to see the results of your influence. You've taught your last class but in a very real way have never stopped teaching. And others must now carry the Fidler torch as Earth's next communicators of history. I hope that my work would have made you proud, sir, and I so greatly miss your presence.

Born to Explore

1

The Falsest of All

But false prophets also arose among the people, just as there will
be false teachers among you, who will secretly bring in
destructive heresies, even denying the Master who bought them,
bringing upon themselves swift destruction.

—Bible, 2 Peter 2:1–3

Throughout the 1960s, the United States of America faced many challenges
to the fabric of its very existence. Look no further than the plight of Indiana
governor Matthew Welsh. A man in crisis. Citing "tingled" ears, he suggested
that radio broadcasters not play "Louie Louie" because of its "pornographic"
lyrics. Beyond even this, one challenge in particular constituted a self-induced
and expensively dangerous competition with another country. A competition so vast that it extended beyond the confines of our own home planet
and into space.

America's Apollo program, created for the express purpose of landing its
citizenry on the moon, had arisen from U.S. president John Kennedy's public
throwdown during a rousing 1961 speech. He sought to clinch victory in a so-
called space race against the Soviet Union—an ultra-bad rival country filled
with lying communist dogs. Kennedy's challenge hadn't been motivated by
exploration or discovery or science. It was completely political. At the time he
gave his speech, the number of Americans who'd flown in space totaled one.

But the president had inspired his country. Apollo mobilized hundreds of
thousands of talented people in various disciplines from materials technology to life support to celestial navigation to thermostabilizing turkey meat
with cranberry sauce. And within only a few years, Apollo seemed capable of
attaining Kennedy's goal before the decade ended.

Things grew expensive. By 1970 a single flight of Apollo's massive Saturn V
booster rocket cost roughly $375 million. Compare that to the National Science
Foundation's entire 1970 budget of $440 million. And for as much of a burly

geek's heartthrob as it represented, the Saturn V launched as a 363-foot-tall stack of engines and tech, but all you got back in the end was a little three-man capsule the size of a camping tent. That couldn't even be used again! It would be akin, offered naysayers, to flying a brand-new commercial aircraft from New York to London only one time and then abandoning it in a landfill.

Even years ahead of Apollo's first piloted launch, serious cogitation was already underway regarding what might happen in the program's wake. How does anyone follow what is arguably one of the greatest technological undertakings of all time? The country planned to continue visiting space, which for the sake of ongoing public support really demanded that Apollo's successor embrace reusability. Not $375 million throwaways. Determining exactly what that entailed would consume years and quiescently prove that when respected professionals convene to decide their industry's next major evolution, agreeing on what to do is hard.

The machinery began accelerating in early 1964. With floodwaters of money draining into Apollo, Kennedy successor Lyndon Johnson mandated an investigation of what lay ahead for the country in space. His directive led to a semiformal "Future Programs Task Group" headed by an industry engineer. Its report impacted the president's desk no later than January 1965. It was overdue by several months.

Although totally discomfited by the report's overall approach—specifying long-term goals without any political support yet behind them—National Aeronautics and Space Administration (NASA) administrator James Webb nonetheless called it "a timely and valuable working document for use within NASA, and other agencies, as a foundation for further analysis and discussion." Created in October 1958, NASA governed the country's space activities from behind a charter defining them as civilian and peaceful.

Naturally, America also harbored military space objectives. Like spying. The country's entire space program could have easily been run as a branch of the U.S. Army, which controlled most early flights. But continuing to do so carried the risk of provoking other nations. Space could end up a battlefield to occupy, arm, and contest. NASA intended to separate, by definition, the country's space activities from its military ones.

Conventional wisdom suggests that Webb, with his glowing comments on the Future Programs' report, was fooling even himself. The tome represented vanilla reporting—an overwrought attempt to dodge fundamental questions at hand while technically completing the assignment. Still, it merits attention because the report put into play a formal dialogue about America's cosmic future. Most sections of it concentrated on piloted flight. Three optimistic paragraphs were devoted to a hypothetical Jupiter mission, blasting away on unproven booster rockets with special upper stages. Or perhaps atop the mammoth Saturn V itself. Over and over between its uninspired front and back covers, the Future Programs' report anticipated a continuing use of expendable rockets for Earth-orbiting as well as planetary missions.

That said, the report's authors did advocate for the development of what they called "a recoverable orbital transport." Its basic premise: a large mother ship with a smaller vehicle riding piggyback. Both using wings. Both flown by pilots. After lifting off in conventional fashion from a runway, the mother ship would accelerate to extreme velocities. At high altitudes, the piggybacking passenger would light up its own engines and dart away into Earth orbit while the mother ship turned back to land on the same runway from where it began. Aboard the orbiting space plane, astronauts would do their thing and commute home at the end of the workweek, also landing on a runway.

Nobody called it a "space shuttle" per se, but that's what it was.

This notion of ground-to-space mass transit gained additional footholds during a key Houston conference in October 1966. NASA's budget had just suffered a $712 million mauling. Congressional back chatter sought further reductions. Apollo hadn't even flown. Don't think of the National Aeronautics and Space Administration, by the way, as some singularly monolithic structure accessorized by a giant parking lot filled with cream-colored Buicks. Comprising the space agency is an aggregated cluster of geographically diverse "centers." Each professes a specialty. For example, the Marshall Space Flight Center in Alabama developed the Saturn boosters for Apollo. Overseeing these centers are administrative officials at NASA Headquarters in Washington DC. And for this 1966 Houston conference, DC administrators gathered with reps from Marshall and Houston's Manned Spacecraft Center.

Ostensibly, this conference would further examine how reusable transport could facilitate the construction and maintenance of a space station. Attendees yapped opinions at each other like neighborhood dogs. At the end of it all, one Marshall participant described this event as "the beginning of the space shuttle as such." He hadn't been part of the Future Programs Task Group.

1. Early concept art for a reusable space transport craft. The larger mother ship would slowly generate controversy over its cost and role. Courtesy NASA.

Neither had Richard Nixon, who clinched the U.S. presidential race in November 1968 and began facing a number of scratchy loose ends from previous administrations. Along with the Vietnam War and widespread racial unrest, Nixon had been gifted Apollo—a popular and clearly defined program racing breathlessly toward the lunar surface on a gilded highway of tax dollars. Nixon wouldn't have to so much as lift a finger to watch it play out successfully during his time in office, sharing glory he didn't earn. Once American boots left their first prints, however, what came next *still* remained sharply in doubt.

A new, thirteen-member group convened with the idea of providing Nixon's administration with some sort of guidance. They called themselves the Task Force on Space and plunked a Nobel Prize winner at the helm. With many professional scientists aboard, the tone of its final report proffered realistic and depoliticized appraisals of how astronauts fit into a comprehensive plan of space exploration. One line read, "We are against any present commitment to the construction of a large space station." The task force thought it unnecessary because an already-underway derivative called Apollo Applications would supposedly address the same thing, albeit at a more modest scale and over a longer time frame.

Although some politicians felt an American flag should be planted on Mars next, "a great majority of the task force" opposed a "commitment to a manned Mars landing" at that time. The insanely high cost of sending people there would have further jostled an already-imbalanced relationship between piloted and robotic exploration. Why commit the money when comparable scientific research could be done with much less expensive remote-control machinery?

"We do not know precisely what may be the proper or most useful functions of man in space."

Also within the report came a blatant call to action: "The U.S. program for planetary exploration by instrumented probes needs to be strengthened and funds for such probes increased appreciably." The task force noted that the technology to do this already existed "and will greatly increase the scope and depth of human knowledge and perspective."

One particular task force member by the name of James Van Allen no doubt represented a major influence on that report's prevailing attitude. He once argued, "The progressive loss of U.S. leadership in space science can be attributed, I believe, largely to our excessive emphasis on manned space flight. And on vaguely-perceived, poorly-founded goals of a highly speculative nature."

This physics professor, from the University of Iowa, had been one of the first to study cosmic radiation in Earth's upper atmosphere. (Admittedly, it's kind of a niche.) He used surplus rockets, picked up for a song, that carried specialized and hand-built instruments in their noses. Throughout the 1950s, Van Allen found himself on a personal and never-ending quest to study that esoteric radiation at its purest. To hear from space itself with real answers. His quest led to pioneering experiments aboard the earliest American satellites and planetary probes. Cosmic radiation aside, understand this: Van Allen was one of the very first to genuinely *explore space*.

Over time he rose to become a preeminent researcher in the space science community. Commensurate with that was his midwestern, even-keeled rationale for what mattered. Throughout Van Allen's career, he remained vocally suspicious of the real value and overall impact of human spaceflight. During the operation of an experiment, the mere presence of a human could jeopardize everything a scientist had spent years preparing. "An astronaut's sneeze

2. A space scientist in his natural habitat—the classroom. James Van Allen works with a student during the early 1970s. Courtesy University of Iowa James A. Van Allen Collection.

could wreck a sensitive experiment in a microgravitational field," he postu-lated. "Clouds of gas, or droplets from thrusters of the spacecraft, or from dumps of water or urine, ruin the local vacuum and optical observing con-ditions. And complex magnetic and electric fields associated with manned spacecraft preclude certain kinds of radio observations." Say what you want, but Van Allen usually made unassailable points. To understand the real space

THE FALSEST OF ALL | 7

environment, to produce clean true data, often required humans to be absent from that environment.

"Apart from serving the spirit of adventure, there is little reason for sending people into space."

But Van Allen certainly held no final authority over policy, and mixed messages emanated from NASA's top ranks. In a February 1969 public speech, Acting Administrator Tom Paine specifically called for "the establishment of that projected major research laboratory in the sky, the permanent U.S. space station, accessed by a low-cost space shuttle." Understand that NASA already had a space station in the queue—a temporary "workshop" one, using a repurposed Saturn fuel tank already in orbit. For missions lasting maybe a month. Paine wanted long-term habitation and seemed ready to die on this hill.

President Richard Nixon had been in office less than a month when, on February 13, 1969, he dashed off a memo to multiple peeps: his vice president, his secretary of defense, his science adviser, and Tom Paine. In one lean paragraph Nixon simply explained, "It is necessary for me to have in the near future definitive recommendation on the direction which the U.S. space program should take in the post-Apollo period." Yet another new collective sprouted—the Space Task Group, with Vice President Spiro Agnew in charge. And whatever the group's final recommendations, Nixon demanded them by September.

Wherever the country went in space, the trip would always involve launching off Earth. If it went only as far as an orbiting space station, well, you'd arrive shortly. If people or probes headed for Venus, Mars, someplace farther, then the same heavy-lift vehicle could (probably) be used. And so the appeal spread for a reusable, airline-style transport vehicle. Make something capable of servicing most any satellite. Capable of launching planetary spacecraft. Or bringing crippled satellites back home, even! Avoid one-shot boosters in any capacity. "Fully reusable or near fully reusable systems offer the maximum potential for an economic and versatile space shuttle system," read a line from one report.

Proudly helming the task group, Vice President Agnew still had a few years left until he'd disgracefully resign for a series of extraordinarily poor choices (such as taking bribes). In the meantime, *he* wanted Americans on Mars and wasn't ashamed to say so. On the record. He considered the effort a worthy Apollo encore. "I think we shouldn't be too timid to say by the end of this century we're going to put a man on Mars," suggested the vice president. He was in Florida for *Apollo 11*'s blastoff to the first human lunar landing and acknowledged his opinion as being somewhat in the minority.

The following month, Agnew's task group released its final draft. Eagerly anticipated by American space industry officials, the publication had been equally anticipated by their Soviet counterparts—who, up to that point, had failed to land people on the moon or to return lunar samples using robotic machinery. "What a historical paradox!" derided a key space engineer from the Union of Soviet Socialist Republics. "In the USSR, our lunar program crisis was a result of failures and disasters. The Americans' crisis—at the moment of their greatest triumph—was because they hadn't yet decided what to do next."

Referencing a supposed "strong and wide-spread personal identification with the manned flight program," the report's opening gambit unsurprisingly parroted Spiro Agnew: "We conclude that a manned Mars mission should be accepted as a long-range goal for the space program." Such a mission could pop as early as 1981, claimed the supporting paragraphs, though repeatedly classifying this effort as likely taking place somewhere down the road.

One later section did recognize a "documented interest in the Space Transportation System," calling for a reusable and economical Earth-orbiting ferry as part of the country's space infrastructure. It should be able to, in the report's words, "carry passengers, supplies, rocket fuel, other spacecraft, equipment, or additional rocket stages to and from orbit on a routine aircraft-like basis."

In response, Nixon staffers cornered Agnew with frank news: The president would never support that Martian fantasy, as he didn't want his administration associated with another NASA money hole. The intervention finally got Agnew to back down.

With some of these affairs occurring publicly but others privately, Richard Nixon approached the podium on March 7, 1970, to inform the country of its future in space. Two lunar landings had happened; *Apollo 13* would leave the ground only a month later.

After leading with the recent Apollo successes, Nixon's remarks shifted. "We must now define new goals which make sense for the Seventies," he loftily began. "But we must also recognize that many critical problems here on this planet make high-priority demands on our attention and our resources." Nixon then outlined six major objectives in space, including a continuation of the Apollo program and robotic planetary exploration. He remained noncommittal about sending people to Mars.

But the president did mention efforts to reduce the cost of accessing space: "We are currently examining in greater detail the feasibility of reusable space shuttles as one way of achieving this objective."

Most everyone still pictured this shuttle thing as two piloted and fully reusable vehicles working together: mother ship and orbiter. It remained the preferred configuration into 1971, albeit with a little gotcha: Almost $10 billion were projected as necessary to advance this thing from paper to operational machinery. Nixon gagged.

The more people worked the various configuration options, the more of a wild card the mother ship became. Its design had to be ideal over four completely different stages of flight: low-speed takeoff and climb, acceleration, high-altitude operations, then return approach and landing. All under vastly different conditions of weight and balance. The mother ship was projected to consume the bulk of development funds, without ever being the gorgeous thing in orbit. It wouldn't carry any astronauts. It was just a taxicab driving the starlet to the ball; all eyes would be on her. Who cared about taxicabs?

The impasse broke in May 1971 by way of a two-man organization that had the advantage of being emotionally uninvested in any space proceedings to date. Employing neither aerospace engineers nor politicians, but economists, the Mathematica company had been contracted by NASA Headquarters to evaluate shuttle configurations from an *economic* perspective. Mathematica ignored hurdles such as creating high-powered new engines or exotic reusable heat shielding, assuming that technical challenges would sort themselves out. It also ignored logistical bugaboos such as quickly rehabbing launch pads or turning around landed shuttles to go again. Mathematica focused purely on the greenbacks.

Starting from the generally accepted estimate of $10 billion to $14 billion in nonrecurring development costs, Mathematica factored in a baseline flight itinerary encompassing NASA, the Defense Department, and a broad spectrum of commercial interests. And provided that the aforementioned clients needed at least the same amount of launch services as had been required for the last eight years, a reusable shuttle made *economic* sense. The numbers told them so. But numbers didn't have to invent human-rated aerospace technology from scratch.

After considering approximately two hundred(!) different configurations from industrial heavy hitters such as McDonnell Douglas, Boeing, Grumman, and Lockheed, Mathematica also zeroed in on the wide-body mother ship. It alone would require billions in development costs and decades to pay off. A mother ship *failed* to make economic sense. Okay, was there something more cost-effective?

Ultimately, Mathematica settled on a concept it named Thrust Assisted Orbiter Shuttle. Instead of a mother ship, the shuttle would ride atop an incongruous arrangement of heavy-lift rockets and jettisonable fuel tanks. "This eliminates the need to develop a large manned, reusable booster," went Mathematica's argument. And in doing so, the stratospheric up-front costs could practically be halved. Industry contractors already knew how to whip up fuel tanks and big dumb boosters; it was just a matter of bolting everything together and then clipping a shuttle on it somehow. "The cost per launch," argued the economists, "can be as low as $6 million or even less."

NASA leaders wasted little time capitalizing on Mathematica's findings. The space agency's relatively new administrator, James C. "Jim" Fletcher, penned a slanted opinion meant to convince even the most steadfast holdout. In a section titled "Why the Space Shuttle?" Fletcher tossed out a few ambitious oversimplifications such as how the machine "lands on a runway, ready for its next use. And it will do this so economically that, if necessary, it can provide transportation to and from space each week." He predicted between thirty and fifty flights a year, arguing further that possession of a shuttle "will give us a quick reaction time and the ability to fly ad hoc military missions whenever they are necessary."

As ultimately agreed to by clients, contractors, and increasingly excited members of the White House, the shuttle's fifteen-by-sixty-foot payload bay would carry up to sixty-five thousand pounds of whatever America wanted in space. Those specs had been heavily influenced by the wishful U.S. Air Force and its window-peeping spy satellites, both existing and planned. The air force's attraction puzzled individuals such as Van Allen, who noted that military space activities were already being fulfilled and in more cost-effective ways. Why in the world did people now want a shuttle?

Per the engineering drawings, this new shuttle orbiter sported three currently nonexistent, high-pressure engines but no fuel tanks for them because Mathematica's "economically viable" hardware configuration prioritized cargo space. *Don't carry empty tanks around during a mission*, went the rationale. As such, the shuttle would have to piggyback onto a large, externally mounted fuel tank and parasitically gulp from it on the way up to orbit. Also, even though the three new engines offered an unprecedented power-to-weight ratio, they still couldn't lift the very orbiter to which they were bolted. So the tank, in turn, would be held between two massive booster rockets. The choice of solid fuel to run these boosters made things simpler to build and therefore cheaper up front. Parachuting both down to the ocean for recovery also made them reusable. Such boosters certainly existed but had never before been certified for launching humans.

When James Fletcher bestowed upon his president a model space shuttle, Nixon came alive, because for unto him had been born in the city of Houston a savior—'twas the shuttle to space. It promised a very different angle on spaceflight. Yet one that was justified if you squinted a bit and one that was endorsed by cascading waves of government panels.

At this exact moment, while ceremonially revealing a final design, the program could've received a name. Something majestic. Something to give it a little pluck and dash. Names help sell a program and retain public interest. Apollo had been named after Greek mythology's inspirational god of light and knowledge. Nixon could've done no worse in calling his new program Hermes. A fellow Greek god, the athletic Hermes possessed wings in his function as a messenger to Olympus. It aligned well with the concept of a shuttling orbiter and certainly fit better than most other ancient deities.

But the president left any name off his new program. It didn't have a name at all. It was just called the Space Transportation System.

Richard Nixon hadn't created Apollo. He'd inherited it, didn't want it, and worked madly to distinguish his own administration from that of Kennedy's. Nixon saw no need to "win" anything in space. He shied away from such over-the-top grandiosity as piloted flights to Mars or on-orbit stations. That stuff was state fair crazy. America had enough problems on the ground. Nevertheless, his administration could neither ignore the emerging relevance of working in space nor refuse to do anything with it. All along Nixon had recognized the critical role of making some kind of contribution. A modest next step to be remembered for. And this would be it. Leaving the program untitled seemed to help it *remain* modest.

"An economical space plane," celebrated one industry advocate. "Capable of putting a fresh egg, every morning, on the table of every crew member of a space station circling the globe."

2

Intermarriage

If we go ahead with a Shuttle, can this cure one sick child? Can it provide housing for one indigent family? Can it educate one illiterate youngster or provide for one hungry household?

—Senator William Proxmire (D-WI) arguing against the investment in a space shuttle, April 9, 1973

Money flowed into the space shuttle at near-relativistic speeds. Program costs surpassed $3 billion, then $4 billion. By the spring of 1979, the whole affair had become a money pit. The shuttle's insanely powerful new engines? Its magically reusable, heat-blocking protective tiles? Both almost hopelessly behind schedule. White House budget wranglers whispered to U.S. president Jimmy Carter that he should scuttle the entire program.

But the president kept shuttle money flowing. For certain Carter endorsed, and celebrated, NASA's comparatively unglamorous robotic planetary missions. Earnestly he'd contributed an optimistic message of hope and future galactic community for inclusion on twin space probes called *Voyager*, at that time hopscotching their way past the outer planets on a "Grand Tour" as it was known. Often Carter tuned in during mission milestones. "We spent a couple of hours watching the Voyager pass Saturn. This has been one of our best scientific efforts." Carter recorded this in his diary on November 11, 1980. But human spaceflight? Eh, not really his thing.

Why, then, would Carter shovel money into a black hole?

"I was not enthusiastic about sending humans on missions to Mars or outer space," he claimed, years later. "But I thought the shuttle was a good way to continue the good work of NASA. I didn't want to waste the money already invested."

For all the windy, up-front dialogue about shuttle capabilities, and all the back-bending engineering breakthroughs, many missions demanded even more. See, America's potent creation would be able to fling seven smiles and bunk beds and government coffee into a shallow, 186-mile-high Earth orbit. But that was about it. By itself, a shuttle would lack the ability to send probes to even nearby planets. Disposable, one-shot boosters had been doing that for years. Yet all had been phased out in favor of America's designated new first stage—the space shuttle payload bay. Any planetary probe deployed from there would require a supplemental booster tacked onto its backside. One that, of course, took up space and weight in the payload bay, stealing both from bona fide cargo.

Some probes needed *two* boosters because even the supplementals were wimpy. All used solid fuel—which, yeah, is safer than a bunch of liquid sloshing around inside tanks. But solids don't offer the same amount of thrust. They impart less energy to the spacecraft. And are an instant-on, zero-to-one-hundred experience, meaning a higher payload strain and an increased risk of damage.

Well, what about using a Centaur inside the shuttle's payload bay? Available since the mid-1960s, this innovative upper stage could lift 35–40 percent more payload than any other (based on takeoff weight). A Centaur was incredibly light—using the pressure of the fuel in its tanks to carry structural loads. It was simple—using pressurized helium, not just pumps, to move fluids. It was efficient—piping cold fuel past hot parts so the fuel expanded and spun turbines for free. Its engines also could be restarted in space multiple times. And it was "smart"—recognizing during one probe launch a critically low trajectory and automatically thrusting for several extra seconds to regain course.

Aerospace professionals loved these things. Centaur's blow-you-away performance increased spacecraft size, instrument count, and mission duration. It put *Surveyor* landers on the moon and thrusted twin *Pioneer* probes out past Jupiter. An international organization working to implement global communications relied on Centaur to haul up its transponder-laden satellites. America's Department of Defense wanted its own on-orbit data relay system (because of course it did) and wisely specified the use of Centaurs.

But never, ever had a Centaur occupied the same launch pad alongside live, human pilots. The stage furnished what it did largely because it employed an incredibly powerful yet devilishly hard-to-handle fuel. A fuel useful only at 423 degrees below zero Fahrenheit, where ordinary metals turn super brittle, and things can go to hell in a hurry. The Centaur was a beast that breathed liquid hydrogen. And if disaster fell, would bleed it too.

So what? went the counterargument. The Saturn V used liquid hydrogen in its second stage *and* carried astronauts. What was the big deal? No other upper stage compared. Deployed from a shuttle payload bay, Centaur could hurl 5,600 pounds of spacecraft to nearly anywhere in the solar system. It could surgically place more than 10,600 pounds into geosynchronous orbit, even if the payload was forty feet long. If a shuttle mission had to abort, Centaur engineers promised they could drain its tanks out the back within five minutes.

Many astronauts fretted. Centaurs were spooky. Carrying such a volatile beast inside their tested-but-not-really-tested shuttle hadn't exactly been spelled out in the job description. Regardless, in early January 1981, program administrators decided that the risks of liquid hydrogen outweighed the negatives. Centaur and the shuttle would make a love connection as America's next-gen launch platform.

By the time a shuttle first left Earth that April, three technicians had already died in one. It'd happened only weeks prior to the launch and had nothing to do with Centaur. After a countdown test, these techs had entered the shuttle's aft compartment and were overcome by a lack of oxygen. Beyond this tragedy, the program as a whole had rung up $5.974 billion worth of costs, not including gratuities. This represented a 17 percent overage from the original estimate. Not too shabby, actually, as far as national space programs go.

However, the first couple shuttles out the door, *Columbia* and *Challenger*, each struggled to lift even fifty-seven thousand of the originally promised sixty-five thousand pounds. Three future orbiters would supposedly be able to pick up the slack, although the exact tonnage still varied. The capabilities of any given mission would have to be determined on a case-by-case basis. And of course, not all hardware was completely reusable. The half-million individual parts composing each external fuel tank would rain down as trash in the ocean.

"I think the Shuttle is a great mistake. It's taken all the funding that should have gone to planetary exploration." Such was the wailing of an American biologist who pleaded with NASA policymakers to focus more on science. "The only really exciting thing that I can think of now, that's coming up, is the Galileo mission."

Naturally, James Van Allen had something to say about where things were headed. "The Shuttle's contribution to science has been modest, and its contribution to utilitarian applications of space technology has been insignificant," he wrote in an opinion piece. "At the end of the day, I ask myself whether the huge national commitment of technical talent to human spaceflight, and the ever-present potential for the loss of precious human life, are really justifiable."

None of these decades-old discussions and decisions about how to build a space shuttle, or what to put in one, or even whether to make one in the first place, ever solicited the opinions of a man named John Casani.

On a butt-ass cold Tuesday morning in January 1986, he joined a dozen-odd people in a second-floor meeting room at Cape Kennedy's Operations and Checkout Building. It was relatively early, around eight o'clock. Everybody wore suits and ties and uncomfortable dress shoes. A few miles away in a clean-room hangar, crews labored over semifinal preparations of the *Galileo* spacecraft—in line to ride a shuttle that May. Following its deployment in low Earth orbit, the van-size *Galileo* would rocket away to spend its operational life orbiting Jupiter in pursuit of noble scientific discovery. The Centaur-in-shuttle arrangement constituted the only direct flight to Jupiter and would occur within a reasonable two- to three-year time frame. *Galileo* absolutely needed Centaur. With it, the outer solar system would still be reachable even if departing from an Earth orbit as low as 105 nautical miles. The Jupiter machine also signified Centaur's very first use inside a space shuttle. No shuttle had even test-flown an empty Centaur. Its presence had the Cape on edge.

A twice-married and middle-aged father of four, John Casani had already invested over eight years of his professional life heading the Galileo team as its project manager. Within that whole entire span, the spacecraft and mission had slowly come of age, albeit in the shuttle's shadow. Originally planned to launch some five years earlier, by January 1986 *Galileo* had endured a nightmarish four major redesigns completely unrelated to its own problems. These redesigns occurred because the shuttle's development had lagged behind schedule, or because it could no longer carry the amount of payload originally quoted, or because the high-energy Centaur stage had first not been allowed, then approved, then killed by Congress, then resurrected again. At one point in development, the *Galileo* spacecraft had actually been cleaved in two—each requiring its own separate launch—because no shuttle could lift the entire thing in one go. New launch dates meant new trajectories to calculate, new course-correction maneuvers to plan, new sequences of science observations to build and coordinate.

For worker bees such as Casani, sending things up to orbit on a shut-

3. How everything would've been carried inside the shuttle's payload bay. Folded-up *Galileo* is leftmost. The large tapering cylinder is Centaur. Its aft body and engines are housed in a cradle that would tilt up for deployment. Courtesy NASA/JPL-Caltech.

tle created enormous headaches in places where there didn't have to be any. "We knew that the JSC people were *so* predisposed against the Centaur," he claimed of higher-ups at NASA's Johnson Space Flight Center in Houston—a place that, until this Jupiter mission, dealt exclusively in crewed flight. "They didn't *like* the idea that we had a Centaur. They didn't *like* the idea that they had [to deal with] *Galileo*, you know?"

Until then, shuttle people had never crossed paths with the planetary spacecraft crowd. Although both went to space, they lived in completely separate worlds with separate rules. One perfect example was scheduling. Centaur engineer Joe Nieberding later recollected one of many awkward conversations with shuttle management. "Hey," he began one discussion, "if we launch on *this* day, you guys have to deploy us at *this* time."

"Oh, no, we can't do that," argued the shuttle manager. "The crew's asleep at that time."

"Well, like, why don't you wake them up?"

"That's their manifested sleep time," came the retort. "They're not going to do anything else."

What a bunch of stuffed-shirt pomposity. "So," concluded Nieberding, his voice tinged with frustration and disappointment, "we tried to help them understand the constraints that we were dealing with for a planetary launch out of a shuttle. Which they'd never encountered. They had no interest in learning."

The meeting room John Casani stood in faced north, and its big windows permitted mostly clear viewing of the shuttle's launch pads roughly seven miles away. On the northernmost Pad 39B stood *Challenger* with its external tank and boosters, fresh from yesterday's scrubbed launch. Another attempt would supposedly happen sometime this morning. But the meeting room didn't have a TV, and nobody could hear the public-address system at the visitor complex so far away.

The morning's gathering had been summoned by Nieberding's coworker Larry Ross, who managed the Centaur program out of NASA's Lewis Research Center in Cleveland, Ohio. One day prior, on the twenty-seventh, he'd ticked through nineteen to-dos, fielded a late-morning teleconference, then caught a mid-afternoon United flight into Orlando. Ross looked forward to his meeting and orchestrated similar get-togethers approximately once a quarter. "On all my major projects, I—I would always convene a group of people that were generally at my level in the organization. That is to say, the bosses of the project managers," detailed Ross. He called it the Senior Manager's Panel. One attendee came from Honeywell, which provided Centaur's guidance system. Another trekked over from the Johnson Space Center's dedicated Shuttle/Centaur office. Defense contractor General Dynamics actually built the Centaurs, in San Diego, and had sent its own guy as well. An impressive collection of talent and experience ringed the table that morning as Ross pulled out a spiral notebook.

"And he had me because I was a senior representative of the major user of it," mentioned John Casani of his presence at the Centaur meeting. It'd been nominally scheduled for a day and a half. "Everybody had to, you know, say what their issues were, what their problems were. And his idea was, you know, if you had the high-enough attention, and there was an obstacle that anybody was having, you know, that was an environment that it could be resolved in. It was a good, good system."

Casani worked out of a separate NASA-funded facility—the Jet Propulsion Laboratory in Pasadena, California. JPL's specialty was, and is, robotic planetary spacecraft of *Galileo*'s type. The huge machine had been designed and built at JPL under Casani's management and direction. When it finally left Earth, the Lab would also serve as its center of operations.

People grabbed seats. Ross flipped to a clean page in his notebook and penned everyone's name across the top margin. Seventeen names in total. He got things underway. "I wanted to know status and problems, and *what are you doing about the problems? And do you know what he's doing over there?*" During his twenty-three years at Lewis, Ross had enjoyed great success with this approach; every participant seemed to appreciate it. "For them to come together and see their peers paying attention to the program, and the progress their peers are making, and the efforts they're making? That stimulated them to do the same thing. And it worked beautifully."

The General Dynamics guy spoke up about parts availability. He was having some problems getting certain things. Another chimed in about old parts; they could still be used but the specs kept tightening—to some degree because of this new direction with Centaur. It elevated already-high standards even higher.

During the years leading up to this meeting, more than sixty firings of Centaur's base configuration had occurred. Larry Ross therefore felt supreme confidence in the shuttle variant known as Centaur G-Prime. *Galileo*'s mission-specific, flight-grade Centaur had rolled out of the General Dynamics plant in August and subsequently transferred to the Cape after three weeks of testing. Once on-site, it moved to a converted launch pad for additional work. Only a few days prior, the stage had aced a critical test of filling and draining its fuel tanks. "Major milestone!" celebrated Ross. "We may have been off by a couple days, but it was very, very close to what we had been planning for this major test. And here we are in January, getting ready for our main launch!" *Galileo*'s actual shuttle, *Atlantis*, had been specially modified for the unique payload, including new fill/drain spigots in the fuselage, and ports for venting.

His meeting wended on, Ross jotting down the evolving discussion in a

4. Larry Ross, speaking just a few years after the 1986 Galileo meeting.
"We knew that there was no optimum stage currently in hand for the cargo bay,"
he commented. "But neither Marshall nor Johnson Space Flight Center
wanted us in the game of having anything in the Shuttle." Courtesy NASA.

mostly readable half cursive. The Honeywell guy commented on what he felt to be a tight delivery schedule for retrofitting both G-Primes with updated guidance systems. (At this point, only two finished stages existed: one for *Galileo* and the other for a solar probe called *Ulysses*.) Ross flipped to a new page in his spiral notebook. The rep for the engine supplier reported a minor engine deficiency it'd been chasing. The Ulysses guy regarded things on his end as being in "good shape," which Ross noted.

The wall clock pressed toward noon, and someone in the know suggested they pause because *Challenger* would be launching shortly. What a spectacle! Any aspect of spaceflight is dangerous, but liftoff is perhaps most so, owing to the huge amounts of explosive fuel involved. Many thousands of components must flawlessly interact to ensure success. On this cold Florida morning, a few of those would let everyone down.

"Let's go, everybody, stand at the window and watch," prompted the guy.

Up from his chair, Casani was herded into a nearby corner office that promised better viewing angles. "And we stood at the window and watched." Thence came a sad sigh while recollecting what he personally witnessed some seventy-three seconds into flight as *Challenger* throttled up its main engines. They did this for what would be the very last time because one of the shuttle's leaking booster rockets cracked away from its lower mount and triggered a complete structural failure of the external tank. Instantly this created a rapidly burning, near explosion of fuel that enveloped and completely destabilized the winged orbiter. It broke apart and the crew died.

Ladies and gentlemen, America's space shuttle. Welcomed into the world as some celebrated new member of the family, a supernatural savior almost, the lying mask of this creation had finally been torn away in revelation of the false prophet it really was.

Silence fell over that meeting room. From high over the Atlantic, pieces began raining down.

Jesus, thought John Casani to himself, while watching the situation unfold in real time. *There's seven people in that goddamn thing.*

3

No Plan, No Mentor, No Guidance

We had no way whatsoever to get *Galileo* to Jupiter.
In fact, the only trajectory we knew would work was one
to the Smithsonian Institution.

—JPL engineer and Galileo team member Bill O'Neil

People crowded the office windows, shoulder to shoulder. Glued.

John Casani remembered, "There wasn't a sound, not a word uttered, for at least fifteen or twenty seconds."

Then someone rasped, "Oh my god."

Next to Casani, squeezed up with everyone else, Larry Ross knew plenty about shuttle contingencies and said to himself, "Oh my god." Then he thought, *They're gonna perform a return-to-launch-site abort.* This split-second maneuver, available only within a very narrow window of time during ascent, was regarded by most astronauts as impossible—on par with a straight-up miracle.

But perhaps *Challenger's* crew would pull it off?

Larry Ross went monotone. "And of course, they did not."

More pieces drifted from the sky. Ross's people drifted away from the windows and back toward their seats but only to collect belongings. Everyone knew the meeting had ended the same time *Challenger* had. At the tail of his notes, Ross dashed, "Meeting terminated after 51-L tragedy." He used NASA's official designation for the flight. Today we know it simply as *Challenger*. Just say that one word, and most anybody who was alive at the time knows what you mean.

Most of those same people can also recollect exactly where they were when the *Challenger* accident happened. Certainly that was the case for Don Gurnett. "I was at the Jet Propulsion Laboratory for the flyby," he indicated of his whereabouts. "I used to have to travel very often out there for meetings." A physics professor at the University of Iowa, Gurnett had much more going on than just his plasma wave experiment awaiting launch within *Galileo's*

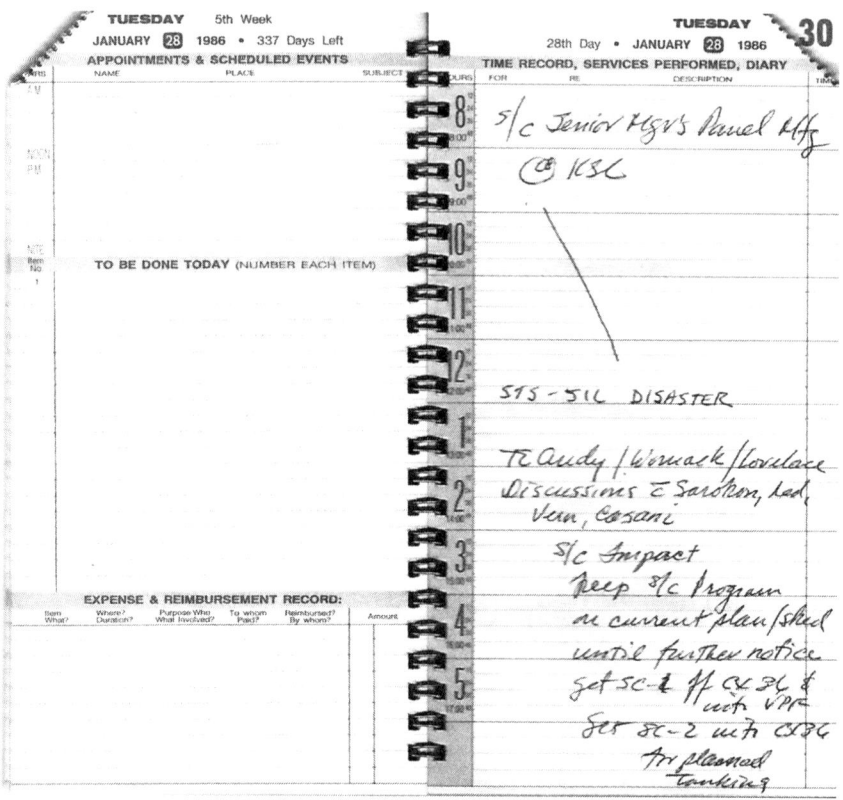

5. Larry Ross's original Day-Timer page from January 28, 1986. At the bottom
he added to-dos signifying an intent to proceed with Centaur readiness.
"SC-1" and "SC-2" refer to individual Centaur stages themselves, while "CX36"
is Complex 36, where Centaur tests were occurring. Courtesy Larry Ross.

belly. A similar creation of his rode aboard the *Voyager 2* spacecraft and had just made its closest approach to Uranus only a few days prior. "The flyby."

Another bizarre day for Gurnett, another example of how his whole relationship with Voyager had been off-axis from the start. He'd initially been rejected from that mission as an experimenter, only to gain admittance after a do-or-die appeal that saw him join forces with an adversary. In the name of science had Gurnett done so, burying the hatchet while agreeing to share credit for any Grand Tour program discoveries.

"We split those evenly between the two of us," Gurnett explained, concerning who'd be the senior author for each of their published papers. "Alternated.

He was first at Jupiter, I was first at Saturn. And I think we had agreed that I would be first at Uranus." Gurnett roundly looked forward to his upcoming turn at academic top billing—absolutely unaware that his once-reviled partner would not live to see Neptune. Or of what was just about to happen to the shuttle.

"And I'm in a Voyager meeting, and—and they have a TV in there, and somebody switches it onto the right channel to watch the *Challenger* launch. And we're all sitting there. We watch." Practically speaking, Gurnett's meeting also ended with the fireball, and pretty quickly nobody much felt like talking about Voyager or Uranus.

Down at the Cape, Larry Ross huddled with four stragglers—Casani included—to quickly brainstorm what would happen next. "Our mindset was, we don't know what this is all about," offered Ross of the disaster's possible cause and its long-term impacts on Galileo. "We don't know what's going to happen. The best we can do is preserve all options by not letting time get away from us." In his Day-Timer, Ross scribbled a few notes about keeping the program on its current plan and schedule until further notice—prescient thinking, indeed. Despite those chaotic first moments, he'd realized that without solid guidance from the top echelon at Lewis, each of the program's various subordinate groups may well develop its own approach for dealing with the fallout. The groups needed unity. Synchronized plans. That would take time. Until meeting with all the other section heads, he wasn't about to lead Shuttle/Centaur in some unannounced direction and therefore stuck to the script.

"We need to press on," argued Ross to the group, and everyone agreed.

"Don't change anything."

Glumly Casani adjourned to visit *Galileo*'s nearby prep hangar and check in with his JPL team, which had also paused to watch the launch. Larry Ross stuffed the Day-Timer into his belongings. Then he, too, filed from the room.

Word came down in late February that both Ulysses and Galileo had been postponed for at least thirteen months. The announcement included nothing about when shuttles might fly again. But launching *Galileo* over a year down the road introduced new hassles because Jupiter would've moved, as all planets do. Leading to fuel problems, trajectory problems. *Ulysses* also needed Jupiter in a certain spot for stealing a gravity boost on its way to orbiting the poles of the sun.

Crews pressed forward as though these announcements hadn't really occurred. *Galileo* transferred in mid-May to a different building at the Cape where it was finally mated with Centaur G-Prime #1. But then came a hard stop.

NASA CANCELS JUPITER, SUN ROCKET

June 20, 1986. WASHINGTON—The National Aeronautics and Space Administration Thursday canceled development of a modified Centaur rocket that it had planned to carry into orbit aboard the space shuttle and then use to fire scientific payloads to Jupiter and the sun.

NASA Administrator James C. Fletcher said the Centaur "would not meet safety criteria being applied to other cargo or elements of the space shuttle system." His decision came after urgent NASA and congressional investigations of potential safety problems following the Jan. 28 destruction of the shuttle Challenger 73 seconds after launch.

The move was a stinging blow for the U.S. planetary exploration program.

Stopping a project like Shuttle/Centaur is about as straightforward as stopping a glacier. "It takes place over a period of months," Joe Nieberding explained. "There was a team of people laying out the plan for the actual stoppage of work." Multiple contracts had to be canceled: engines from Pratt & Whitney, computers from Teledyne, navigation systems from Honeywell. "There's termination costs and schedules, and you have to, you know, get rid of some hardware. Preserve others. There was a whole plan worked out to do that." Over two hundred stop-work orders went out in total. To paraphrase the general language of each: *You are hereby directed to take those necessary steps for an orderly shutdown and cessation of all work.* Most contractors had the bulk of things buttoned up by mid-September, with the understanding that all tasks had to be completed by the end of the year. All engineering paperwork went to a central location.

Completely lacking any ride to Jupiter, Team Galileo was dead in the water. Had the spacecraft been slated for a conventional booster, it would've launched in early 1982 and already been well into its prime mission.

Instead, the *Galileo* spacecraft sat in a room at the Cape with the lights off.

The unprecedented and demanding role of Galileo project manager had been bestowed upon John Casani years beforehand, with no advance notice, while riding in a car being driven by his boss. The pair were in Florida alongside a broad contingent from the Jet Propulsion Laboratory. Everyone's job there was to complete launch prep on both *Voyager* probes destined for the outer solar system. At the time, Casani's hands were basically overloaded as Voyager's project manager. He sat atop the pyramid, with everyone else on that project ultimately reporting to him.

Voyager launch operations occurred within a sprawling facility called the Cape Canaveral Air Force Station. It was adjacent to yet separate from the somewhat more recognizable Kennedy Space Center complex used for Apollo launches. Voyager occupied Complex 41, laying more up on the north end. The facility also contained offices and prep hangars and basically everything Voyager's team needed to get its birds off the ground. A day or so after the second one left on September 5, 1977, people were clearing out.

Casani's boss, a longtime JPL guy named Bob Parks, steered the rental car down a miles-long road leading away from Complex 41. Parks was high up the chain and reported straight to Lab director Bruce Murray. Casani remembered the events in the car pretty clearly. "We were driving off the Cape, back to our motel rooms in Cocoa Beach. And we were gonna pack up and grab our bags and get on an airplane."

After a bit Parks divulged, "Well, we got the J-O-P project approved in the Congress just last week."

Casani knew a little about this. The abbreviation stood for Jupiter Orbiter with Probe, which represented a decent chunk of the recent appropriations bill. Missions often start out with dreary names because it's enough work as it is just to get them going. With any luck at all, the giant Jupiter machine, in whatever final form it took, would be anointed with something more inspirational. But things were finally a go! How exciting! "That was a big topic at JPL," Casani later said of the Jupiter mission, "because it had been quite a difficult struggle to get the Congress to approve it."

Parks kept talking while driving. He told Casani, "And so when we get back, I want you to head up the project office." Casani wondered aloud if he'd be giving up the Voyager reins. But Parks told him no. A new outer-planet department would be created at the Lab with Casani heading it up. Voyager management would transfer there, with J-O-P as a new roomie. It sounded like a step up. Casani thought he was being promoted.

Reinforced Parks, "This is gonna be your next assignment."

By way of informal conversation, John Richard Casani—a Pennsylvania expat who'd ended up at JPL almost completely by chance after road-tripping to California on a total whim—was now in supreme command of building the most expensive and complicated interplanetary spacecraft envisioned thus far. How funny then that Casani himself, born on September 17, 1932, grew up in a household that lacked a common piece of technology many others had—a TV. "My father didn't believe in it," explained Casani, stifling laughter at the bizarre philosophy. Every story about his father—John Charles "Jack" Casani—always came out in a noticeably fond manner. "Whenever we all wanted a TV, he says, 'No!' He says, 'We're waitin' for color!' And then when color came out, you know, he was waitin' for 3D! So we never had a television set in the house the whole time." As years went by, this patriarchal barricade, standing between young John and such things as broadcast entertainment, would continually influence his son in perhaps unintentional ways.

To discuss life and family and his career as one of the most important people who ever worked in solar system exploration, the younger John Casani had agreed to meet—but only at a specific restaurant befitting his palate. Casani arrived almost precisely on time. And the combination of dark sport coat and blue button-down he chose to wear made for something of a nonverbal cue that every meeting seemed worthy of dressing up for.

With bread on the table and wine orders underway, Casani's guest swung open a three-ring binder to the first of many pages. In rough chronological order, they contained discussion topics written long in advance. But the initial questions flopped. Badly.

Why stay at JPL for so long? For the thrill of space exploration? Casani responded, "I don't think there's anything about JPL, or work that I've done, or work that I *do*, that thrills me. I mean, that's not the word I would use."

I see. Hmm. How would you describe your legacy? "You know, my legacy is gone," he deadpanned. "We probably should throw a lot of what my legacy is away. And start over with a clean sheet of paper."

O-kay, well, I've been reading about your management practices. Let's talk about them. "What you're probably learning from me is I'm very undisciplined, unorganized, and not very structured in terms of what a good management practice would anticipate," came the response, with no laughter in sight to stifle. "And besides that, you know, my wife has been dead now for, I don't know, ten years? So you know, things aren't the same."

This meetup had been in the works for months but was imploding like an old Las Vegas hotel. The table went quiet. Lying near the bread basket, a small audio recorder had nothing to record.

Then tell me about growing up in Philadelphia. Pay dirt!

Casani's paternal grandmother had been born in Cuba of European parents. After reaching the United States, she ended up at Mount St. Joseph Catholic high school in the Philadelphia suburbs. And from there, attended Trinity College in Washington DC. How did Casani remember all this with such ease? "The woman that my dad married, strangely enough, went to the same high school, probably twenty or thirty years later, and to the same college. And then my dad had one daughter, and *she* went to the same two schools."

However great all this education may seem, it actually caused a rift within the otherwise stable family at large. A rift centering on John's father, the one and only, not-to-be-argued-with John "Jack" Casani. A man born in 1898 who went on to serve during World War I. A man who, after his release from the army, wanted to attend college and be all academically inclined as his smart Cuban mother was. But the elders told him, "No. You're going into the family business." This referred to the Casani Candy Company, home of "quality treats since 1865." It sat on the distribution end of the confectionary chain, buying from manufacturers in bulk and then wholesaling to local retailers. Today the company's website claims to have handled initial distribution of the Hershey's chocolate bar.

Candy isn't exactly an essential item, which made things tight during the American Great Depression and World War II. "People might like their candy," Jack Casani once told his son, "but they *have* to have toilet paper. And they *have* to have toothpaste. And they *have* to have soap." When the economy's in a downturn, a candy company might not be the wisest choice for employment.

Jack had rebelled against any sort of personal involvement with Casani Candy, even though the guy basically had no choice. "*His* problem was that he didn't want to be in the business in the first place," detailed the younger Casani. "He didn't like the idea that he had to go into the business. He wanted to go to college! And his interests were always the classics, you know. Literature. He read everything."

At the time, young John remained largely oblivious to all this. He certainly had no interest in business life during his formative years. Never paid attention to it. "I was always interested in tinkering," Casani admitted. "Playing around with mechanical stuff." His eyes lit up a little when he said that.

6. In this family photo from the 1940s, Jack and Julia Casani have taken the kids on a beach vacation to Cape May, New Jersey. Young John is just to the right of his father. Directly in front of John is his brother Kane, who'd also end up at JPL. Said John of Kane, "He was just a pesky little brother, and I was just kind of a pain-in-the-ass older brother." Sister Anita stands at far left. Two more boys were still to come! Courtesy the Casani Family.

Jack Casani and his wife, Julia, plus their five kids, lived in the familial familiar suburbs of Philadelphia. Eldest child John experienced a typical Pennsylvania Catholic upbringing. He went to Norwood Academy in Chestnut Hill, as did his brothers, while sister Anita went to Fontbonne. Both schools were run by the Sisters of Saint Joseph. Five years John's junior, Bill Cosby simultaneously grew up only a handful of miles—but worlds away—in the city proper.

As a youth, John acquired a passing interest in the scientific world. It was minor, but present, and happened in part because his seventh- and eighth-grade teachers occasionally handed out a little freebie science newsletter. It contained various tidbits: stories of a telescope under construction at Mount Palomar in California or fun factoids about new types of research rockets used by a place called the Jet Propulsion Laboratory. So much high tech out there.

During this same period, John also discovered an old high school textbook of his dad's titled *Physics*. This accidental find became one of those serendipitous moments that redirect major life choices. "It was full of levers and pulleys and all of that kind of stuff. And I *devoured* that book. I loved that. And I thought that I wanted to grow up to be a physicist."

But adolescent John, who'd just transitioned to the Jesuit-run St. Joseph's Preparatory School, found himself in the wrong kind of program. "It was a *very* liberal arts education, and they had two curriculum: a classic and a regular. And I took the classic, I think, because my father wanted me to. Because that's where *his* interests were. And that meant I had Greek and Latin and English." A little trigonometry did come his way. But practically nothing in the hard sciences.

Thankfully, this style of coursework failed to affect his core interests. "All through high school I was always building stuff and tinkering with stuff." He joined the swim team for a little physical activity, racing the 440 and the individual medley. One day, the family had an electrician out. This anonymous man departed without likely knowing the impact his service call had on adolescent John. "I watched him, and I decided that I could do that. And so I did a lot of *that* stuff around the house and for my friends."

John revamped an electric train set, adding little flashing signs and other accouterments. His parents were quizzed about potential household enhancements. "My mom wanted something that would turn the lights off after she left the pantry. And I worked out something like that. You know, I built it around a fluorescent bulb tube starter, you know, and it had a little time delay feature in it. And I was able to do that."

The kid was just getting started. "I built a garage door opener before you could buy one!" That in particular delighted father Jack, who excitedly operated it via push button directly from the garage, which sat at the back of the property. The John R. Casani pièce de résistance, however, would have to be what he did with the basement toilet. "I fixed it up so that when you opened the door, the toilet would flush," he pridefully boasted.

"You're just a Rube Goldberg," his father told him.

The phrase became a common refrain around the house. "That's all I ever heard," Casani alleged. Everything young John did? Nothing but *tinkering*. That word became a derogatory term. "I never heard *engineering*. Never *ever* heard *engineering*." At the time, he had no idea the discipline of engineering even existed. It never came up during his classical high school education. Nobody he knew studied engineering or worked in this field. "The concept of an engineering degree, or an engineering curriculum, or that somebody could have a profession as an engineer *never* occurred to me. And I never *knew* anybody that was an engineer." He'd watched that electrician work in his house and failed to make the connection. Maybe electricity involved physics? On downtown outings, he'd gaze up at the imposing Lincoln-Liberty Building on South Broad, right there in Center City—home to Wanamaker's Department Store and directly across the street from Philadelphia City Hall. Never made a connection to engineering. Maybe construction involved physics?

After graduating high school, John learned what the plan was when his dad came up and said, "You're going to Penn." Sans consultation, he'd been signed up for the University of Pennsylvania. It waited nearby. John hadn't been recruited for swimming but showed up poolside not long after first semester began and told someone, "Hey, I wanna swim."

"Fine," came the response, and just like that he joined the team. According to Casani, that year marked the first in which the Amateur Athletic Union (AAU) had offered the 150-yard individual medley event at the collegiate level. Casani swam it at the first meet and won. This made him the overall AAU champion for that event—a distinction lasting only until another meet the following weekend, when he lost.

"I never regained it," he sighed of the championship. "I wasn't that good."

One thing Casani *was* a natural at—being social. He joined the Phi Kappa Psi fraternity and saw his circle of friends explode. And oh yeah, there was also the matter of classes. He met with an adviser, who tossed out the inevitable question: "What do you wanna do?"

John didn't really know for certain, so he gave his default answer: "I wanna be a physicist."

"Well!" exclaimed the adviser. "You gotta take a *lot* of mathematics if you wanna be a physicist!"

That advice prompted John to sign up for numerous math classes. "But then I also had to take all the other stuff. Which was, you know, economics and sociology and history and French and English. And when it came time to choose a major, physics wasn't one of the ones that was offered." Decades later at the restaurant table, Casani kind of adjusted his jaw while contemplating what to say next about this unforeseen barrier. "And I'm not sure at that point that I wanted to be a physicist. I knew then that physicists weren't interested in doing the stuff that was in the book that I was so turned on about." Fulfilling needs—that's where his heart lay. Solving problems. Arranging levers and pulleys to *do* things.

If he couldn't do physics, well, then *what*? He looked down the list of options for majors based on his two initial years of liberal arts coursework. "None of those things was what I wanted to do." This sucked. In a pickle, John called home.

"Dad," he confessed over the phone line, in vulnerable tones, "I can't pick one of these things. I—I don't know what the hell I wanna do."

Jack kept quiet for a bit to let the boy talk.

Now came the pitch. Big breath. Suggested John, attempting to be casual, "I should go in the air force for a couple years and, you know, mature a little bit. And after two years I'll probably know what I want to do."

At this point, as many fathers will surely attest, various potential responses come to mind. John figured his dad would say something inspirational at best and clichéd at worst. But neither happened. "Instead of saying, 'Sure, son, go in the air force,' he offered to kick the shit out of me."

His hugely ticked-off father levitated with rage. Ordering, "You go back there! And don't you come home until you've picked something!" The phone call ended right about there.

John hung up and tried to process what just transpired. That'd been some rough sledding. "He was bound that *all* of his kids were gonna become professionals of one kind or another," Casani later justified his father's attitude, putting the overly harsh response into context. "He thought, if you have a professional education, you're not—what was his expression?—'subject to the vicissitudes of the business world,' or life, or something like that."

Today this approach makes perfect sense to the longtime JPL engineer. But back then, as a sophomore in college languishing with no compass, it only threw his future into question. *Welp*, he thought, *guess the air force is off the table*. He had no idea what in the hell to do next.

Not long after the distressing phone call, John bemoaned his predicament to a fraternity brother, who suggested, "Well, why don't you try engineering?"

"What's that?"

"Penn has an engineering school," his friend continued. "Why don't you come down there and see what's goin' on?" John's fraternity brothers always seemed to have crazy ideas. Another one of them, Louie, had always joked about moving to California and seemed half-serious about it.

John met with Penn's dean of electrical engineering, who agreed that this wayward pupil could start the program the following year as a sophomore. John already *was* a sophomore. Also, as the dean pointed out, summer classes would be necessary to fulfill some missing requirements. *"And,"* Casani elaborated, looking back on the despairing situation, that was "provided that I went to the Drexel Institute of Technology, which was just a couple of blocks away, at the next semester. And take a drawing course or a drafting course or some damn thing like that which I missed as a freshman."

A leap away from liberal arts, yahoo. The next painful step would be selling this paradigm shift to his dad. A man already on his guard. "I wasn't sure whether he'd go for it because it meant one more year of college now, because I'd be entering as a sophomore instead of a junior." Naturally, John worried that his parents might freak out—what with five kids queued up for an expensive educational pipeline and some of them unexpectedly taking longer than planned. "So I wasn't sure what my dad's reaction would be."

As it happened, Jack simply declared, "Fine. Go and do it."

John knew darn well that under no circumstances, even if he were facing a destitute life on the streets, should he ever ask about going to work for Casani Candy. His father would've combusted. "He didn't want *any* of us in the business," Casani warned. "And he made that clear, right from the beginning, that there was not gonna be any way we were gonna get into the business. We were gonna do something else altogether. And I wasn't particularly interested in the business either, you know?"

John felt confident in the new direction partly because a buddy of his had already started at Penn as an engineering major and would be in the same program. "Aw, this is great!" his buddy said. "We'll both be sophomores in electrical engineering!" Academic codependency—what a thrill. "I can help you! We'll do our homework together! I can help you with some of the stuff you missed, and this'll be really great!"

But that fall, the guy disappeared. "He never showed up," Casani reported. Total ghosting. "He'd flunked his freshman engineering exams, and they kicked him out. And so I was on my own. And I went through three years of electrical engineering and graduated in 1955."

Years down the road, when Penn faculty conferred upon him an honorary PhD, John Casani's senior-year project had remained legendary enough to justify formal mention during the ceremony. "The Robot Car," as everyone called it, performed its sole public demonstration in an on-campus parking lot one Saturday prior to the semester's end. The effort saw John and another electrical engineering senior collaborating with two seniors from mechanical engineering. Together they accessorized a professor's 1950 two-door coupe so that it could be driven by remote control. Fundamentally this consisted of two custom radios paired in a transmit-receive configuration. The transmitter stayed outside the car. Eight commands could be sent from it wirelessly: shift into drive, accelerate, steer left, brake, and so on.

Once received in-car and decoded, electrical signals triggered an arrangement of physical actuators and linkages crafted by the two mechanical engineering students. "They had the more difficult part of the project," Casani advised, "including providing the professor confidence we weren't going to have any fatalities during the demonstration." To hedge against this, extra credit was promised to an undergrad volunteer who naively occupied the car's back seat. Around his wrist went one end of a rope. His single instruction: If things go bad, pull. Its other end led up front to a knot around the car's emergency brake handle.

"It was pretty rudimentary," admitted Casani. "But it worked! And the kid in the back seat never had to jerk on the brake rope. I never did find out if he got extra credit."

Despite the elation of discovering a program that truly fit his interests, and graduating from it in good standing even without the promised assistance of his missing buddy, John Casani had no clue where to go. "I didn't have a plan," he admitted for life after graduation. "I didn't have any mentor. I didn't have any guidance. I didn't have anybody that could suggest, you know, what I might do. I just floundered."

4

The Kind of Person We Need

We didn't know how to do what we were supposed to do.
We were too dumb to know that what they were
askin' us to do would really be hard.

—John Casani on his early days at JPL

The Rome Air Development Center lay in upstate New York, more or less between Syracuse and Albany. It's where Casani nabbed his first postsecondary employment. The center did a lot of electronics-based R&D work for, with some irony, the U.S. Air Force.

What'd you think, John? "I really didn't like it there very much." During Casani's first winter in Rome, an entire week went by where local temperatures never climbed above ten below zero. Philly wasn't so bad, but Casani hated living in a world colder than his own kitchen freezer.

"And I was sayin', 'This is *bullshit*.'"

Not long after, and quite unexpectedly, his old frat brother Louie happened to ring up Casani at work in subzero Rome.

"John, do you really wanna go to California?"

"Hell yes."

"Well," Louie told him, "I've got a job offer from a refinery out there. Why don't you come home, and we'll go!"

Repeatedly had the two discussed such a thing. The most memorable time occurred on Saturday, the tenth of October 1953. Still in college, the pair had attended a Penn Quakers football game against the Golden Bears of the University of California, Berkeley. "They beat the crap out of us," remembered Casani. Rampant anguish had coursed through the stands of Franklin Field.

"California's Golden Bears unleashed a host of lightning fast backs, a line that Swiss-cheesed the Red and Blue forward wall, and a passer par excellence in Paul Larson, to trounce a surprisingly inept Quaker squad, 40–0." So went an uncomfortably candid postgame summary from the local paper.

A skunking wasn't the only thing Casani remembered. "They had the most beautiful cheerleaders," he gushed of the opposition, "and all that kind of stuff. It was sort of mesmerizing to a couple of Philadelphia kids." The idea of moving to California bubbled up right there in the middle of the lopsided game as both men intently scrutinized the bobbing line of smiley, well-coiffed damsels prancing across the field like springtime lambs. Aches of primal masculine lust flared inside the men.

In frozen Rome, Louie's call was enough for Casani, who gave notice and went home to pack. He called his father, saying, "Guess what, Dad? I'm going to California!" And then lied: "I got a job out there!" Casani doesn't recall much objection from his parents about the idea of moving west. "If they had any adverse reaction to it, they never expressed it to me, and I didn't sense any. Maybe just a reflection on my lack of sensitivity or gratitude. But I don't think so."

The guys planned on making the trip in Louie's '52 Ford and loaded up, as Casani reminisced, "everything we owned." Neither had been an adult long enough to acquire much, and so the car filled, more or less, with what two young men might gather for a weeklong vacation. "But for my buddy, it also included his 78-RPM record player and record collection. As well as a case of motor oil his father insisted we take just to be safe. I had no idea why he thought we would need that, but it was typical of the 'wilderness' impression many entrenched easterners had about leaving the East Coast and going west." For his part, Jack Casani is not known to have offered any automotive maintenance tools or supplies.

What an adventure for an already-exciting summer of 1956. Elvis Presley had broken into the charts with "Heartbreak Hotel." Mickey Mantle reigned at the baseball plate. Golda Meir was transitioning into her new role as Israel's prime minister. The Melbourne Olympics were underway, albeit with the equestrian events taking place in Stockholm because Australia's tight animal quarantine laws did not permit bringing in foreign horses.

To all that hullabaloo add two young men in a car, who spent upward of fifty-six days zigzagging westward across North America while overnighting at any college that was home to a chapter of their beloved Phi Kappa Psi fraternity.

"We had a great time doin' that," grinned Casani, the joy evident in his face. "I think the two of us had in our mind that, you know, this was nothing permanent. This is just going to be, to go out there and maybe spend a couple of years there. Have a great time, you know, and come back to Phila-delphia, where we belong, and have something to talk about at cocktail par-ties for the next two years."

Finally the guys hit California and beelined for Phi Kappa Psi's house at the University of California, Los Angeles. There they could crash temporarily. But Louie's job awaited him at Braun over in Alhambra, a good twenty-five miles to the east. After one night, they repacked and drove to the University of Southern California (USC) frat house, which made for a (slightly) better commute. Louie started at Braun, and John Casani was once again alone.

"Well, shit!" Casani told himself, two long weeks after their arrival. "I don't have a job!"

In wandering desperation, he found the USC student union and approached a couple girls working behind its front desk.

"Hey," he started, "I graduated last year. I don't have a job."

They said, "Oh." Adding, "That's no good."

The women directed him to a nearby on-campus office, where all manner of questions came up about what might interest him. "What's your degree?" they wanted to know. Casani told them it was in electrical engineering. But finances were at the point where he'd take any job at all.

"They didn't realize that I was not a USC graduate," Casani pointed out, long after this all happened. "I wasn't trying to be duplicitous at the time. It said *student union*, so I figured all the colleges worked together somehow." He chuckles at the naivete of this now, but the tactic netted him job interviews. Considering his degree, the USC student union contacted a few of the many aerospace engineering firms in the area. Casani visited Northrop Corpora-tion, home of such aircraft as the venerable P-61 Black Widow, a World War II night fighter. He also interviewed at North American Aviation, which at that time was developing the exotic rocket-powered X-15 aircraft.

And then Casani got sent up to Pasadena for an interview at the Jet Pro-pulsion Laboratory. That name rang a long-dormant bell in the back of his mind. Sounded like something from eighth grade. JPL did lots of secretive high tech for the military. Such as missiles and ways to steer them.

The USC employment counselors told him it would be a good place to work, a plum. But disclaimers abounded. "It may not be all that easy to get a job there," they confided, more than once. "Most of the people that work there have either master's or PhDs."

Neither were in hand, but he still really, really needed a paycheck, so Casani decided to go anyway. He committed to an appointment. Behind the wheel of a secondhand '47 Plymouth, Casani made his way up Oak Grove Drive in Pasadena to the Lab's main entrance. Slight issue: The JPL guy who was supposed to interview him had gone on vacation. Instead, Casani met with the guy's boss—a lean and well-dressed, no-nonsense manager named Jack James who visually resembled a Secret Service agent. They hit it off; James's wife just happened to be from Philadelphia. "He wanted to know what things I had worked on," Casani recalled. "And I described them to him. And you know, he seemed interested in that. He asked me a couple questions. I guess he figured out I wasn't bullshitting."

The two did not discuss Casani's auto-flushing toilet breakthrough. Overall, the meeting seemed to go well. Casani retreated to the USC frat house, and in July, North American invited him to come work on its cruise missile program. For the opportunity to touch evil Soviet communists with Uncle Sam's patriotic nuclear devastation, John Casani would be paid $405 a month.

It sounded intriguing. "A nice amount of money at that time," he snuck in, as kind of an aside. North American's offer tumbled around inside Casani's head. Then Jack James called USC's Phi Kappa Psi house with an offer of $395 a month to work on guided missiles. Also intriguing. Less money. "But I liked Jack," Casani reasoned. "A lot." Jack James started courting the prospective employee with the relentless intensity of an obsessed stalker. The Phi Kappa Psi house had a pay phone in its lobby, and that thing rang every day: "Jack James for John Casani, please." James once even telegrammed. Certainly more interest than North American had shown.

"So I decided, what the hell," Casani finally told himself. "This is a nobrainer. I'm gonna take the job at JPL. It's a more prestigious place." Even though he didn't have a concrete idea of what all they really did. Even though it should've been called the *Rocket* Propulsion Laboratory. But Jack James had kind of sold him on the place.

"It's because I liked Jack that I took the job." The calendar said 1956. Only a year later, North American's missile project would be canceled, and they'd lay off over ten thousand people.

Had Jack James not been the substitute JPL interviewer that day—had Casani instead met with whom he was originally supposed to—things may well have turned out rather differently. It came down to attitude. "I was kinda

7. JPL's front entrance signage, 1957. Established in the late 1930s by Caltech faculty and students, the Lab's original purpose was to enhance the performance of jet aircraft by way of rockets. Courtesy NASA/JPL-Caltech.

loose, you know, and I was more oriented to havin' fun," Casani disclosed. He categorized the original interviewer as smart and all, but "a very straitlaced guy." The kind who wouldn't have mixed well with Casani's going temperament. "I was too flighty, I think, for him."

At long last, John Casani had landed a job in a field that interested him and at a location boasting temperatures warmer than Philly's. Didn't seem to be a bad gig. Maybe he'd stay a while?

Across the breadth of its confidential campus, the Lab operated on military contracts. Such as the one for Juno. To reach its target, the intermediate-range Juno missile needed a working control system—sort of an electronic guide dog—to lead the works to its destination. And as with any new piece of high technology, Juno had tallied many, many problems in its early development. One of them had been assigned to new hire John Richard Casani.

He joined a team working on Juno's backup guidance system, known as Codorec. Most of its development work had been finished by the time Casani arrived. "But it had never flown," he explained. "It was all laboratory stuff." The system badly needed genuine in-flight testing. But Juno as a whole, or even as a part, wasn't yet airworthy.

In an effort to make progress, some JPL higher-ups wanted to stuff Codorec into Juno's smaller cousin, the Corporal missile, and test using that instead because Corporals actually worked. Being operational, they could stand in as Codorec surrogates. "Fly it in a Corporal missile," elaborated Casani of Codorec, "using the Corporal guidance system to actually steer the missile but telemetering back to the ground what the Codorec system *would've* done had it been in control. Okay?" This testing scenario made sense, and plans to implement it had been drawn up. But nobody had yet done the work. It went to the new guy. "Figure out how to get all this stuff into a Corporal missile. Wire it up, get it, uh, you know, put the antennas on and get the power to it, and get the commands and make it work with the ground system. And so that was my job." Shades of his robot car.

The education commensurate with a bachelor's degree in electrical engineering had brought John Casani to a certain point mentally. It got him thinking in specific directions. But nothing in the world equals actual on-the-job experience, which Casani began slowly acquiring inside Building 18 on the JPL campus. He felt intrigued by the task at hand. What a challenge! He had to learn great chunks as he went along. As part of the effort, Casani stuffed his head with everything he could about the Corporal missile—an intricate bit of technology several orders of magnitude removed from his father's garage door opener.

Today Building 18 is confusingly sandwiched between Buildings 259 and 288. The whole campus is out of order like this because a building's number corresponds to the order in which it was constructed. And beyond the obvious usefulness of consecutively numbered buildings, JPL's undulating grounds further lack the sensibility of a grid system. New Lab employees get lost all the time and have to ask for help.

Although not lost per se, John Casani lacked the ability to tackle Codorec all by himself from start to finish. So as everyone else did on campus, he formed a network of consultants, peer reviewers, and others off of whom he could bounce the latest situations. "What about *this*?" Casani would toss out to a colleague regarding his problem of the moment. It could be anything: an unexpected test result. A wiring issue.

Luckily the culture of the Lab was one where everybody always jumped in. "Here's what you have to keep in mind," people would say to Casani. "Here's one way to approach it." Or, "This is what you might want to try next." Little nudges always got him going again. When he needed that help, it materialized. And any helper knew that before long, the guy he'd just helped would be coming back around for a need of his own.

"There were a lot of smart guys there that were very willing to help and mentor me," Casani proclaimed with noticeable fondness. "There were a lotta, *lotta* people like that, that I worked with."

One anecdote from this late 1950s period, of particular thrill to Casani, had to do with a masking system for the Corporal missile's radio. Although the missile had been operationally deployed in Europe, it had weak spots. One was electronic countermeasures. If Soviet operators intercepted any command signals between the Corporal and its ground controllers, those signals would just resemble ordinary noise. "And they wouldn't be able to jam it," Casani indicated of the enemy.

But for all the planning of this weapon, an obscure detail made it through the cracks: The Corporal's radio system needed to synchronize its masking technology with ground control, which became apparent in testing. "I got to develop that," Casani mentioned. "So that was a very small part of it. But that's something they sort of forgot about until the last minute, you know, and said, 'Hey, we need this.'" From scratch, he invented a little circuit that went into production on the Corporal. Elegantly it solved the problem, enabling the missile's radio masking system to work as intended.

"It was a great opportunity for me," Casani pleasantly recalled of his time on Codorec. "I was designing and building stuff that had to work." Cool stuff indeed! Military stuff! Confidential! John Casani didn't know it at the time, but he would spend the remainder of his professional life at the Jet Propulsion Laboratory. So would, of all people, his younger brother Kane. Nearly three years separated them in age. Both had gone to the same high school, where Kane initiated a trend of following most every path laid down by his sibling. Kane had swum for St. Joseph's, studied (civil) engineering at Penn,

8. An army Corporal missile on a mobile launch platform, late 1950s. It burned liquid fuel, flew surface to surface with a range of seventy-five miles, and could carry a conventional or nuclear warhead. Courtesy NASA/JPL-Caltech.

joined the same fraternity, moved to the same state, and then began work at the same place. The brothers even lived together—in a south Pasadena apartment along with another engineer.

At one point while the Codorec effort was underway, Jack James came to visit the man he'd hired. "I just wanna tell you something," James informed Casani, after taking him aside. "You know, I really admire the way you're goin' about this. You've got things under control."

Casani, head deeply immersed in the task at hand and preoccupied with everything still to do, wasn't quite sure how to respond. *Where was this coming from?* He didn't work directly for James. But here were accolades all the same.

Jack James told him, "You're the kind of person that we need around here."

Within a year, this secretive military facility would head in even newer directions than that of radio-control missiles. It happened because of an inanimate object—an otherwise unremarkable metallic ball that another country had endeavored to mount atop its own missile. A much larger missile, capable of reaching space.

5

The Chair Dares

The less you say about my first marriage and my first wife, the happier
I would be with your end product.

—John Casani to his guest at the Italian restaurant

In a great month for movies, October 1957 offered London cinemagoers *The Bridge on the River Kwai*. It came out on the second day of the month, eventually won Best Picture at the Oscars, and today is regarded as one of the finest films ever made.

If World War II epics didn't seem attractive, tickets could also be purchased for the perhaps not so highly regarded *The Brain from Planet Arous* with its story of parasitic alien invasions and worldly domination. What—not sounding great? Okay, well, maybe try the edge-of-your-seat *Devil's Hairpin* and its focus on the evils of hot-rod racing and premarital sex. But if that wasn't going to cut it either, then wait for November's *Eighteen and Anxious*—a tantalizingly conspicuous and rather self-explanatory title.

The exact same day *Devil's Hairpin* reached theaters, a diligent group of engineers inside the Soviet Union were not in line to see it. Pursuing other priorities, they instead managed to launch a reconfigured military rocket toward space. It carried a beeping ball of electronics now known as Sputnik 1, which holds the distinction of beating all other human-made objects into orbit. Sputnik denoted Earth's first "artificial satellite," as the Reds called it.

America's response finally came via a lean cylinder named Explorer 1 that orbited in February 1958, long after Soviet engineers had also lofted their jaw-dropping encore. Sputnik 2 launched only a month after its predecessor, carrying a luckless stray dog on a sad, isolated, one-way trip into Earth orbit and death by heat stroke and historical immortality. Continuing their work of integrating complex missile systems, Jet Propulsion Lab employees overcame two key hurdles on the way to Explorer 1's success: They encapsulated its science payload with a functionally protective enclosure of their own design and crafted the booster rocket's solid-fuel upper stages.

Keep in mind that JPL played only one part, overall. A separate team of army engineers supplied the booster's imposing first stage. The Cape Canaveral Air Force Station provided launch facilities. And a small roster of students under the University of Iowa's James Van Allen supplied Explorer's key science instrument—a cosmic radiation detector. The device became the first in a series that ultimately revealed the presence of trapped radiation "belts" orbiting our planet. Over a year after Sputnik 1, in October 1958, scientific satellites came under the purview of the newly created National Aeronautics and Space Administration and its pledge to use space peacefully.

Earth orbit had been conquered. Reaching the moon (or at least the vicinity of it) represented the next tangible goal, and both superpowers began pivoting their space efforts toward lunar probes. The Jet Propulsion Laboratory transitioned from military projects to solar system exploration. John Casani loved his work. Father Jack approved. "I think he took great delight in the fact that my interests were in something other than business!"

Settling in at the Italian restaurant, a much older John Casani adjusted his sport coat to hang more comfortably. Took a quick slurp of water to lube the throat. After finishing Codorec, he transferred to a disjointed new venture called Pioneer, which was caught in bureaucratic consolidation from multiple military programs down to one civilian space agency. All *Pioneers* were not created equal. Their first couple launches occurred under the air force, using air force missile boosters. But Pioneer's next two went by way of the army and atop Junos, with JPL managing a payload that looked and operated differently from that of the air force.

All *Pioneers* flew instruments involving the University of Iowa and James Van Allen. He justified the inclusion in a characteristically simple way: "After we discovered the radiation belts, I proposed we send missions past the moon to see if it was magnetized." Unlike today's ponderous approval processes, Van Allen was able to green-light his experiments by submitting a glorified memo to a science board. It was forwarded to responsible parties at the air force and army, who were all but begging for science payloads to justify launching into space.

"A buncha little different pieces" is how John Casani described the army's *Pioneer 3* payload upon greeting it for the first time. "There was a little radio transmitter. There was a battery pack. There was something that controlled the power, and there was some relays in there, and there was two instruments." Others at JPL had scratch-built the radio system and batteries. Casani's role was to integrate, test, and deliver a finished payload. So he had to mate Iowa's instrument, a rudimentary camera, and the various supporting electronics with their custom-machined mounts. Connect everything to the batteries and radio. Test. Wedge it all inside a cone-shaped fiberglass body small enough to hold in one hand. Then test again.

Imagine tackling such a project—a new, expensive, self-contained, miniature automatic science lab upon which people's careers are depending. Based on the engineering drawings, every component should fit together and work together. But the first week of integration might prove that certain mounting holes don't line up, or an electronic part needs replacing, or other assemblies don't fit after all. It's not a model kit. There's no picture on a box to 100 percent show what it'll look like when complete. Indeed, there *is no box*.

"We had to package all this stuff," explained Casani of *Pioneer*'s multitudinous elements, "and make sure it would fit to the upper stage of the booster stage, and so that was my job! To get it all integrated, tested, go through environmental testing. Make sure that it could take the temperature and the vibration and what have you." When finished the thing weighed about thirteen pounds— less than most bowling balls. Gently Casani cradled it as if it were an infant.

Pioneer 3 then headed for the next waypoint on its journey—America's heartland, the state of Iowa. Casani's baby passed into the hands of James Van Allen, "'cause he was gonna calibrate it after it was all put together. And make sure that he could, you know, properly interpret what the spacecraft was actually seeing with the signals that were coming back to the ground." A custom shipping container had been made for the gilded plumb bob, but Casani didn't quite know how to handle the logistics. This called for visiting his boss, Gene.

"Okay, I've got this thing," Casani started, referring to the *Pioneer*'s ungainly shipping container. Its form factor resembled that of a living-room ottoman. "I got to get it on an airplane."

Gene just stood there, kind of indifferent, monitoring where this went. Casani babbled on about every last stress point. "I don't like the idea of putting this in the baggage compartment. I don't know how it would be handled, or anything about it." He was terrified to let the probe out of his sight.

"Don't—don't even think about that," Gene jumped in.

"What are you talkin' about?"

"Just buy another ticket. Strap it in a chair."

Casani froze as the simple brilliance of this soaked into his spongy brain. As he later recalled, "And the idea that I would be authorized to buy another ticket to put this box on the seat next to me and strap it in, you know, ah, was, uh, *wow*!"

He told his boss, "Okay!" Then laughed. "That's a good idea. Why didn't I think of that?"

Pioneer 3 went up in early December 1958. Unfortunately its booster shut down a few seconds early—enough to kill the lunar flyby. One day later, *Pioneer 3* rebounded earthward, but Van Allen actually welcomed it. The probe's return plunge blazed a different path through the radiation belts, sending data the entire time, and further refined our understanding of how the belts are shaped. Cheered Van Allen, "From our point of view, the flight was much more valuable than it would have been if it had flown to the moon!"

Late 1958 marked a time of transition. Only days before *Pioneer 3* flew, U.S. president Dwight Eisenhower signed an executive order transferring JPL from army control to NASA. Casani and colleagues welcomed such news as the Lab would be permanently shifted away from military endeavors. What a seismic move. Explained Lab director Bill Pickering, "JPL argued for, and received, a charter to develop the deep space missions. As a personal aside, I was delighted to hold a contract that said in essence, 'Go out and explore the depths of the solar system.'"

Casani himself experienced a transition of his own, away from bachelorhood. He married in 1959. Marie was ten years his senior and came to the relationship with three children already. "Everybody was telling me, 'It's not a good idea,'" he remembered. "I was getting all kinds of advice like that. But she was a lot of fun!" And Casani liked kids anyway. He moved away from brother Kane and into Marie's Altadena house. For her birthday, John bought his new wife a new Mustang.

Debugging its design along the way, the JPL team had gotten one *Pioneer* out the door. What had been learned from building it? How did *Pioneer 3*'s internal temperatures during flight compare to what was predicted? What was its actual center of gravity? How did the electronics hold up? "We were just starting to use transistors," pointed out Casani of the period in which these *Pioneers* were flying. He spoke highly of a colleague named Dean, "the only guy that really knew how to design circuits with transistors." Dean would come to him with laundry lists of questions: How much power do I have to work with? What functions need to happen at the same time? And Casani would help tease out every detail.

On a day-to-day basis while integrating *Pioneer 4*, Dean labored alongside a tech who usually handled the tools more than Casani himself did. They were supposed to be equals. But Casani had a better grasp on procedures and the overall big picture, and frequently redirected his colleague with statements such as, "Hey, we gotta do *this*." In a tactful yet candid way. And as this arrangement settled into normalcy, Casani began putting down his managerial roots.

"I never thought of it as management at the time," he pointed out. "We were just workin' together, you know?" But more and more coworkers began gravitating toward him for answers and guidance because he always had time to help. Why was John Casani such a person for others?

Mere days into 1959, Soviet news outlets announced their country's launch of the "cosmic rocket"—a bulbous amalgamation of metal and batteries that today is called *Luna 1*. It hurtled past the moon with only four thousand miles of separation and demonstrated a supposed lead in a supposed race through space.

Couldn't America speed things up for *Pioneer 4*? Laughed James Van Allen, "You wouldn't believe how much work goes into making a little box full of equipment like that. Sides just packed full of transistors and very sophisticated circuits!" He listed several indignities of testing. "Shake it on the shake machine. And put it in the vacuum, to be sure it doesn't spark over the high voltage." At some point in the process, a previously unknown weakness would sometimes unmask itself and ruin everyone's day. Van Allen tossed his hands open in a small gesture. "Then you try to figure out, *What do I do about that?* Well, that goes on and on for weeks!"

Eventually, like a U.S. Navy SEAL on graduation day, *Pioneer 4* emerged from testing after having survived every inhumanity thrown at it. Casani bought another round of plane tickets to Iowa for its final calibration. And from there things advanced to the Cape. *Pioneer 4* finally went up on March 3—two long months after *Luna 1*.

In addition to becoming the first U.S. probe to almost completely escape Earth's gravity, *Pioneer 4* passed within thirty-seven thousand miles of the

9. *From left*: Wernher von Braun, John Casani, and James Van Allen pose with *Pioneer 4* during final tests in early March 1959. Their circular ID badges indicate this picture was taken in Florida at a prep hangar near the launch pad. Courtesy NASA/JPL-Caltech.

lunar surface—although still too far away to trigger its camera. The fiberglass cone then spun off into a lazy orbit of the sun; its batteries slowly exhaled. Well, what about the original question? Did it determine whether the moon was magnetized? Per Van Allen, "No effect of the moon was detected."

America's hodgepodge of lunar probes morphed into an expansive project called Ranger. It intended to crash-land probes onto the moon while collecting all sorts of readings on the local environment. Both John and Kane Casani worked on it, although never side by side or even at the same time. Early in Ranger, John assumed a wide-ranging systems engineering role. Later, Kane received the focused task of rough-landing simple experiments on the surface.

If Ranger proved anything, it was that NASA and JPL weren't seeing eye to eye on how to run complex technology endeavors. Six Ranger flights in a row all went down the tubes before one finally succeeded in sending home pictures as it crashed. These doings commingled with two parallel efforts, one successful, to fly by Venus using similar hardware. All the while, Soviet engineers lobbed forth their own responses bound for the moon, Venus, and Mars.

Casani and his wife welcomed a baby son, John Charles, or "Charlie." Ranger failures painfully continued into the early sixties. Things got so bad that JPL administrators ended up in front of Congress to explain why all these multimillion-dollar flights kept going belly-up and embarrassing the hell out of everyone. The Lab *did* successfully manage to fly past Mars in the summer of 1965. And took notice as the Soviets managed to almost reach the surface of Venus—by way of an impressively evolving series of armored-up washtubs called Venera. This shadowy solar system we knew so little about only a decade prior? Beginning to lift the veil!

All the while, Casani inched his way forward and up. Come the late 1960s, he served on a doomed paper project called Voyager Mars. It aimed to set enormous landers onto the Red Planet and test for the presence of life. Most of its budget disappeared into overblown concept studies. Voyager Mars exposed the risks of thinking too big for planetary missions and failed to survive Congress.

Then, not at all suddenly, in 1967—as humans approached the frenzied peak of racing to land people on the moon—an otherwise diminutive NASA advisory board recommended from its low-tier backroom digs that the time had now come to start planning visits to Jupiter. "The next logical step, you see," orated James Van Allen, quite calmly, of his agreement with this approach. "We knew how to do it, and we *could* do it. And time is ripe to get on to this, to the outer planets."

Strictly speaking, anybody could've said that . . . and so what? How easy for some idealistic scientist to lean back in his battered leather swivel chair and sip cheap instant coffee while making such sweeping pronouncements! Sheesh. Well, had someone else called for such action, the resulting flow of history may have charted a different path altogether.

But James Alfred Van Allen was, by this time, practically in his own category, having leveraged a basic inquisitiveness and polite midwestern style to become a prime mover in the direction of American space exploration. He made its compass needle point in specific directions and stands today as one of the very first to give it purpose. A favorite topic continued to be cosmic radiation—extremely common particles reaching Earth from beyond our solar system. "There are a few per minute hitting your body every day, sittin' right there," he once told a visitor to his office. "You don't know it."

After the United States launched that first Explorer in 1958, Van Allen had defaulted onto the second, third, then the fourth satellite launches. And the University of Iowa's Department of Physics and Astronomy found itself pretty much a mainstay. It rode the various Pioneer iterations. Iowa gear went out on the first flights past Venus and Mars. "Workin' my way through the solar system," Van Allen said of his baseline strategy. He joined various advisory panels, collaborating with fellow space scientists to recommend a sensible order of planetary exploration because governments never had a great sense of what to do next.

No question, a certain huge planet with stripey clouds and a raging red storm "piqued my interest fairly early on," as Van Allen put it, due largely to "a famous paper by a couple of colleagues about the radio emissions from Jupiter." But the prospect of visiting this intriguing world would have to wait until technology, popular opinion, and the agreed-upon timeline made such a thing possible. He wanted to study Jupiter about as much as he did cosmic radiation. Or planetary magnetic fields. Or other fundamental properties of our solar system. Plasma waves, anyone?

Serious consideration of Jupiter entered congressional minds as early as March 1963, during multiple hearings before the "Committee on Science and Astronautics," as Congress termed it. Already at that time, President Kennedy's moon challenge dominated the typical person on the street's thoughts about space. If exploring it meant something else, uh, what might that be? Wasn't an important race to the moon underway?

But during his own presentation at these committee hearings, NASA staff scientist Homer Newell calmly folded planetary exploration into the mix. He tended to think more as Van Allen did. "Another topic of great interest is to continue to explore the origin of the solar system," Newell advocated to the congresspeople. "This means that we will want to investigate the planet Mercury, and the planets Jupiter and Saturn, in the course of our program."

The topic persisted into the periphery of more hearings, which occurred the following month. Additional NASA staffers outlined the possibility of stacking modified Saturn V booster rockets with a new upper stage running on nuclear propulsion. (Completely untested but whatever.) Some real heavyweight trips could then be flown, they pointed out, specifically referencing "Jupiter probe missions, which could be accomplished with this nuclear Saturn V."

These trips could be great. But they weren't going to happen any time soon—not with the lion's share of NASA money going toward the man-on-the-moon goal. Even previously science-oriented programs such as Ranger had been co-opted for support of Apollo's lunar landings. Van Allen picked up his own narrative: "Meanwhile, I undertook to take every advantage of going to any planet I could." He tried not to let the situation buckle his enthusiasm. He supported the moon race and all, but recognized it as the political program that it was.

One thing the U.S. government funds well is study contracts. All the while Voyager Mars simmered and bubbled and slowly boiled over, oodles of study contracts had been issued regarding outer-planet missions. The defense contractor Lockheed had already popped out one response, back in September 1964—a progress report on its investigations of a possible flight through the asteroid belt and on past Jupiter. One major goal: "measurement of Jupiter's environment." Hang on, that was it? Well, nothing had been sent that far out, and nobody knew what Jupiter was really like.

The report concluded such a trip to be realistic—while noting the diminished solar energy at Jovian distances and the resulting need for alternative methods of powering the ship. Instead of solar, Lockheed preferred a decently well-understood technology based on simple physics: When two dissimilar metals in contact are subjected to opposing temperature extremes, a small amount of electricity is generated. The "thermoelectric effect," it's called. Inefficient but stable. If bunches of these metallic, wishbone-shaped "thermocouples" were strung together, they could power a spacecraft all the way out at Jupiter. Lockheed's report contained sample line drawings of how the major parts needed might fit within a launch rocket. The parts included a large folding antenna for communication at such ridiculous distances but no insight on how the antenna would be unfolded.

So this was already possible way back in 1964? Sweet! Then why not throw some money at it? Push forward and go? To Jupiter! It was *huge*—minus the sun, Jupiter contains two-thirds of all material in our solar system. Best estimates put the size of its core alone as somewhere between ten and a hundred Earths. Let's visit that!

It was *weird*—bleeding into space about twice as much heat as it receives from the sun. This internal process, whatever it is, drives complex weather systems and layers upon layers of swirling clouds. With a few planet-size storms thrown in for good measure. Not to mention its bizarre Great Red Spot, maybe the longest-lasting storm that ever was. We're talking nineteen thousand miles across, with its own layer of clouds riding a good five miles higher in altitude than others. *So* mysterious. Let's measure that!

It was *fast*—spinning primordial gas and goop all the way around in less than ten hours. Earth's weather is driven primarily by the sun. But Jupiter's weather is pushed along by its own aggregated internal dynamics. Let's find out why!

Don't forget its plentiful company—with the Jovian system then known as being home to a dozen moons, any one of which could be just as crazy as Jupiter. Some might have water. Some might have life. Of course, none of them might have much of anything, but nobody knew. The whole planet—the whole neighborhood, for that matter—was all so poorly understood. This faraway and hostile system . . . why not go there next?

Recalled Van Allen, years later, "I thought, 'Gee, it was just the logical progression of the advancement of the field, was to go to Jupiter!'" But politically? No chance. Not at this point in the late 1960s, with the moon race about to climax.

Research still cranked away in the background. Multiple studies (and philosophical opinion papers) attempted to sort through what was truly known of the outer planets and break down what likely would involve reaching them. "We are expending a great deal of effort on only a small fraction of the solar system," scolded one paper from January 1967. It went on to assert, right up front in its abstract, that "scientific investigation of the solar system is shown to be much less difficult than is commonly believed." Speculation ensued about various outer-planet experiment types and weights, specifically regarding a certain huge place with a big red thing on it. "Jupiter merits particular attention," asserted the author, echoing a recent conclusion by the Space Science Board. Both had the same reasons as everyone for studying it: crazy weather, crazy radiation levels, and an even crazier magnetic field. The paper continued: "The planet may have retained much of its original material. This material could be indicative of the fundamental composition of the planetary system." Just look at the rich mission profile, it argued. Look at the easy use of gravity slingshots, enabling various flight paths through the Jovian system and close swing-bys past all those moons.

The paper, crafted by prominent Lockheed engineer Maxwell Hunter, pointed out the unresolved nature of whether gas giants such as Jupiter even had solid surfaces that could be landed on. But he did try to be specific where possible. Carry the uppermost booster stage all the way to Jupiter, advised Hunter, and you'll have fuel for braking into orbit. You'll need a larger antenna dish than what'll fit in the launch rocket, so use an inflatable type. One fascinating possibility came up in his writing that has yet to be realized more than fifty-eight years later: "For a large planet like Jupiter, the primary mission might be that of landing on its satellites."

When Hunter referenced the Space Science Board, he called attention to one of America's most influential entities—the National Academy of Sciences. This place has been around a while. It came to be during the nation's Civil War, in 1863, and functions to this day as a nonpartisan source of advice to the government on all matters involving science. It is a nonprofit and, despite its creation by an act of Congress, nongovernmental organization as well.

The Space Science Board itself, as might be imagined, was not part of the academy's original charter. Ongoing processes of discovery result in the creation of new scientific disciplines, resultantly expanding the fields in which the academy can proffer advice. For example, space. The 1957–58 period had witnessed a worldwide program of coordinated scientific research known as the International Geophysical Year. It started even before the first Sputniks and led to this new Space Science Board joining what the academy had already been doing for almost a hundred years. At its core, the board comprised twenty space scientists (with Van Allen on its roster). NASA operated a separate advisory group, the Lunar and Planetary Missions Board, which formed the same year as Max Hunter's report and functioned as something of a complement to the Space Science Board.

Two other publications from May 1967 explored various aspects of long-duration missions to the far solar system: How big of a spacecraft could be launched with the current boosters? How to power the different types? What to put on them? How to get from A to B? How to stay in contact over extreme distances? "A spacecraft weight of less than 1000 lbs. for 85 lbs. of experiments appears reasonable," mentioned one. Another section went so deep as to address the regulation of temperatures inside hypothetical spacecraft. An entirely separate document, produced for NASA Headquarters, anticipated the technical hazards of a small probe transmitting data to its mother ship while descending into Jupiter's atmosphere.

By the time these studies hit the street, a rare and attractive situation had been identified. Making their way through space, the four largest outer planets would—starting in the late 1970s—all briefly align on the same side of the sun. Close enough for a single ship to visit all four, using gravity to leapfrog from one to the next. Advocates called it the Grand Tour. A regal name for certain—yeah, great—but the project's undesigned spacecraft only had about ten years to get off the ground. And the planetary alignment offered no advantage for orbiting Jupiter.

These sorts of studies furthered the idea of outer-planet missions, demonstrated viability, and made for decent progress. But not the kind of progress equating to a launch date or startup funding or anything even close to some tangible plan. Visions of possible visitation dogged a certain physicist in Iowa. It ate at his brain. "A special appeal of Jupiter was its huge radiation belt," continued James Van Allen, seemingly unable to stop gushing about every cool thing the place had on offer. If any planet ever had a talent agent, Jupiter's was Van Allen. While discussing it, he'd randomly announce such things as, "And extraordinary atmospheric dynamics on a very large scale!" By gaw, didn't the at-large public see the value that he did in flying out there? Wouldn't Congress kick in just a few bucks? He kind of got into a funk about it. Dragging himself to one group/board/task force/etcetera meeting after another, James Van Allen felt himself starting to lose patience. The genteel midwestern approach wasn't cutting it. He shifted to *badgering*—his word.

At the next meeting, up popped Van Allen. "How about Jupiter? How about Saturn?" he'd say to the person next to him, to the room, to whomever. And at the meeting after that: "How about Jupiter? How about Saturn? How about the outer planets?" He made it a personal priority to get in everyone's face. To anyone listening—or for that matter, even if they weren't—he'd blab away: "How about the outer planets? They have much more interesting physics than Mars does."

Nobody escaped.

"Come on, guys, don't forget we haven't even come close to any of the outer planets yet!"

Finally one guy had enough. He was an important guy in a position to facilitate—the guy who chaired NASA's Lunar and Planetary Missions Board, which included Van Allen as its current agitator in chief. "Okay, Van," dared the chair, "why don't you develop your rationale and see what you can do with it?"

Van Allen stopped complaining and leaped into action. "So I got together a group of sort of kindred spirits, and we *developed these rationales*. And that was the basis for the *Pioneer 10* and *11* missions." The stated objective read simply, "To fly through the asteroid belt and reach the environment of Jupiter." It actually picked up funding. Van Allen had earned his wish. Official invitations to the scientific community went out on June 10, 1968—even though NASA HQ had not yet officially approved the mission. Some seventy-five Jupiter proposals came back, in competition for eleven open experiment slots aboard the unbuilt spacecraft. Iowa learned it got one of them on March 24, 1969, a month after formal NASA approval, and Van Allen would finally be flying past Jupiter.

"I was greatly relieved and delighted," he smiled.

6

Misdirection at the Brigadoon

We did it. We can report to Voyager: Come on through,
the rings are clear.

—A. Thomas Young, deputy director of Ames Research Center,
following *Pioneer 11*'s encounter with Saturn

At places such as JPL, the work doesn't always end when the workday ends, and what used to help bridge the gap between work and home was a bar on North Fair Oaks Avenue in Altadena called the Brigadoon. Frequently John Casani huddled there after-hours with colleagues. "And we would continue the design process, you know, over cocktail napkins and stuff like that."

Marriage hadn't panned out so well. The age difference, in particular, had proven to be huge. When in the company of Marie's friends, John—despite being a born socializer—had found it tough to maintain conversation with a whole group of people at least a decade older than himself. Same problem when she came along to get-togethers with his own friends. Ultimately, "we went our separate ways," John acknowledged. They'd owned a house in Altadena, so he found an apartment nearby and shared custody of the preschool-age Charlie. Marie kept the Mustang. John mostly focused on work.

One late Friday afternoon at the Brigadoon, a group of JPLers were standing near the door and about to leave. Someone asked John if he'd be up for getting together on Saturday. But that wouldn't work because he was already signed up for rodeo cowboy lessons that day. It required forty-five minutes of driving to the ranch. Overheard by the Brigadoon's new cocktail waitress, Lynn Seitz, that odd little exchange piqued her interest. Over the ensuing weeks, she kept an eye out for this man who apparently was advanced enough to work at secretive aerospace facilities yet sufficiently primal enough to learn the ways of a cowboy. How interesting.

Not long after, Casani returned to the Brigadoon with a young lady in tow. The couple shared drinks at the bar. A scotch on the rocks for Casani. He saw the owner and told him, "Sam, I think we'll have dinner here tonight." Sam delegated the task to Lynn, who scurried around for a bit.

Finally she approached Casani. "Sir, your place settings are ready."

He looked around but didn't see any table for two. "Okay, where?"

Lynn gestured to a lone place setting at the bar. "Well, there's one here." Then she pointed to another at the Brigadoon's farthest corner. "And . . . there's one over there."

In minor confusion, Casani reappraised what she'd done, then turned his attention back to Lynn. "She was so beautiful," he later marveled of the Brazilian-born woman. And something of a wiseacre to go so far as to configure geographically isolated place settings. That night marked the last date Casani had with anyone else. From then on, it was all Lynn, all the time, as the two became an instantly inseparable pair—not to mention a rather ideal one. She moved in with him even before exchanging vows. If any single person on Earth perfectly complemented John Richard Casani and possessed the wherewithal to balance his world, it was Lynn. "Before we got married, she told me that she wanted to have six kids," he laughed. "She had three right off the bat!"

All were boys. In fact, babies were coming so fast that the couple had to improvise. John removed the front passenger seat from their '63 Volkswagen Beetle and replaced it with a laundry basket. "And, uh, we carried three babies around that way. And as each one got a little bigger, they'd move from the basket to the back seat with my wife, and the new baby would go in the basket." The family operated that way for years. Older brothers were often tasked with carrying younger brothers and placing *them* in the basket. One of the Casani boys later remembered being told, "Put him in feet forward, because that is safer." Once all the kids were old enough, John reinstalled the passenger seat and used the Volkswagen for commuting to work.

After a moment of open-mouth silence, the inevitable question came: *John, was the laundry basket fastened to the floor?*

"Nope. Nope."

Together, John and Lynn would raise four boys, adopt a fifth, and travel the world on adventures one after another. They'd host memorable parties— some lasting more than twenty-four hours. And decades later, long after a fabled Jupiter orbiter had flown and given itself to science, John would hold Lynn's hand as she died.

10. John and Lynn on their wedding day, December 13, 1969. Years later, Lynn would advocate moving back the official year of their marriage so as to be more inclusive of John's son from his previous marriage. Courtesy the Casani Family.

11. Lynn and John at a social occasion in 1969. "You can tell from the facial expression that she wasn't buying much of what I was trying to explain to her," he commented of the photo. "She had a pretty good BS detector!" Courtesy the Casani Family.

The slow progression toward a Jupiter orbiter mission subtly churned in the background like a saltwater taffy machine. Even more study contracts. More reports. Some of them discussed the fields and particles kinds of experiments James Van Allen wanted to run: cosmic radiation, magnetism. They work best when swept through space, mounted on the outstretched arms of a gently spinning craft. It's akin to waving your hands around in a dark room to find what's there. To boot, this spinning approach stabilizes the spacecraft and helps reduce its complexity.

Many reports called attention to the unprecedented imaging opportunities offered by the vantage point of an orbiting spacecraft. The cameras will be right in Jupiter's back yard! But what camera takes good pictures when it's spinning in circles? Really, the ship needed to be gyroscopically stabilized across its three axes of motion. The proper exploration of Jupiter would thus seem to require two fundamentally different types of spacecraft.

A milestone analysis from September 1970 acknowledged the impending Pioneer Jupiter flights, offered nods to the hypothetical four-planet Grand Tour survey, then settled back to assess where the state of exploration would be, assuming both projects flew. "Altogether these missions constitute a launch endeavor of six flyby spacecraft to Jupiter in the 1970's," it summarized—factoring in the Grand Tour's then-projected use of four separate craft. What might carry the torch of all these flybys? "Clearly, now is the time to move ahead with advanced planning and analysis of more comprehensive Jupiter orbiter and atmospheric probe missions which, hopefully, will closely follow or perhaps even mesh with the Grand Tour."

The analysis then sequenced the incredibly vast and nebulous topic of Jupiter exploration into definable and manageable nuggets. At the topmost level sat broad "regimes" such as *atmosphere* and *interior. Atmosphere* broke down into *composition* and *dynamics,* which further cleaved into the identification of such objectives as *elements present* or *amounts of particulates.* At the lowest and most detailed level came *measurables,* involving actual evidence of those objectives. Supplemental figures and tables laid out different types of measurements to be made at the Jovian system. What kinds of instruments would do this work? How much might they be expected to weigh? What were the power requirements? Data rates?

Subsequent paragraphs found common ground with earlier studies, detailing the need for spinning *and* self-stabilized crafts. No way to combine those two requirements on one ship! The author speculated whether a modified *Pioneer* could fulfill the parameters that needed spinning. Additional language mentioned the possibility of launching two completely separate ships on one booster. A key recommendation: Send a detachable experiment package to drop into Jupiter's thick atmosphere.

"No single orbit would satisfy all science objective and instrument requirements," the analysis pointed out. Small tight orbits would be great for studying the planet itself. But the most revelatory information about Jupiter's moons would be gathered by varying the spacecraft's orbital parameters, in a somewhat irregular pattern, so as to fling the craft past the most intriguing moons. This would also help mitigate radiation exposure. Once every science objective—and safeguard—had been fleshed out into a detailed flight plan, no single orbit would likely be the same.

For anyone who read the report cover to cover, a robust Jupiter mission could be accomplished within the state of the art. No fantasy equipment or techniques required. No magic. It just needed a financial commitment.

Dainty *Pioneer 10* left Earth in March 1972, with its near twin following just over a year later. Both worked wonderfully, giving humans their first close-up Jovian look-see as *Pioneer 10* whipped past the planet going some eighty-two thousand miles an hour. *Pioneer 11* later darted past Saturn. Iowa had instruments aboard both. (According to Van Allen, a flight path had been identified that would've taken *Pioneer 11* on to Uranus. But concerns over the craft lasting through its Jupiter–Saturn trip, plus an unfavorable angle past the ringed planet, saw the idea "quietly abandoned.")

The ships were incredibly simple, spinning for stability, with all data transmitted home in real time. Iowa's instrument utilized two commands only: "power on" and "power off." Not long after *Pioneer 11*'s launch, its experimenters campaigned to send a third, flight-worthy test spacecraft on an altogether different outer-planet path to make its own unique observations. It'd been built, so why not fly it?

"Our case fell on deaf ears at NASA Headquarters," mourned Van Allen with twinges of disgust. "And the spare spacecraft now hangs in the main gallery of the National Air and Space Museum." He himself had perceived each *Pioneer* as something of a sacrificial lamb—going out to confront the harsh radioactive environment and potentially be killed by it. "Several of the experiments on *Pioneer*," he'd predicted in a 1971 memo, "will probably be disabled in real-time by charged-particle radiation." During the Jupiter encounters, both spacecraft tried to act on false commands generated by high radiation. *Pioneer 10*'s imaging system dealt with at least ten and consequently lost a few pictures. But neither craft was totally cooked. They continued operating and returning valuable readings. Future visitors could indeed be made to survive this wicked environment.

In one congratulatory letter to a Pioneer colleague, James Van Allen optimistically mentioned, "I hope that we have the opportunity to work together again in the future—hopefully on Pioneer Jupiter Orbiter." Maybe by saying it would the thing become just a touch more real.

Van Allen decided against sitting back to wait and see what might happen next. He sought to define that new step to the outer planets, to *reach* it, and he attacked from multiple fronts. He joined yet another panel—the Outer Planets Science Advisory Group. Not ringing a bell? This union of space science experts, a calico crew from multiple institutions, sought to define broad exploratory programs across multiple disciplines. They'd operate throughout the early 1970s. Van Allen ended up chairing it, along with the similarly named Outer Planets Science Working Group—itself a subset of the advisory group.

This subgroup audited already-planned experiments under the mindset that any science package headed for space should be as up-to-date as possible. They asked themselves: How have new discoveries impacted this experiment since it was originally proposed? Is what it wants to examine still relevant? What aspects should be updated or maybe deleted? After careful reviews, the working group would recommend specific changes.

From his overcluttered top-floor office inside the University of Iowa's Van Allen Hall (same guy, not his cousin), Van Allen explained his role in terms that might have sounded boastful if they weren't true. "And I was, you know, the leading advocate for missions to the outer planets. And I was made chairman of both those bodies. And I wrote the advocacy document for missions to Jupiter and Saturn. So I was in the front lines there, carrying the flag for the outer planets." He couldn't turn off what turned him on—those giant marbles of swirling gas and clouds. Both practically begging visitors to release their secrets. "I was mostly interested in getting to Jupiter, because Jupiter's a huge radio emitter and obviously a magnetized planet, had large radiation belts, and I was just dying to get there to investigate those!"

A bonfire's worth of mission design studies overlapped each other into the early and mid-1970s. One after another assessed the latest in everything from propulsion and navigation to reduced instrument sizes. A study from June 1972 focused on possibilities with a so-called space tug—fat tanks of fuel sending probes to distant planets. Other studies addressed questionably overambitious topics. What about a 1986 launch to Uranus and Neptune using an untested propulsion system? Or how about intercepting an asteroid in 1989? Maybe a retooled *Pioneer* to Saturn, continuing on to drop a probe into the clouds of Uranus? Just about everything seemed to be on the table.

But celestial mechanics have a way of intruding, no matter where people might *want* to go. An early February 1973 report for NASA HQ compared three upcoming "opportunities" for a Jupiter orbiter: launch in 1980, during late '81/early '82, or wait until '83. No question, the planetary alignment during the '81–'82 window offered the greatest payload capacity by far.

As 1973 played out, virtual buffets of mission combos were hashed through in one laboriously produced report after another: A 1980 launch to Jupiter and Uranus. An '81 Saturn-only trip. An '82 Saturn–Uranus. All merely fly-bys. All written in dry, tax-code prose able to render someone asleep faster than all-you-can-eat pasta. But every one of those options could actually be flown! All had calculations. Weight budgets for fuel, the heat shield, the science. What made the most sense to do?

Sort of but not really paying attention, John Casani basically kept his nose to his work and gradually moved up the chain at JPL. He took on a major role for an ambitious Earth-to-Venus-to-Mercury flight called *Mariner 10*. "I was the spacecraft system manager for it," he indicated, shouldering a decidedly higher-level position on this one. The ship flew past Venus in February 1974 and used that planet's gravity to angle on toward Mercury. Just as planned. The flight path sent it near our innermost planet during late March, tracing a slow arc back around for a second Mercury pass to happen in late September. Before that occurred, though, some news came through the *Los Angeles Times* on Sunday, May 19, 1974:

> John R. Casani, spacecraft system manager for the Mariner 10 Venus-Mercury project, has been named manager of Jet Propulsion Laboratory's Guidance and Control Division.
>
> The 41-year-old Casani succeeds Garth E. Sweetnam, who died last March. Casani has held managerial positions for more than a decade

in the design and development of spacecraft which have explored Mars, Venus and Mercury.

In his new post, he will lead JPL's development of spacecraft power, guidance, control, and celestial sensor subsystems.

Somewhat poetically, John Casani had just taken over from the guy who would've originally interviewed him, likely *not have hired him*, but had happened to be out of town that day John first visited JPL. In their previous season, Casani's beloved Penn Quakers had lost to both Harvard and Yale on the gridiron. But he himself was riding a lucky winning streak of professional advancement.

Casani detailed how the promotion occurred. There was no review panel or anything even remotely formal. His previous boss, Gene, simply came over to talk privately but fumbled the lead-in. "You know what? Yer gonna, we're gonna . . . you're gonna go take over a division."

This was big. A JPL division is the root organizational level of a specific discipline. Examples are instrumentation and communications. It would involve managing many, many people.

Looking at his boss, Casani said, "What?" And just stood there. "I didn't think I was ready for that."

"Nah, you're ready."

Reliving the exchange many years later, Casani chuckled. "So they were thinkin' about me. They were watchin' me, and they saw something. I don't know what the hell it was. I've never thought of myself as a good manager. I thought I was more of a leader. I could get people to do things, you know, and get people to work together and get people to feel committed to what they were doin'. And maybe it was that?"

Maybe it was.

7

Shortsighted and Penny Foolish

It is apparent that the reconnaissance mode of study of
the inner solar system is completed.

—Representative Don Fuqua (D-FL) during his pitch
for NASA's fiscal year 1978 budget

When the Grand Tour scored funding in May 1972, JPL put its biggest-ever bun in the oven. Two unmatched, full-featured, and headline-making spacecraft called *Voyager* began final design and construction. Launch goal: late 1977. The Lab would manage the project from start to finish. Officially, the money went only as far as Saturn. Unofficially, the design team wanted Neptune and worked to preclude any limitations that might prohibit them from flying on to Uranus at least, should funding be extended. As grand as it was, the four-planet Grand Tour did not include Jupiter orbiters or little drop probes into the clouds of Uranus or other such extras.

Uranus popped up during yet another workshop—at Ames Research Center in May 1974—only days after the news article ran about John Casani's promotion. Unlike other workshops, which tended to microscopically labor over science objectives, this one brought together contractors and JPLers and straight-up NASA employees, plus researchers from all over the States. General Electric had people there. Langley Research Center. The Kitt Peak National Observatory. Hughes Aircraft. Stanford University. TRW Systems. As host, Ames certainly had its own fair share of attendees. Since a chunk of the subject matter dealt with atmospheric entry probes—an Ames specialty—it only made sense.

He didn't attend this conference, but John Casani understood the developing schism between Ames and JPL as relating to outer-planet exploration. Both facilities competed for NASA contracts to pay the bills. When choosing an overall mission profile, you first had to choose your baseline platform, and there were two: the spinning Pioneer from Ames and JPL's stabilized Mariner.

As of late, the Ames crowd had been gathering praise in real time for *Pioneer 10*'s flyby, which utterly validated the whole idea of a spinner. It took the world to Jupiter! The design could be straightforwardly upgraded for orbital operations there. "A simple spinning spacecraft with some fields and particle instruments on it—*only. And* an atmospheric probe," noted Casani. Made total sense to him. After detaching from the orbiter to commence its one-way trip, the probe would need to spin for stability anyway. Basing a Jupiter orbiter on Pioneer signified a potentially easier trip through Congress. If it ever got to Congress!

But the big trade-off was photography. *Pioneer 10* and *11* sent back images of Jupiter and Saturn that barely exceeded the quality of those available on Earth. Strictly speaking, the ships didn't even have cameras. Instead, each *Pioneer* used a specialized telescope mounted at the outer edge of its circular antenna dish. The telescope peeked through a small notch. As the ship spun during its voyage, the telescope scanned narrow swaths of the local surroundings— some three-hundredths of a degree each time. It measured the *properties* of light: reflectivity, luminance, scattering.

Lucky for us visual creatures, this data could also be processed into recognizable imagery by essentially laying the scan strips alongside one another. But what work! A single *Pioneer* image required detailed advance planning and could take thirty minutes for the ship to build. Multiple automated (and time-critical) commands had to precisely advance the telescope's position as the spacecraft flew by.

Next, factor in wonkiness. High radiation levels briefly bullied *Pioneer 11*'s telescope into randomly skipping around. This led to a challenging process of reassembling the strips. Occasionally, data from an individual swipe never came home at all; an otherwise great shot of Jupiter might suffer from an empty strip of black screaming right down the middle, as if that part had been censored. Or the sensitivity of the telescope would vary, generating light and dark bands across a reconstructed image that had to be leveled out.

But even with things working properly, spatial pitfalls could never be avoided. Two planetary features from the same longitude, for example, might appear skewed in the final photo because they'd been captured over time. Also, otherwise intact raw imagery suffered from being initially displayed as warped—as if the planet was made of Silly Putty and had been stretched by giant hands. The geometry had to be fixed before things looked right. No picture taken in these conditions would ever be as good as one from a real photographic sensor mounted on a platform stabilized across three axes of motion. A platform such as Mariner's, which had one JPL guy at the Ames conference scrambling in defense.

"I wanted to raise some points where I think the Mariner mission has really not been well understood by this group," he mentioned to a few colleagues. Didn't people want great photos? Won't happen with a camera that's whipping around at nearly five RPM, right? And hey, what about all of you wanting to point spectrometers at Jupiter? Those would need to be on a fixed platform too. And don't forget the power advantage! Mariner's antenna design offered extra gain over the stock Pioneer, meaning higher data rates. Everything comes home faster. You scientists are going to be happier. Your results get out to the world more rapidly. Don't you want that?

The guy also had something to say to those who'd been overheard spreading bad info about pairing a Mariner with a spinning descent probe. "Now, there were some remarks, too, that puzzled me about whether or not we knew we could deploy a spinner from a three-axis stabilized spacecraft. Certainly we can! There are a couple of very good designs," he fired back.

As the conference formally got underway, which exact planet they should strive for became something of a hot potato. *Pioneer 10* had screamed past Jupiter only six months prior. Nobody knew how tough it might be to send a probe *into* those swirling clouds. "My understanding is that the preliminary results indicate that the entry problem there is not quite so severe as we once thought it was," announced the workshop chair, as he brought the meeting to order. "I understand you are going to be looking at probes for other missions to the outer planets."

Next came the introduction of a key figure who held in his hands the future of everyone sitting in that room. That man was Dan Herman of NASA Headquarters, who dwelled specifically within its Office of Space Science and Applications. His key job description involved putting science experiments into space. (And according to his introduction that day, he'd parked illegally in order to arrive on time.)

Herman kicked off his comments by describing what HQ *had* been planning, up until recently, and what'd already been pitched to Congress. A nearby projector screen lit up with slides depicting a typed grid of mission profiles. The top read OCTOBER PLAN, with calendar years ticking away to the right from 1979 through '89. "Currently, this plan is in the process of being changed, because our thinking with respect to the outer planet probe missions has changed." Herman referenced one of the advisory groups Van Allen chaired and its recommendation of three mission finalists—all of which, perhaps not surprisingly, used the guts of Pioneer:

> 1979: a Jupiter and Uranus flyby, with a descent probe into Uranus
> 1980: a Saturn flyby
> 1981: a Saturn and Uranus flyby, with another descent probe into Uranus

Supposing the first two missions went as intended, the 1981 outing would instead send its probe into Titan, that fascinating moon of Saturn. Titan possessed an atmosphere rich in the complex organic compound methane. It's common enough on Earth but not so much on other worlds. "Why is there any methane left today? How is it resupplied?" a report on Titan would later ask. "We have no answers to these simple, basic questions."

Dan Herman continued with his deconstruction of the October Plan. "The scenario had a couple of weaknesses in it," he conceded to the group. "The major one of which was exposed at the Titan workshop held here at Ames about a year or so ago." That gathering had questioned the prevailing strategy of using one universal entry probe design no matter where the destination. People didn't like that anymore; unique worlds demanded unique designs and unique science. Life might dwell beneath Titan's heavy haze, demanding a whole regimen of quarantine restrictions that'd be completely unnecessary elsewhere. Everyone needed to consider the *specificity* of atmospheric descent probes. One for Jupiter hadn't been on the table because, until Pioneer, nobody had enough data to even design a testing setup for such a probe.

"So, for several reasons, our thinking has changed," continued Dan Herman. The nearby projector clacked to a new slide—a much busier grid this time, showcasing multiple families of potential missions running year after year into the early nineties. His OUTER PLANETS section alone comprised nine distinct and fairly reasonable options. The INNER PLANETS section contained, in hindsight, several overly farfetched nonstarters, including the VENUS LARGE LANDER penciled for dual launches in 1989.

Herman took a minute to describe how two upcoming Mars Viking orbiter/

lander missions were suffering cost overruns. The situation rippled through the entire bottom line of American planetary exploration. "Since our overall budget does not increase, funds for planning for new missions is from the *same* funding that has to accommodate overruns. We, therefore, had to alter our thinking and decide which missions we wanted to do as scheduled, and which missions would have to be deferred."

One subpanel meeting that week would conclude that Uranus, based on models and predictions, offered a relatively forgiving atmosphere. "The logical first choice for outer planet entry missions," came its recommendation.

"The total heat load," a colleague echoed, "is considerably less than what it would be for Jupiter!"

A Uranus probe did remain on the list—switching platforms from Pioneer to Mariner, "which we want to launch in 1979," outlined Dan Herman, as he continued with HQ's latest thinking. "As far as a Jupiter entry probe is concerned, we are discussing a cooperative program with ESRO at the present time, using *Pioneer H* to do an orbiter mission in the 1980 opportunity." The European Space Research Organization had been looking to collaborate with NASA on space science projects; maybe this one would pop.

Herman offered more details on the Pioneer-based Jupiter scenario and even had concept art to show off: Against the blackness of space, a ship looking very similar to *Pioneer 10* approached Jupiter, with one thin oval tracing its flight path about the planet. The descent probe followed a separate dashed line heading straight in for impact. "The Probe would be released before orbit capture, and the spacecraft would serve as a relay for the Probe during entry. Then the spacecraft would be captured, and would achieve a highly elliptical orbit about the planet." *Pioneer H* referred to the extra spacecraft Van Allen had lamented not using.

After Herman relinquished the podium, the conference proper got underway, and everyone broke off into groups. The viability of Jupiter no longer seemed up in the air; it had plenty of fans. In counterargument, "we know so little about Uranus," groused one attendee. Reaching it would be troublesome because the planet's definitive location in space hadn't yet been locked down. Improved approximations would have to be made, paid for with money that didn't yet exist.

Uncorked debate soon raged. They got into it with the spinning versus stabilized approaches. One guy resurrected the alleged problems of releasing a spinning probe off the stable Mariner. "There are a lot of things that have not been investigated," he fretted.

Arguments flared about how to proceed. They argued about whether the antenna on the mother ship would have to move as it received signals from a descent probe. They argued about probe heat shields—about how high the entry temperatures would build up on the various types. "All we have looked at are a small family of blunt cones and Apollo shapes," complained one guy. "Stability, of course, is an important problem! We want to know what orientation the probe is in at all times."

"Put the question the other way around," barked someone else. "If you could optimize the heat shield design to go to Jupiter, do so. And then ask yourself: What science could you take along with that?" Best case, any Jupiter probe would experience twenty seconds of life in the upper atmosphere before dying. That's it. Building something takes years. How many years are you willing to put in to have twenty seconds at Jupiter?

They also argued about whether stuff would last for a seven-year journey through space. "Most of our instruments are ready to fly, but they are not necessarily ready to fly all the way to Uranus," noted someone from Ames. "It is going to take a while for us to be sure that after seven years of sitting around on a spacecraft, or on the shelf, these things will operate in a way in which we can understand them."

Someone else brought up the idea of a probe before *the* probe. Similar to a junior precursor. The Fisher-Price® My First Descent Probe. "Throw it off of some vehicle that happens to be flying by there," said the guy, referencing Jupiter. "And not expect too damn much of it. Just use it for a learning experience." He asked for comments on this "piggyback experiment," to use his term, and certainly got them.

"With a probe you have a fifty-fifty chance of getting nothing," rebutted an Ames guy. Wouldn't it make more sense to put this pre-probe money and effort into testing?

"If it fails you *will* get something!" shot back a different Ames guy, in a spontaneous Ames versus Ames cage match. "You will know that your design was inadequate!"

As the days ticked by and the event started winding down, a longtime staffer at NASA HQ took his turn. "If I may, I'm going to deviate a little bit from the chairman's admonition to 'stick to Uranus.'" Instead, Jupiter really seemed like the place to visit first. Wasn't this what so many already wanted—to study that crazy environment of magnetism and radiation? Without any "will it last for seven years to Uranus" business?

Dan Herman's prodigious option charts had listed two identical 1979 launches to fly by Jupiter and Uranus. Next would come that European collaboration to orbit Jupiter using the leftover *Pioneer*. Hopefully in 1980. Two Mariner-based Jupiter orbiters would then go during that golden 1981 heavy-lift launch window. Followed by a couple of relatively stripped-down, Pioneer-based Jupiter entry probes, happening years down the line in 1984. "This decision was made with the advice of the scientific community," he noted of the ordering. Uranus before Jupiter. "Not because it ranked below the other planets in terms of science interest, but on the basis of when it was estimated that we'd have the technological capability to do it."

Well, things had now changed because of *Pioneer 10*. It reported more favorable Jovian conditions than previously reckoned. A descent probe could totally survive the heat of reentry; they could build it. All in all, dainty *Pioneer 10*, according to this man from NASA Headquarters, "in some respects opened a Pandora's Box. Which should be opened. There is no complaint about that. But undoubtedly we are going to get pressure to bring a Jupiter-probe mission off sooner." To the group he posed a hypothetical: Knowing what we do, can the Saturn/Uranus probe concept be used at Jupiter? "If we can, then I am sure many people will want to do a Jupiter probe mission sooner."

The conference drew to a close. "We had contractors talking to each other," praised one of the lead speakers in his summation. "So we have had contractors and we have had Headquarters people and Center people and scientists all communicating with each other. To me, the whole thing has been very much worthwhile."

Momentum slowly built on some form of Jupiter orbiter with an accompanying descent probe. An early 1975 study, incorporating European space scientists, demonstrated international support for a mission of this caliber. But the dueling platforms had yet to complete their showdown. Advocates took sides. You could argue for Ames Center and its simplified Pioneer. Good for fields and particles! Cosmic radiation! Magnetic fields! Or you could stand with JPL's stabilized Mariner. Best for spectroscopy! Best for photos! More precise targeting and tracking of the probe!

Land's sakes, which one should they go with?

That September, two nondescript, backroom JPL engineers unleashed a paradigm-changing idea at a conference in St. Louis. It dwelled within their "reference mission concept," based on the stabilized Mariner, and included a Jovian descent probe. But they also included something else—in their words, "a daughter satellite, dedicated to particle and fields measurements, ejected into an independent orbit about Jupiter." Essentially, the men were proposing to fly both designs at the same time.

That might not have been the first domino to fall against the proven track record of a totally spinning ship. But it made for a big one. A flagship Jupiter orbiter, based on the Mariner platform, coalesced into JPL's "marketing plan," as some called it, around the tail end of 1975. A major next step would naturally be the convening of yet another working group, or board, or committee, to determine the major science objectives and payload requirements.

Brand-new JPL director Bruce Murray took office on April 1, 1976, and his proclamations that day were no joke. Voyager's project manager, in the thick of it and still aiming to leave Earth in two years' time, got pulled off the field. "For multiple reasons," the guy was told. "Budgets are in flux. Less money for planets. Shuttle costs are killing all of us. JPL must diversify," explained Director Murray. Voyager's dedicated, invested, and highly capable project manager would now be overseeing a new effort on renewable energy. And into the number 1 Voyager seat went . . . John Casani, who'd been running the Guidance and Control Division for little more than a year.

He didn't come in cold; already guidance and control had its hands full supporting Voyager. As Casani explained of his ex-division, "It provided the control system, for controlling the attitude of the spacecraft. *And* the power subsystem." Guidance and control was deep in the loop. "So I was very, very close to the Voyager project."

He was also close with Voyager's outgoing project manager. "I really shouldn't speak for him, because he's never said this," Casani remarked about the man he replaced—a veteran of the Ranger debacles who stoically shuffled over to the new energy effort and tried hard not to look back. "He was probably the right guy to do it, ah, but I'm sure it really tore at his heartstrings to move off." They overlapped by three months so Casani could get the hang of things.

Despite his busy new role, Casani at one point found himself up at Ames

12. This 1970s artist's sketch depicts the J-O-P concept as it was then understood. Dominating the craft is a gigantic folding antenna dish, required for long-distance communication at Jupiter. Nuclear power sources extend on boom arms. The probe has just separated from the orbiter, falling away. A small probe relay antenna is hunkered behind the main antenna. Courtesy NASA/JPL-Caltech.

Center for a workshop on science requirements and options for the Jupiter orbiter. Guess who headed up the workshop? Yeah, Van Allen again. Casani felt out of place. "I don't know why I was up there for this thing, 'cause it was before the launch of *Voyager*. And the two, the imaging community and the fields and particle community, were arguing" about the now-classic catch-22 of "how to add a camera to the mission without screwing up the fields and particle science." JPL had been working on some ideas, but none looked promising.

Suddenly Van Allen wanted to speak. "I was there when he said it," Casani recalled. "He says, 'Well, why don't we make it a *despun* spacecraft? Part of it's spinning, and part of it's not spinning?'"

Everyone froze. Someone asked, "What are you talking about?"

"Well, Hughes does that," Van Allen told them, referencing California's mega-ogre aerospace contractor. He said it wasn't a new idea at all. True, Hughes had been using dual-spin technology since the 1960s on their Syncom satellites. Each one depended on a spin-stabilized kick stage to reach high orbit. The cylindrical body used wraparound solar cells for constant sunlight from the slow rotation. The opposite end of a Syncom remained locked in orientation for precise antenna pointing or whatever its mission required. Hughes called it the Gyrostat System. "A fully stabilized platform, while maintaining features of well-proven spinner technology." So proclaimed a 1967 internal company summary. "The unique feature of Gyrostat is that the platform is *virtually unlimited in size.*"

Was this for real? The workshop attendees all looked around at one another. Van Allen shrugged. "Hughes does it all the time."

The room erupted with compromise. "Yeah!"

At least, that's what Casani recalled. He wrapped the story by saying that Van Allen "asked if that might work. I said we'd take a look at it. That was the germ of the idea."

Dual spin could be the aspirin for so many headaches. It would unlock the use of science experiments from both categories. Anything could go. From this latest workshop, Van Allen's group delivered to NASA management a proposed roster of investigations and mission requirements—based on a dual-spin Jupiter orbiter with accompanying descent probe into the planet's atmosphere.

Uncertainties of mission mode and science types now reconciled, the secondary points began to jell. In January 1982 the whole ball of wax would leave the belly of an orbiting space shuttle and light up four stages of auxiliary rockets to embark on a direct, thousand-day road trip to Jupiter. This "J-O-P," as people started calling it, would mark the very first use of a shuttle for an interplanetary launch. Regardless, by 1982 the shuttle would no doubt have dozens of missions under its belt with every single bug worked out. Should be good to go! Fifty-seven days ahead of reaching Jupiter, J-O-P would release its tagalong plumb bob into the depths of the host planet's complicated atmosphere. The probe would report continuously until crushed by the intense pressures within those swirling clouds of this night-sky jewel.

Next, J-O-P would fire its massive braking rocket to slow down and be captured into Jovian orbit. Its framework mission profile ensured a dynamic tour of the busy Jovian system. The orbiting laboratory would use gravity from its host planet and moons to fly at least eleven evolving ellipses that would gradually send it past each of the major satellites over the course of twenty months. In this way, J-O-P would visit the entire local system.

The course promised to gather between ten and a hundred times more detail than what was anticipated for either *Voyager* flyby. One publication called J-O-P "a scientific descendant" of Voyager. All in all, the new mission promised to generate many firsts. First planetary flight from the shuttle. First orbiter at an outer planet. First entry probe into the atmosphere of an outer planet. Hell yeah!

One thing stuck in the back of John Casani's mind. The dual-spin satellites Hughes made went only as high as Earth orbit. All they did was relay phone calls and teletype messages. But what about operating complex science instruments all the way out at Jupiter? At that distance? At those radiation levels?

Casani sighed. "Nobody had really looked yet at what it was gonna take to despin a spacecraft."

Change had been occurring on multiple fronts in Casani's life. He and Lynn and the boys had recently vacated their undersized Altadena post-and-beam house for an expansive, five-thousand-plus-square-foot mini-mansion on South Orange Grove Boulevard in West Pasadena. It was old enough to have a porte cochere near the front entrance and a carriage house in the back. What an upgrade from Casani's post-divorce apartment—a place not only awkwardly small but also an unfortunate hangout for neighborhood skunks, which gained entrance through the cat door. Lynn wasn't sure what to make of the place when she moved in. The landlady—an ancient Russian woman, survivor of the revolution—would push the skunks around with a broom until minimally out of her way. John and Lynn adored the Russian woman but not the crazy skunks, and they'd sworn to flee as soon as the right place came around. Their next couple houses weren't quite the ticket either. But a Casani family forever home eventually materialized on South Orange Grove.

The house was fantastic, only a few miles south of the famous Rose Bowl football stadium, and the family enjoyed its suburban coziness. As the last week of December approached, a neighbor asked John and Lynn what their strategy might be for the upcoming Tournament of Roses Parade—a massively attended, jam-packed affair broadcast live to the world. It occurred on the morning of New Year's Day, prior to the even more massive Rose Bowl college football game played that afternoon at the stadium.

"We didn't realize that we had moved into a place that was *on* the parade route," John later explained. He thanked his neighbor for visiting and requested any kind of advice.

"There's gonna be hundreds of people on the street," warned the neighbor. "They always, you know, park on your lawn and everything else. The rest of us all put fences up to keep people off." With strong intonation, he recommended the Casanis act similarly. The family followed through by temporarily surrounding their property with tall and imposing chain-link fencing to hopefully ward off trouble. But recognizing the social opportunities, the Casanis also invited several friends of the boys, plus their parents, and staged a teensy New Year's party. Out in the streets, music bands came and went. Somebody even showed up with a live elephant and was giving people rides.

New Year gatherings would become a Casani tradition. "It got to be a pretty good-sized party! Every year it got bigger."

Something else also happens every year: Congress reviews and debates future NASA money in an excruciatingly drawn-out way. It's almost worse than watching rocks weather. Anybody who ever considered the Soviet Union a model bureaucracy need only have attended one of the twenty-two NASA budget hearings held between September 1976 and February 1977. Members of the U.S. Congress conferred with professionals from every corner of the aerospace industry. These people's livelihoods would be affected by the final dollar allocations: people from NASA field centers and key contractors and headquarters itself. People from the U.S. Air Force to the European Space Agency to the outlying scientific and industrial communities. The aggregate results of these discussions formed NASA's budget request.

But it was only a *request*. People ask for plenty of things they never get. Lord Humungus in *The Road Warrior* asked those refinery people to "just walk away," but that didn't happen, now did it? Same exact thing here. Nothing in NASA's proposed budget enjoyed any kind of guarantee just because it showed up on the initial version. Before *any* new money became available, detailed funding breakdowns had to survive independent gauntlets in both houses of Congress. These were almost torture-level reviews, real medieval, going line by line by line. Anything still breathing next endured a presidential washing machine of revisions, though nothing of substance typically changed. The final decisions could then become law and disbursement checks written.

On February 24, 1977, numerous public bills were introduced on the House floor. One of them, sponsored by a Texas representative, carried this full official title: "A bill to authorize appropriations to the National Aeronautics and Space Administration for research and development, construction of facilities, and research and program management, and for other purposes." Say hello to House Resolution 4088—NASA's entire budget for fiscal year 1978. It requested some $4 billion for everything from space shuttle facility construction (the biggest line item by a mile) to modifications on a chilled water system at the Lewis Research Center ($860,000) to initial start-up costs on a certain "Jupiter Orbiter with Probe." Seven months remained in which to agree on every last thing 4088 contained before the new fiscal year took effect on October 1.

NASA's wants came up for House floor discussion on March 17 and began in a congressionally traditional way. "The gentleman from Florida will be recognized for thirty minutes," proclaimed the House chairman. He nodded to the chair of the Subcommittee on Space Science and Applications, who approached the mic. "The budget for fiscal year 1978, which we are recommending today, is a modest increase over last year's of 9.7 percent, which will only hold the line on the buying power of NASA. Even under these circumstances, NASA has been able to propose sound programs for the remainder of this decade."

He saluted the man who originally introduced the bill. Lauded his colleagues. Mentioned that his parent committee generally liked what it had here but wanted fourteen tweaks to the funding allocations. Plus three other changes. Overall, it wanted to tack on about $19 million more than the original submission. So as far as congressional things tend to go, they were close.

By an order of magnitude would money continue flowing toward the shuttle—over $1.3 billion worth in this current round, as requested by NASA itself. "Your committee reviewed the progress of the Space Shuttle program and found it to be within cost and on schedule," noted the gentleman from Florida. The funding request would be preserved.

"However," he continued, in what might've initially sounded like bad news, "the committee observes that a highly complex program with a demanding schedule such as the Shuttle could effectively use increased funding for both the design, development, test, and evaluation phase, *and* initial Shuttle Orbiter production phase of the program." Wait, was that thing getting a bonus? Yes. The House wanted to give the shuttle more money, introducing a proposed brand-new section to the bill that added $95 million atop the existing request—but only in the event that money somehow did not get spent elsewhere.

As his thirty-minute timer unforgivingly ticked down, the rep spent a few minutes apiece on major budget aspects, including facility maintenance and energy technology. A quick discussion of lunar and planetary efforts came last, with a particular something coming dead last.

"The next logical step in our exploration of the outer planets is an orbiter and probe mission to Jupiter," he recited. "A new start request this year." In educated, impressively detailed, and persuasive terms, the rep spoke about the significance of the Jovian system, the scientific community's interest, and the importance of exploring it. He knew about J-O-P's general mission and flight profiles. He knew about the proposed experiments currently under review. He knew about Jupiter's atmospheric storms. He even knew about theorized properties of the Galilean satellites. In sum, he *knew what was not known* about the Jovian system.

"This extremely rewarding mission has been enthusiastically supported by every science advisory group that has examined it," came the assertion. "Therefore, I am urging my colleagues to support the Jupiter orbiter/probe mission as the next logical and appropriate step in our exploration of the outer solar system." He wound down with some final comments on unrelated planetary missions. And with that, the gentleman from Florida completed his pitch: "I urge passage of H.R. 4088," he said and yielded the floor.

The response was broadly positive, with at least thirteen members of Congress rising to explicitly support the bill. They thanked the gentleman from Florida for such a thorough job. They noted the wisdom of NASA's four proposed new starts—one being J-O-P. "These programs will provide invaluable data to maintain a sound and growing knowledge base," remarked one member in a noticeably polished sound bite.

"The NASA budget is a cost-effective budget," opined another, after rising in praise.

Hopeful statements cascaded forth, one after the next, filled with talk of alternative energy, the increasing role of satellites, and the significance of the impending era. "The Shuttle will be the DC-3 of our space program—cheap, reusable, and efficient," offered a rep, after coming to his feet in support of the NASA bill. "Routine and dependable access to near space."

"This budget should be viewed as an investment—as opposed to just spending money on space," chimed in yet another representative. "This is true because these investments will grow or multiply to create more economic benefit in the future than was invested in the present." The circle-jerking dialogue continued, in generally upbeat tones, with the only real controversy being whether any NASA funds would go toward the development of a Concorde-like supersonic airplane.

They voted on the whole package: 389–0 in favor. It then went on the Senate calendar.

Four days later, outgoing NASA administrator Jim Fletcher joined another round of hearings by sitting with a Senate subcommittee to further discuss NASA's money wishes. Many minutes were devoted to the shuttle's supposedly glowing progress. "The first of the six planned orbital flight tests remain scheduled for 1979, and we expect to reach operational status in 1980," Fletcher testified. He stressed the relative importance of a five-shuttle fleet and talked up the necessity of having both East and West Coast launch facilities. Then he snuck in mention of "a next major step in the systematic exploration of the solar system. This is the Jupiter Orbiter Probe mission."

No question, these months were hectic for the space biz. One of the Voyager Grand Tour ships arrived in Florida on April 11 to begin its final prelaunch processing cycle. A second left Pasadena on April 21, bound for Florida also, to join its twin at the Cape Canaveral Air Force Station. Just to the north, crews were hard at work reconfiguring two Apollo launch pads for shuttle operations and may not have noticed the second *Voyager* quietly arrive on April 25. John Casani juggled these Voyager milestones with a personal one—the April 2 death of his father, Jack.

Disaster struck in early May: The House subcommittee tasked with NASA funding, in an attempt to slim things down just a little, cut every J-O-P dollar from the budget. JPL director Bruce Murray categorized what happened as an "ambush" and complained how the deletion occurred "without warning." Years later, the guy would still be pissed off about it.

The funding deletion occurred partly due to NASA's ongoing assertion that its favorite new start, the belle of the ball, remained an orbiting astronomical telescope. The National Academy of Sciences ranked it their number 1 thing too. Such telescopic love had chilled the House subcommittee's excitement for NASA's three other proposed new starts. Besides, with the shuttle nearing completion, and intelligence officials practically crying with Beatlesque anticipation over its critical national security role, shuttle moola needed preservation. One congressman—the same congressman who'd just pulled the plug on J-O-P—called the shuttle a sacred cow.

Why should Jupiter command an artificial deadline? "Not every project that the scientific community wants can have first priority," reminded one key Congress member involved in the decision. He was speaking to colleagues on the House floor. "And that is why we made the budget priority choice—to provide the Space Telescope—but we denied the Jupiter Orbiter Probe. Jupiter will be there five or ten or fifteen years from now when this project can be reinstated."

Alarms rang throughout the administrative halls of JPL. As Director Bruce Murray warned, "The key Mars/Jupiter alignment was very favorable in 1982, only partly so in 1984, and not at all after that." Delaying could kill the whole damn mission.

Meanwhile, NASA's overall budget slogged through the molasses swamp of governmental review. Its turn (finally) came up for Senate discussion on May 13, with one member specifically noting the relevance of J-O-P. "The $20.7 million authorized for this mission will also maintain the continuity needed for systematic and economical exploration of the solar system," he applauded. After three laborious readings, the bill passed that same day. It passed in amended form with Jupiter still on the rolls, because whatever the Senate did with bills was completely independent of the House—and the Senate hadn't deleted J-O-P.

Worker bees at JPL remained busy on multiple fronts. May's third week saw plenty of activity at the Cape's Launch Complex 41. For pad wring-out and countdown tests, one *Voyager* went atop a powerful, reliable, mature, comparatively low-cost, fully tested, and very much single-use Titan-Centaur booster. A spare third spacecraft arrived on May 23 in case the others needed parts or a complete swap. Funded and nearly ready, the *Voyagers* would still go regardless of whatever monetary crises the next generation had to face.

Summer unspooled. Mima Jaušovec won the French Open women's singles championship. Would-be do-gooder Anita Bryant successfully crusaded against a Florida law banning discrimination against a person's sexual orientation. Simultaneously, JPL fought for J-O-P's retention in House Resolution 4088. By this time, space program advocates had pressed for *any* amount of money to begin formal work on the Jupiter mission. Even half the origi-

13. Bruce Murray (*right*) discusses Voyager during a 1977 Lab visit by Charles, prince of Wales. A scale *Voyager* model in front of them mimics a full-scale model looming above the men. "Bruce was very aggressive, and he always wanted more," reported John Casani of Murray's management style. "He had people diving for cover, you know? They were scared to death of Bruce." Courtesy NASA/JPL-Caltech.

nal request would still be a green light. Until both sides of Congress agreed on the disposition of J-O-P, "a telephone and letter network spontaneously sprang to life, building its own momentum and reaching some Congressmen from home-district constituents," as Bruce Murray put it. Some of the more impassioned advocates, including James Van Allen, personally visited members of Congress. The *Washington Post* even ran an op-ed on the situation and endorsed J-O-P.

July 19 saw the final showdown commence on the House floor. Why kill J-O-P? "Let me tell you why," began the head congressman in opposition. "In denying the Jupiter Orbiter Probe, we were and are continuing to make an effort to strike a budget priority choice. This year's budget includes $36 million to initiate funding of the space telescope, which will ultimately cost the American taxpayer at least $800 million." He again spotlighted the preference of the scientific community for the telescope, despite real congressional concern about its timing and cost, and closed with a funding reality check. Something had to give, he intoned, and that something is Jupiter—which, of course, would still be out there doing its thing for the conceivable future. Go later, NASA.

"Mr. Speaker, I rise to express my opposition," proclaimed one congressman, upset about the idea of deleting funds. He highlighted the relevance of exploring Jupiter's mini solar system. Plus the critical 1982 launch window.

Others rose in solidarity. A California representative noted the upcoming Voyager launches with melancholy. "Unfortunately, however, no further missions are currently funded, and the nation's planetary exploration program is near stagnation levels." Eagerly he supported the Jupiter flight.

"We will never again have the same opportunity—in our lifetimes, at any rate—to be able to send a probe to Jupiter as cheaply as we can do it today." So went the opinion of a congresswoman from Maryland. "Shortsighted and penny foolish!" she scolded. The planned mission would cost less than $1.50 per person.

"Jupiter is literally pregnant with potential information relating to all aspects of science," opined another congressman, employing a previously unused adjective.

Multiple viewpoints welled up. "I believe that the project is worthwhile; the project does have scientific value," began one congressman from Michigan after being allotted three minutes to speak. "The problem is, where do we get all the money to do all the things we want to do?" He talked about the eventual sum cost of J-O-P by the time the mission ended. The cost of a possible rendezvous with Halley's comet. The cost of a separate mission to orbit and map Venus with top-quality radar. The cost of a lunar polar orbiter. All told, these very high-buck missions might collectively top a billion, easily. "I am not telling the members each one of these is proper or improper," the guy continued. "What I am saying to them is: we have looked at the Jupiter Orbiter Probe. We have said, on the basis of the money that we believe is available now and will be available in the next couple of years, this is a place we can forego." His three minutes were about up. "This is not a turndown to NASA. It is a forced examination of spending choices." He sat down.

Someone counterargued that canceling J-O-P would mean a disintegration of the "highly developed team of scientific, business, and industrial personnel" who were currently intact and organized for the project. You'd let that kind of synergy just dissolve?

The original guy wanting to kill J-O-P shot back, referring to the recent flood of phone calls from supporters across the country. "We *asked* people in the field of astronomy what priority they would attach to the large space telescope and the Jupiter Orbiter Probe missions," he said, half exasperated. "Without exception, they indicated that they would prefer the large space telescope." He pointed out that the majority of those on the Committee on Science and Technology *also* preferred the telescope over J-O-P. NASA costs tend to rise, he observed, and the shuttle's $5.1 billion "slide rule" projection from 1971 had already swelled to approximately $12 billion to $14 billion. Talk about pregnant!

"But that does not bother NASA," alleged the congressman, apparently hell-bent on being Van Allen's killjoy. "They are always talking about the cost of the program in 1971 dollars. Well, you cannot argue both ways! *We* are talking about the cost of the Jupiter Orbiter program in 1977 dollars and *that* cost comes to around $320 million."

Lest anyone think this guy hated NASA, space, or exploration in general, he was Representative Eddie Boland (D-MA), who claimed to the assemblage that his subcommittee overseeing space budgets had "dealt rather generously with NASA over the years." Even before NASA officially came to be. He believed in the cause. He was a fan. "But the problem is that NASA is lining up all of these programs year after year." Yeah, everyone including Butters the pet bunny was calling a Jupiter orbiter "the next logical step," but every year it was something different. What the hell would NASA want next year?

"If taking $20.7 million out of this budget is going to really harm or injure NASA, then there is something wrong with NASA and there is something wrong with the space program."

After another member stood to offer even more warnings about the 1982 launch window, Eddie Boland shot back. He lambasted the rude propaganda being ladled onto his colleagues by outside forces. "I know the members are receiving all kinds of letters indicating that this cut would cause all kinds of irreparable damage to the U.S. space program. That is a lot of nonsense," charged Boland. What about the $1.7 billion, roughly, spent to date on eighteen trips to other *next logical steps* like Venus or Mars or the moon or even Jupiter and Saturn already with those Pioneer thingies?

"I have no objection to them. I have been voting *for* them for over 22 years." There comes a time for many when lines are drawn in the sand. And for Eddie Boland, his line was a mission to orbit Jupiter.

One of his fellow naysayers took up the charge. "It is said that we need something for people to do out at the Jet Propulsion Laboratory in California. That is a heck of an argument, that we have to start a program just to give people something to do." Maintain their payroll for what—graphs of radiation levels? "Is it going to be worthwhile?" he asked the room. "Because we are not funding the Jupiter Orbiter Probe this year does not mean we are terminating our space program. We approved the other new starts."

More back and forth, with those for and against, until the Speaker retook control by announcing, "All time has expired." Boland moved to vote. They did, but it was a voice vote—which to everyone's ears sounded as if it was a loss for Boland.

"Mr. Speaker," he jumped in, "I object to the vote on the ground that a quorum is not present." The sergeant at arms rounded everybody up for proper electronic voting, which commenced immediately. Boland lost again, 280–131, with 22 not tending a vote.

President Carter signed House Resolution 4088 on July 30, and it became Public Law 95-76. Money for J-O-P became officially official, the eventual news of which thrilled John Casani as he and his boss departed the Cape. Their just-launched *Voyagers*, awesome as they were, would only fly past Jupiter. But J-O-P would *remain there*, in orbit for years, as the planet's longtime companion.

The very first thing worth doing on his new project, realized John Casani, would be to give it a properly majestic name.

8

Spin-Up, Dual Spin, Despun

Since fiscal year 1978, the total project costs estimate has increased by
an average of $106 million annually.

—U.S. General Accounting Office report on the Galileo project

As the pair drove off, their sideview mirrors displayed the receding facilities of the Cape Canaveral Air Force Station. At the wheel: Bob Parks. Riding shotgun: John Casani.

"In my position as head of the flight projects, I would be instrumental in selecting the project manager," Parks later explained of his role at the time. He picked candidates for many positions, with the review process rippling upward through multiple levels. But "project manager" occupied a category all its own. "That one went all the way to the Director for approval. But I never got turned down."

Casani, having just been informed of J-O-P's funding approval, now received Parks's unexpected anointing as chief banana for both J-O-P *and* Voyager. An overworked Casani numbly responded, "Okay." He needed alcohol. The recent days had been nerve-racking. He couldn't think of much else to say at the time—especially with the contrails of both launches practically still hanging in Florida's salty air.

They still had some time in the car, and Casani's enthusiasm grew. "I felt like, here I was, getting a new project! I was gonna be the project development manager right from the start!" At the hotel, both men loaded out and buzzed to the airport and boarded the flight to Los Angeles and home. Despite general weariness, Casani appreciated the way his career seemed to be shaping up. "This was gonna be one that I would have from start to finish. And in addition, I was gonna get to keep Voyager!"

But nobody reckoned on the hurricane-force, post-launch trouble that came blowing through the Voyager project. "What was happening?" Thirty-nine years later, Casani voiced that question to himself, then wordlessly took

14. A *Voyager* in a prep hangar at the Cape. Folded boom arm (*left*) holds various science instruments. Below *Voyager*, struts connect to a "kick stage" for additional thrust post-launch. The dark vertical cylinder at right of kick stage is the nuclear power source, also folded for launch. Courtesy NASA/JPL-Caltech.

a moment to organize the history in his head. Two wizened eyes glanced into the Pasadena sky while assembling his response. As if answers danced up there in the January clouds, waiting for him to pick them out.

Casani finally said it came down to this: Even with both *Voyagers* off the ground, and two lengthy years until Jupiter, they could easily blow the mission and the whole Grand Tour. "We still had a lot of work to do to prepare the sequences to be loaded into the spacecraft, to do the science observations as the spacecraft flew by Jupiter. There was a *tremendous* amount of work to be done. And that was—and we had a schedule laid out, and the resources, you know, set up to do that. And we started to get behind schedule on those activities." A hundred people joined the project to try and get on top of things. But there were almost too many things to get on top of. Hardware problems threatened to plow them under. Neither ship was operating as expected.

"I couldn't do both," admitted Casani of his predicament at the time. He had to be honest with himself. "I wasn't able, really, to devote enough attention to the Voyager problem *and* also manage this Galileo thing. Which was developing, you know. Which was becoming its own major activity."

Sensing a possible train wreck just over the horizon, an observant Bob Parks soon told Casani to abandon Voyager and concentrate solely on bootstrapping the new Jupiter mission. Casani did as told. "Now, if I was gonna feel bad, *that* would've been the time I would've felt bad." He sipped from a just-refilled glass of red wine and slid into a little impersonation of Bob Parks: "'I'm gonna take Voyager away from you. And I'm gonna take it over.' So he took over my role as primary responsibility for Voyager. I pressed on with Galileo."

While retelling this aspect of the story, Casani referred to the Jupiter mission as "Galileo." But at the time, no name had been chosen, and everyone used "J-O-P." It corresponded to all the funding documents and congressional paperwork. A descriptive name, yet bland. And unintentionally confusing. Were people supposed to say the letters individually? Or run them together as the pseudo-word *jop*? That wasn't far from "jalopy," which doesn't exactly call to mind a lot of high-tech, space-age workmanship. This project needed a real name. Something distinguished. Something to excite people. How could JPL do that?

"Ah, we ran a contest," explained Casani, seemingly a tad embarrassed by the approach. He readjusted in his chair a little. "Well, I mean we asked everybody on the project, you know, to submit their candidate names." Pressed on that, he clarified that *everybody* really only meant the dozen-odd people reporting to him. Compared to other suggestions, "Galileo" stood out for its obvious connection to the man who first discovered moons circling around another planet—the very moons and planet this spacecraft would be visiting.

Casani went on to detail the elaborate name approval process: "And uh, then we went back to headquarters and said, 'This is the name that we propose.' And they said okay, and that was it." Boom: Galileo! Order up new letterhead and change the door signage.

But when Vicki Melikan learned of what happened, she rang up Casani, spraying white-hot fury into the hapless phone. In her role as JPL's congressional liaison, Melikan had a vested interest in how Congress perceived Lab operations.

"She had been a lobbyist for General Motors when she came to JPL. And she knew everybody on the Hill," added John Casani. According to him, Melikan had been instrumental in orchestrating J-O-P's funding approval. Casani called her work masterful. So it would stand to reason that at this earliest stage of the mission, Vicki Melikan had much more emotional investment than did Casani, who was then in the process of getting rage-phoned.

"She was furious," came his vivid recollection.

She spat, "You're changin' the name?!" And roared into a you-better-not kind of modulated Melikan monologue on how confused everyone in Congress would be if the change took hold. *Those people have so much going on that they won't know what you're talkin' about, John! They just finished voting for something called Jupiter Orbiter with Probe.* "Now somebody's gonna ask 'em to vote on money for *Galileo*! And they won't know what the hell *Galileo* is!"

Tongue-lashings aside, the new name fit way too perfectly to discard. It became final—as announced by the NASA administrator on January 23, 1978.

Having finished his recollection, Casani sipped casually before resting his wine glass on the round marble table. Then gently folded his hands together. So much time going by permitted him the luxury of levity. Congress wouldn't realize "J-O-P" and "Galileo" meant the same thing?

"That was her concern. But it turned out to be okay."

Soon after Galileo's approval, the hard part began of completing its final design—a deceptively straightforward-sounding task. Oh sure, bunches of things had already been sussed out and committed to paper; JPL had published its "Orbiter Description Document" a year before J-O-P even got approved. This 189-page booklet represented an advanced, "pre-project" study of both the mission and the spacecraft. Considering this concept hadn't even been sold, the description document brimmed with surprising amounts of detail: how big of a ship, how heavy, how much power, how its general hardware would be configured.

Serving as the orbiter's core would be an unromantically practical, eight-sided chassis—"the bus," they called it. *Galileo*'s master control systems would live there. Literally everything else glommed onto it: power supplies, antenna, propulsion, instrument booms, the descent probe.

The bus itself would spin, along with the overhead communications antenna and side-mounted power sources. Every fields and particles instrument would mount there also, merry-go-rounding through space as the vehicle slowly rotated a few times a minute. Underneath all this would bolt a framework of tanks feeding a dozen maneuvering thrusters, plus a giant engine to brake the spacecraft into Jovian orbit. "The retro-propulsion module," as described in official docs. "A self-contained, primary load-bearing structure of the *Galileo* spacecraft." Its braking engine only worked at a spin rate of ten RPM or better, owing to centrifugal forces on the tanks.

Nestled underneath the spun section would be what everyone referred to as the despun section—home to instruments requiring precise pointing. Such as those for studying Jupiter's chemical composition. Or the cameras. Anything in the category of "remote sensing." And then down at the bottom, grapefruit-like, would hang the descent probe for entering Jupiter's atmosphere.

Both the spinning and despun sections had to lock together for probe release and stability during engine burns. JPL's big description document covered all this but didn't go much into *how* it would all work. "Those studies were almost always done by people who had never done a project. Don't ask me to explain that, but there were people that seemed to specialize in doing advanced studies." So came a basic frustration from JPL's Bill O'Neil, who at the Galileo project's start worked in mission design. "Well, the result of that, quite often, was the mass of the spacecraft was way *under*estimated. The capability of the launch vehicle was way *over*estimated. And the *cost* was way underestimated." Time and time again he'd seen that pattern, and now here with the Jupiter orbiter came another glowing example. Designers liberally penciled in spacecraft appendages and functions to their heart's content—without so much as a half second of contemplation on how such things might be realized.

Continued O'Neil, "More often than not, you wound up with a bag of stuff that wasn't going to work from the preliminary designers. And that's handed to the guys that had to *make* it work. Who were very typically the guys that had done the *last* project."

Indeed, whatever actually got built had to fit; shuttle astronauts couldn't just slam the payload bay doors as if car camping in a Ford LTD. Out of contention, therefore, was a supersized iteration of Voyager's rigid antenna dish. And what's this on the side of *Galileo*—an arm extending thirty-six feet in length for hanging sensitive gear away from the bus? Major assemblies were going to have to collapse or fold or retract and then flawlessly deploy on cue, similar to a complicated pop-up book in space that nobody could touch.

The denizens of JPL would need to commit themselves to fulfilling what had been promised to Congress, to the planetary science community, and to the world. They weren't ready to *build*. JPL had its description document and four-color artist's renderings, but it couldn't hand those to a lathe operator and expect much. Draftspersons needed to produce detailed shop drawings for every bracket, fitting, mounting stud, cable, and connector. All dimensioned to show every last thickness, bolt hole, and assembly tolerance. The project needed a final design. Its next few years would go toward determining *on paper* every last one of *Galileo*'s pieces and how they fit together and what materials to use, down to the torque spec on the nameplate screws. Only afterward would anyone begin cutting metal. And only after *that*, during assembly and testing, would they find out how off track the original paper design actually was.

Various units within the Lab—design documentation, materials, quality assurance—gradually whirred up to speed. "I had a whole bunch of people that had worked on Voyager, and their job was over," indicated John Casani. In particular, his team needed to get a handle on dual-spin technology. "So we jumped on this and looked at it. And it was gonna be a really, really difficult problem. Required *much* more sophistication than just despinning an antenna." The concept involved rotating metal contacts and carbon graphite and tensioned springs and voltage regulators. Concerns instantly arose. What about electromagnetic interference?

Just because the design called for dual spin didn't mean that's what would eventually fly. "And then Bruce Murray was on my case," recalled Casani of his boss, who seemed to regard the despun section as borderline fantasy and perhaps not worth the trouble. "He wanted to know, why I couldn't just—instead of havin' a, you know, doin' it the way I was doin' it—why couldn't I just put some other kind of platform on the end of a boom or some shit like that. So we had to do a lot of other studies that didn't really lead anywhere."

Once the design phase finished, JPL would finally press ahead with prototyping *Galileo*'s major pieces of superstructure. Among them would sit that retro-propulsion module—a synthesis of maneuvering thrusters and fuel tanks and piping and a giant rocket motor for slowing into Jovian orbit. This $15 million component was slated to come from West Germany. If things failed to operate for whatever reason during the module's forty-five-minute engine burn, *Galileo* would sail right past Jupiter and never perform its orbital experiments. Engineering estimates put the module, including fuel, as weighing nearly half that of the entire ship.

The sacrificial descent probe would come from a facility perhaps better suited than any other in the world to create such a thing—the Ames Research Center. People there knew what they wanted but needed specialists to create it. The actual job of fabrication went to Hughes Aircraft, which got the contract in September 1978. That same month, a NASA Headquarters man assigned to the shuttle program testified before Congress. Mostly they were on schedule with development, he promised, and could maybe fly one of the newfangled space planes around this time next year. But the whole shuttle orbiter was overweight. And they were having trouble debugging its main engines.

Could delays affect Galileo?

"Yes, they *could*," said the guy.

Word had hit the streets on July 1, 1976—before the mission had funding and way before Vicki Melikan's name-change trauma. It came in the form of an "announcement of opportunity" from NASA's Office of Space Science, AO-OSS-3-76: "A solicitation of proposals for scientific investigations of Jupiter." The time had come to choose Galileo's experiments.

One specific member of that aforementioned scientific community, Don Gurnett, would later refer to his entire decades-long career in space science as "this whole thing." The joke came across as unintentional, with Gurnett outwardly unsure of where to begin a discussion of his life's work. Certainly, he never thought twice about proposing an experiment for Galileo. "Absolutely," proclaimed the man with a grandfatherly face and thin smile and even thinner white hair. "I was intent, driven, to go to Jupiter."

"This whole thing" studied by Don Gurnett was a relatively new field of space physics called plasma waves. His strong determination to "ride" *Galileo* backed up against a recent, and sour, experience of producing instruments to study these waves. At the time the J-O-P proposal request went public, Gurnett's babies awaited launch aboard *Voyager 1* and *2*—still some fourteen months away. He appreciated being aboard. But his team's plasma wave hardware was older, having used a design cribbed from another space mission and modified to work on the *Voyagers*. It piggybacked on somebody else's radio system, which was already installed for a completely different experiment. Gurnett had to rely on that guy's antennas—versus one of his own specialized designs based on years of fine-tuning. He wasn't heading the plasma wave team either; a colleague did. Overall, his Voyager experience had been a rushed, last-minute deal with too little time for preparing anything more mature.

And of course, each *Voyager* would only fly past Jupiter. That meant a one-shot feeding frenzy, grabbing everything possible during an inhumanely short period of close encounter.

But announcement of opportunity number OSS-3 had brought forth the promise of *orbiting* Jupiter. His number 1 place! Gurnett thumbed the announcement. In sparse, almost clinical prose, it defined a number of high-level parameters that roughly framed out the mission's operational requirements and deadlines. The document's tone projected a certain attitude. Reading its imposing list of deliverables smacked of someone barking instructions: Pay attention! You want to ride on J-O-P? The launch period is early 1982! Here are some investigations expected to ride along! Plan on staying out there for twenty-four months! Get your affairs in order! Don't be late submitting!

Over the preceding decade, Gurnett's evolving techniques for studying plasma waves had made for better data. But with every new proposal, he worried about becoming too fussy. Mainly it came down to the antennas: "For the comprehensive design of a plasma wave investigation, such instruments must have one or more electric field antennas. The design of electric antennas is difficult because they almost *always* have a big impact on the spacecraft structure and dynamics."

For all the detail within JPL's thick description document, it didn't specifically provide for Gurnett's own breed of specialized antennas. What if everyone liked his proposal except it messed too much with what'd already been designed? Antennas could easily get in the way of other instruments. Nobody wanted to spend hundreds of millions on Jupiter pictures with an antenna in the middle of every shot. Would JPL realistically change the fundamental spacecraft design just for one guy working in a field with many other researchers? Gurnett wasn't so sure.

Regardless, he scratched up a proposal and reckoned that one key tactic alone might seal the deal. He'd discovered it back during his Voyager tribulations. "You join your competitor," he laughed. "Kind of like what happens in the industry, right?"

Even something as pure and noble as physics research, explained Gurnett, carried with it the occasional necessity of politics. And so to that end, he approached a colleague named Roger Gendrin and suggested the two of them propose Galileo together. In this way, they'd both have a good shot at making the cut and wouldn't have to compete against one another. If you can't beat 'em then join 'em. Right? Yes?

"You know, that's kind of the game," admitted Gurnett. "It's better to work together than to fight and have one of you lose." Final pitches had to be in no later than November 1, 1976, for the descent probe. Those vying for slots on the orbiter had an extra month.

9

What the Pork Chops Said

The scientific focus of flying that instrument or experiment,
and analyzing the data and understanding what
it's telling you, is an *investigation*.

—University of Iowa research scientist Bill Kurth

Galileo's first project review happened a month after the *Voyager* launches, in October 1977, and its design had put on weight like a pregnant blue whale. Orbiter structural enhancements topped 364 pounds. Changes to the descent probe added another 62. And those were only two examples. New weight calculations exceeded the abilities of the originally planned air force stages. If only that Centaur was available, then the shuttle wouldn't need its outlandish-sounding "109-percent thrust" engines. But Centaur wasn't yet on the table. Luckily, trajectory planners had uncovered a method of flying by Mars to pick up a "gravity assist" of free energy for the long outbound haul. Downside: five extra months of flight time. Not ideal, but workable, and retained the same launch period of January 2–12, 1982.

Galileo's actual science investigations had not yet been chosen but were about to be, and John Casani offered a few insights into that world. He described the announcement of opportunity as "sort of like an RFP—a request for proposal. To the science community. And any scientist who wants to can propose." Some 475 individuals from multiple countries had applied for a slot.

"Now," resumed Casani, "these proposals are reviewed by a team of people. And JPL has nothing to do with this review process. Nor would anybody that's building the spacecraft." He went on to say that NASA Headquarters does sponsor the proceedings. But the actual selection is performed by an entirely separate organization, "whose job it is to round up people that are competent to review the proposals and to evaluate them. The whole process is conducted in, really, under the tightest security. I mean, when I was involved in a project, I had no visibility into what this review process was!" Casani added that sometimes Labbies would be asked for comments on a specific proposal, but that was it.

High over the restaurant's outdoor table, the sun peeked out from behind clouds and illuminated the basket of bread. Casani swirled the current round of wine in his glass. "I don't think anybody at JPL could have influenced the selection of these instruments. They would have been evaluated on their merits."

For anything to reach the launch pad, Galileo's money train would have to keep chugging down the tracks. On February 2, 1978, Noel Hinners expectantly sat down across the table from a formidable congressional subcommittee. The meeting occurred in DC's Cannon House Office Building, located just south of the U.S. Capitol and kitty-cornered from the Library of Congress. It was a highfalutin kind of place. Hinners occupied a position of regality in NASA's Office of Space Science. And during this budget review meeting, he requested $78.7 million to continue Galileo development. It fell within a larger $187 million ask for lunar and planetary exploration as a whole.

"The lunar and planetary program has *decreasing* funding requirements for Voyager," Hinners clarified, trying to church it up for the delicate audience. Voyager development was complete, with both missions underway. "The funding increase reflects the buildup in the Jupiter Orbiter Probe development." Side categories like mission operations and data analysis brought the total to what it was. This money, in turn, fell within an even larger request of $513.2 million for the Office of Space Science's entire fiscal 1979 budget. Money for *just* NASA's Office of Space Science.

Lean, mop-headed, and never without glasses on, forty-two-year-old Noel Hinners had inaugurated his space career at Bellcomm—heavily involved in the selection of Apollo landing sites. From there, the geochemistry PhD formally joined NASA in 1972 and began serving as its associate administrator for space science only two years later. To oversimplify things just a bit, JPL's Galileo office reported to NASA's Office of Space Science, which was JPL's customer.

And as Noel Hinners recapped to that subcommittee, "Our first major action after your approval of Galileo was the tentative selection of science investigations, in August 1977." All told, some 115 researchers had received uplifting news about inclusion. Of that total, 29 were for the probe alone. The finalists had staged a kickoff meeting only a month prior to Hinners's testimony, charged with submitting detailed requirements in time for JPL's science confirmation process in October. An approved experiment might be rejected later if it severely impacted the spacecraft's design. No one was aboard for certain until the thing left Earth.

The 115 figure didn't equal 115 individual experiments. Many researchers had banded together on proposals. For the orbiter, scientists would preside over eleven experiments, but that didn't mean eleven identically sized cubbyholes inside *Galileo*. What they got depended on need. The pictures everyone lusted after came with the biggest weight penalty—just under 62 pounds for the camera system, to be supplied by Tucson's Kitt Peak National Observatory. Its electronic imaging chips, explained Hinners, were far superior to the vidicon tubes on *Voyager*. West Germany's dust detector, from the Max Planck Institute for Nuclear Physics in Heidelberg, had also been selected. Hinners mentioned that the formal agreement for the separate West German propulsion module (which, to clarify, was not at all an experiment) had been signed the previous October, with its first joint project meeting held just a couple months earlier in Bonn.

Another experiment intended to study the atmospheres of four moons discovered by Galileo Galilei. Plus Jupiter's own atmosphere. The twin *Voyagers* packed a similar experiment—an ultraviolet spectrometer—but *Galileo*'s more sensitive version would improve upon the returns of its predecessor. "These experiments *all* represent advances beyond the capability of Voyager," assured Hinners. "And in several cases, entirely new kinds of investigations are possible." A specific type of complex hydrocarbon molecule might exist in Jupiter's atmosphere; on Earth it's a building block of life. Would such an important compound also be found way out there beyond the asteroid belt?

"The schedule remains tight," testified Hinners of Galileo's progress. "We did reallocate money to be sure we got off to a good start, but we have had a potential weight problem." A bunch of structural enhancements had had to be tacked on, disrupting everyone's numbers.

One of the congressmen wanted to know, "What does it do to the launch window?"

"It remains the same," promised Noel Hinners. "Mr. Chairman, that completes my presentation."

WESTERN UNION MAILGRAM

1-003917A228005 08/16/77 TLX NASA WSH CDRA

278 WASHDC

DR DONALD A GURNETT
DEPT OF PHYSICS & ASTRONOMY
UNIV OF IOWA
IOWA CITY IA 52242
MSG SL-274

WE ARE PLEASED TO INFORM YOU THAT YOU HAVE BEEN TEN-
TATIVELY SELECTED AS A PRINCIPAL INVESTIGATOR ON THE
ORBITER OF THE JUPITER ORBITER PROBE 1981 MISSION.

/S/ NOEL W HINNERS ASSOCIATE ADMIN FOR SPACE SCIENCE
NASA WASHDC
1122 EST
M3MCOMP MOM

When Don Gurnett received this news, he felt excitement. Truthfully absent from his mind, though, was surprise. "I don't think we had any serious competitors," he smirked—no doubt because of his backstage partnering with one of them.

A printed packet followed the telegram, also from Hinners, with expanded details. His cover letter included language reminding that, as with Mariner Jupiter–Saturn/Voyager, this initial phase of development should be considered probationary: "At the end of the first year, all experiments will be reevaluated in the light of their contribution to the mission and their compatibility with mission constraints before final selection is made."

Gurnett's thoughts then turned to the impending phase of building what'd been proposed. Should be straightforward. Instead of using a third-party contractor, as many others employed, the University of Iowa would fabricate Gurnett's plasma wave instrument right there on its own campus. Gurnett could check in on it every single day if so desired. "There is a huge amount of pressure," he indicated of the process, that maybe isn't so obvious. "You've got to make your instrument *work*. And I can't . . . I can't overestimate that. I—in my early days, I always had the feeling that if I flew an instrument that didn't work, that would be the end of my career." He—rather, Iowa's Department of Physics and Astronomy—had a short few years to get everything together. But they'd done it before.

As 1978 wore on, JPL engineers worked to finish identifying every spacecraft change based on the science roster and to nail down a revised flight path with the Martian detour. Fall arrived, preparing to cap a year of furious JPL work and a colorful parade of worldly events. That year, experts declared small-pox to have been eradicated. Movies about the Vietnam War dominated the box office. In a defeat for enthusiasts of Aqua Net hair spray, Sweden banned aerosol cans. And marking an undeniable low point for the entire history of Earth, newspaper readers were first subjected to the comic strip *Garfield*.

January 1979 brought Galileo's preliminary design review. With Hughes, Ames, and JPL all in the same room, John Casani directed Hughes to add redundancy onto its descent probe. "This had not previously been a requirement," complained one Hughes guy. "A significant impact on a volume-constrained design," he whined. So much progress to date, and now this. "Weight and cost increases!" But the company did it anyway and would go on to survive a follow-up review.

A month before *Voyager 1* flew past Jupiter, in February 1979, Bill Kurth defended his PhD dissertation before the Iowa physics faculty. "And I asked Gurnett if I could kinda hang around and see what came back from *Voyager*," Kurth relayed. He'd helped test Iowa's hardware for it and written some of the documentation. ("I didn't know what I was doing," he'd later admit of the documentation work.) Gurnett told him to pack for California.

In Pasadena, the University of Iowa contingent shacked up at the Saga Motor Hotel on East Colorado Boulevard, directly across the street from Pasadena City College. In its lobby, the Saga featured a huge and nonsensical mosaic of a knight on horseback that could only serve to confuse anyone attempting to identify what the theme of this weird place was actually supposed to be. Iowa's rooms cost $28.89 a night. From a narrow excuse for a parking lot in the back, Kurth and Gurnett could take an immediate right onto North Sierra Bonita Avenue, cross under the 210 bridge, hang a left on Maple, and then merge onto 210 itself, "the Foothill Freeway," for a ten-minute meander northward to JPL and a much nicer parking lot.

Kurth stayed in Pasadena for just over three weeks while attending the sometimes nearly out-of-control festivities of *Voyager 1*'s Jupiter encounter. Days ran incredibly long. He missed his family, ate poorly, and in general slept nowhere near enough. Kurth wouldn't have traded the experience for anything.

When asked, Bill Kurth described how one of his most memorable Voyager moments came during this first planetary encounter—a period of intense

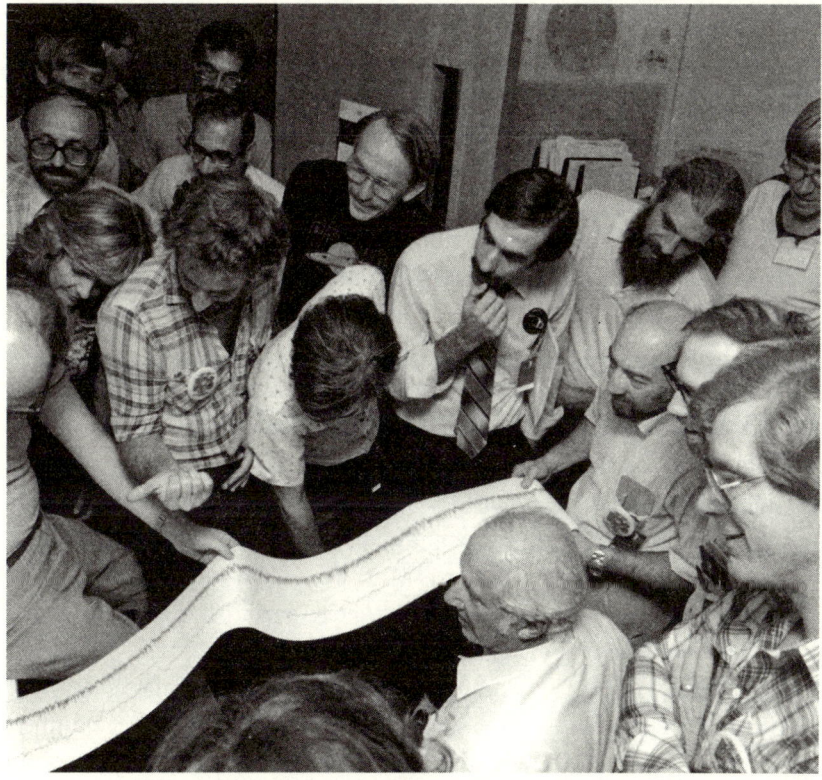

15. A typically hectic scene that often occurred during early *Voyager* encounters. Here, in August 1981, giddy project members ogle a fresh data printout from *Voyager 2*'s Saturn flyby. There was no other way to see it! Courtesy NASA/JPL-Caltech.

activity encompassing late February through early March 1979, when the spacecraft flew closest to Jupiter. Encounters brought everyone to the Lab for a whirlwind of instrument readings, first-look assessments, close-up pictures, and media interviews. With elements of a family reunion mixed in all along the way. Public outreach factored in mightily. Helpful to anyone's cause was a strong scholarly presence when interviewed on television, offering relatable explanations and impressive visuals if possible.

One of the days, Kurth sat alongside Principal Investigator Fred Scarf in the dinky area set aside for the plasma wave team, watching an even smaller black-and-white video monitor. Data from their instrument would intermittently appear during select periods. Otherwise, it showed the latest picture from *Voyager 1*'s camera. "The display was riveting," marveled Kurth. "Each day and each hour, the images became clearer and more detailed, and new worlds were literally being unveiled before our eyes. At the same time, the PWS data were showing evidence of radio and plasma waves never before observed."

Kurth's abbreviation references their plasma wave subsystem. And that subsystem flying way out there—no matter how much of a last-minute compromise it represented—seemed to be doing its job. Days such as these blurred into one another over the course of the encounter. Kurth, Scarf, Gurnett, and others on the plasma wave team spent untold hours quietly focused on their returns dribbling back in real-time from the Jovian system.

"Fred surprised me one day by saying how very sad he was for me. He said that it was so sad that I was experiencing Voyager so early in my career, because it would never be so good again!" exclaimed Bill Kurth. "Of course, what Fred did not know was that the *Voyagers* would visit not only Jupiter and Saturn as was the mission plan, but also Uranus and Neptune." (When the *Voyager 2* plasma wave instrument first arrived at JPL, Kurth had been the one responsible for plastering a "Uranus or Bust" sticker across its shipping container.)

Jupiter saw *Voyager 1* make its closest approach on March 5. Sadly, Bill Kurth didn't know that Fred Scarf would not live to see Neptune, or learn of its Great Dark Spot, or examine the character of plasma waves occurring in the vicinity of that far, faraway place.

Shifting into teacher mode, Don Gurnett now endeavored to explain what plasma waves *are*. He started with the analogy of a performer on stage using a microphone. If that mic is too close to one of the stage speakers, it'll pick up the sound of itself. Which flows back through the audio system, gets amplified, then re-exits the speaker—where it *again* is picked up by the microphone and re-amplified. That's feedback. The whole nasty loop grows in cacophony until some stagehand figures out what's up and moves the speaker. Or yells for the performer to move someplace else.

"Well, in plasma you get a feedback like that," Gurnett segued. "When an electron, like in a radiation belt, is spiraling around the magnetic field, it emits a radio signal." And the frequency of that signal—where you are on the radio dial—depends on how quickly the electron is rotating around said magnetic field. The *frequency* of rotation determines the *frequency* of the radio signal.

"That signal propagates backwards and affects other electrons coming up the field line," he continued, going on to say that the exact same physics are at work in a common laser pointer (only with a laser instead of a plasma).

Any planet or moon with a magnetic field is going to have identical behavior afoot. It means these bodies are themselves such a mighty source of radio emissions that station XEG-AM, in comparison, is a flip phone. And what about Earth? "A tremendously powerful radio source!" Gurnett exclaimed. "But the funny thing is, we never knew!" Natural radio signals from our own planet aren't detectable at ground level because they can't push through our atmosphere's thick layers. Awareness came with the space age. Gurnett's 1971 experiment aboard an Earth-orbiting satellite found our planet emitting a thousand megawatts, in the low-frequency range of a few hundred kilohertz. Gurnett thinks that'd be enough to drown out the entire AM radio spectrum—provided it could get through the atmosphere. "If you were a graduate student in physics *on Mars*," he joked, "you could very well be studying the earth as a radio source."

For a planet to have this feature, it must first have a magnetic field, and not every planet does. But the biggest planet in our solar system surely did, and Gurnett adhered himself to the incoming *Voyager* data streams to learn whether Jupiter might, like Earth, interact with plasma and spawn the curious waves. "Turns out, Jupiter has a similar mechanism!"

MORE HURDLES RISE IN GALILEO PROJECT TO PROBE JUPITER

By Thomas O'Toole, *Washington Post*, August 15, 1979

The Galileo mission the United States plans to launch toward Jupiter in 1982 is running into a rash of unforeseen and threatening problems.

Galileo, the first spacecraft designed to use the Space Shuttle as a launch platform for flight to another planet, is suffering because the Shuttle has run into snags.

The Shuttle engine test program is so far behind that the space agency has had to postpone development of an advanced engine the Shuttle will need to carry *Galileo* into Earth orbit. The advanced engine is called the "109 percent engine," which means it will be burned on takeoff at 109 percent of its rated thrust.

If *Galileo* can't get that push, it will have a fuel problem. That would mean a drastic cut in the mission—instead of making 11 orbits of Jupiter and its four moons, *Galileo* would make only five.

John Casani doesn't remember exactly how he himself got the news. "The most natural way for me to learn would be for me to have received a phone call from NASA Headquarters from my program manager. The guy I reported to back there." Casani also doesn't remember exactly who that person was—an understandable situation with forty years gone. "That was, uh, heh heh heh, I can see his face so clearly now. Wait a minute though. Now it'll, maybe it'll pop into my mind."

Regardless, the call did not bring tidings of joy. Ten years to the day after *Apollo 11*'s historic lunar landing, the word from NASA HQ was sleeting down onto JPL: "We're gonna have to cancel the mission," Casani heard. The shuttle would not be ready in time to launch *Galileo*.

The guy on the other end shuffled right on into a verbal patter that came across as too sales-like and prepared: "We're gonna give you a time to go through an orderly shut-down . . ." But Casani listened only halfway because his mind raced to think of alternate options.

"Wait a second," he interrupted, trying to buy time. His only currency was a plea. "You know, let us work on it. See if we can come up with a solution."

"Okay."

The call ended. Casani stared at the phone. He needed people. Together in a room. Right now.

Shortly thereafter, everyone sat awaiting instruction. Where they went from here came down to a fundamental problem of energy—and what the pork chops told them.

"Let me explain it a little bit," offered a much older John Casani of the situation that faced his project team. More bread landed on the round marble table of the Italian restaurant. "The energy required to get a spacecraft from Earth to Jupiter *varies*. You know? Quite a bit." In high-level terms, he described a concept known as C_3, or "characteristic energy"—essentially, a measure of energy that a launch vehicle can deliver for a payload. Many variables are involved, of course; there's vehicle mass and thrust capability. Plus the critical dates of departure from Earth and of arrival at the chosen destination. The unit of measure for C_3 is kilometers squared per second squared.

"What we generally do, if we wanna get a spacecraft to Jupiter—there's two things you have to consider, you know: when's the arrival date—when do you want it to get there—and when's the launch date. And for any combination of launch and arrival date, there's a certain energy that is required from the launch vehicle for that spacecraft. And you can draw a curve." Where-upon Casani grabbed a blue pen off the table and sketched arcing lines across the back of a square cocktail napkin only slightly moistened by wine spillage. "They're typically, there'll be, the curves are sort of shaped like this. I don't know, you know, sorta like, somethin' like *that*."

With initial curves laid down, he added X and Y axes with "launch date" penned across the horizontal X and "arrival date" climbing up the vertical Y. Talking faster now, Casani explained that his curve stretching through the middle of the graph depicted the minimum energy requirements on the part of the launch rocket.

"But I can't depend on getting launched on *exactly that day*. I have to have what we call a launch period." This refers to a consecutive series of days dur-ing which the spacecraft can leave Earth in possession of enough energy to reach its destination. Understand that "enough energy" breaks down into a balance of rocket propulsion, spacecraft weight, and celestial mechanics. The aiming point is a location in space not where Jupiter *is* but where Jupiter *will be* once the spacecraft arrives. Often the launch period runs twenty days or something similar; it all depends.

Casani wasn't done with his napkin sketching. He shifted into a higher gear, adding new curved lines on the graph in opposition to the original ones, such that the curves joined together to form irregular ovals. "So *now*, you draw a period of time *here* and *here*. And you know, if the curve is flat in *here*," he said, stabbing the napkin, "that means that you gotta design for a spread in arrival dates. Over that period."

Casani threw a question to his tableside guest: "What difference does the arrival date mean? Well, that's the *geometry* that's at the planet when you arrive. 'Cause the planet's moving around the sun!"

While Casani spoke, his eighty-three-year-old arms gesticulated wildly in such a surprisingly agile performance as to virtually draw the solar sys-tem in the air above the table. "And so the angle back to the earth—and the angle of the sun out to the spacecraft—all varies over there. So you have to make sure you design a spacecraft that can accommodate that arrival date. And you have to say, 'Okay, I think that taking into account weather and what the uncertainty, and having the spacecraft ready on time, and having the launch vehicle ready on time . . .' How accurately can you control that? And you know, you've got to allow for a thunderstorm to come through!

Which could delay you for a couple days. So you pick a launch period. Which is a judgment call."

According to Casani, this produces a set of final, graphed, irregular-yet-concentric ovals indicating how much spacecraft can be launched with how much booster energy, and over what period of time, to arrive at the desired destination within a specific range of days. "Knowing what the shape of the C_3 curve is," he elaborated, "you have to design your launch conditions. And these curves are called pork chops."

Casani now applied these concepts to the growing probability of his team not having a launch vehicle ready in time for the 1982 opportunity—regardless of whether *Galileo* itself would be ready. "Okay, so the problem we had, is that the C_3 requirement varies for Jupiter from a *minimum* of about seventy-three kilometers squared per second per something or other, to a maximum of maybe a hundred and five. Something like that. And depending on the year you launch, it can be anywhere in *here*, right?" His finger circled the middle of the example pork chop on the napkin. "It turned out that in 1982, the C_3 required to get to Jupiter was around seventy-five or something. Near the *minimum*," he stressed. "It was low."

But the next opportunity to launch—over a year down the calendar—would need dramatic amounts of energy to send *Galileo* in its current form. How much more? "Was up around eighty-two. It was a *big* increase." And totally beyond what the air force's so-called interim three-stage, solid-fuel booster could provide. "So you talk about contingency. We did not have a contingency that would allow for a slip of one year or two years. Because that would've slipped us out of this minimum energy period."

Helping himself to another serving of bread, Casani's mouth bulged on one side, hamster-like. "Roo yush kanooit!" Loudly he swallowed. "You just can't do it! So the interim upper stage was in that situation. It was not a very high-energy system," relative to something like Centaur. The tacked-together air force stage "could only deliver, you know, seventy-five or seventy-seven kilometers squared per second squared, whatever it was, energy. So once we got moved out of the '82 launch period, the upper stage would *not* do it. *Could not* do it. So we were sayin', 'How in the hell are we gonna get this thing launched?'"

Casani pushed back from the table. He needed a break.

10

Dawn Chorus

*How did my life evolve? It was just things—just like kind of following
my interests more than making any grand plan ahead of time.*

—Don Gurnett

At the time of the seminar, during 1962's spring semester, the University of
Iowa incarcerated its Physics Department within the confines of MacLean
Hall. The building opened in 1912 and is one of five core structures defining
Iowa City's iconic "Pentacrest," with the central one being the state's first
capitol. Okay, so maybe it's iconic only to Iowa Citians. Regardless, flank-
ing the capitol in a rectangular formation are MacLean and three other
massively old, low, off-white stone structures containing departmental and
meeting facilities. All are regal- and humorless-looking buildings, in a category with art museums and national houses of parliament. Most everyone
attending Iowa ends up taking classes in at least one of the four. Within
these buildings, relationships have bloomed and careers have occurred and
dreams have died.

In 1962 Iowa charged electrical engineering students $145 a semester—for
residents anyway. Donald A. Gurnett qualified because he'd grown up only
thirty miles away.

Entering MacLean Hall that afternoon, Gurnett climbed worn stairs to
the top floor and filed inside the southernmost hella-big lecture hall along
with grad students and James Van Allen and somebody on the physics faculty. Slender Gurnett had been a student since late 1957, having begun classes
only weeks before the Soviets terrified greater America shitless by launching
Sputnik, which consisted of God knew what. Years down the road, Gurnett
would deliver his first lecture as a professor in this exact same room.

The visiting speaker, Roger Gallet, had come from the National Bureau of
Standards in Boulder, Colorado. "I'm not sure exactly how he got invited,"
Gurnett acknowledged. But department chair Van Allen would occasionally

16. An aerial image of Iowa City's Pentacrest in autumn 1964, looking southwest. At center is Iowa's original state capitol. The rectangular MacLean Hall is just behind it. Courtesy Frederick W. Kent Collection of Photographs, University of Iowa Libraries, Iowa City, Iowa.

bring in outside speakers on various physics-related topics, and Gurnett reasoned the guy was part of that whole deal.

First up were the introductions and opening statements. "I didn't know what he was going to talk about." Gurnett sized up this man in front: lean and well-dressed with nondescript wire-frame glasses and a somewhat unmanaged hairstyle, which made the visitor look more or less as if he'd been caught by a windblast.

To the people before him, Gallet eagerly introduced a phenomenon he referred to as "natural noise"—a somewhat undefined aural realm of low-frequency garbles. Weird stuff. Coming from space. "Unusual sounds of natural origin," as described by him in a paper written three years before this Iowa visit. A few people were studying it, but not many, and they'd hypothesized answers for some of the sounds. But not all of them.

Presently Gallet turned to the enormous reel-to-reel tape recorder parked nonchalantly off to one side. What existed on its threaded-up tape would change Gurnett's life. He didn't know it at the time, of course, and hadn't known in advance that sounds would even be part of this seminar. Gallet's thumbing of the PLAY button launched twin reels into motion.

The room filled with *shee-e-eew*. *She-eee-ewww* went the sounds, most irregularly. As though cartoon grenades whistled through the air. Looney Tunes effects. The duration of each individual sound varied, lasting up to several seconds in length.

The space physics community pretty much knew what these sounds were already. A flash of lightning someplace on Earth follows our planet's magnetic field line all the way to the other hemisphere. The traveling energy wave then reflects off our ionosphere and races back. While doing so, noises are spontaneously created—for which the simplest description is "whistling." Indeed, the phenomenon has come to be known as whistler-mode waves. These noises occur at extremely low frequencies. Gurnett has no recollection of whether the anecdote came up during Gallet's seminar, but during World War I, German spies accidentally discovered whistlers while trying to listen in on Allied radio transmissions. The Germans had laid out long cables and plugged them into an amplifier and strained to hear anything of value. Unexpectedly, one thing they did hear sounded as if artillery was flying overhead—which may have prompted a few troops to glance upward in concern.

"Actually, you can hear those in your car radio sometimes during a lightning storm," Don Gurnett would later point out. "You ever heard that kind of crash?"

At the front of the seminar hall, Roger Gallet had not quite finished his presentation. There was more. Through the tape machine he next spooled bits of another phenomenon—auroral hiss. And then, as Gurnett described it, "He also played some other really weird things, things that are called *chorus*. And nobody knew what chorus was at all." Just more sounds, also occurring at extremely low frequencies. The tape rolled.

To formulate an idea of what chorus sounds like, try this: Imagine sitting on a tropical beach as dawn arrives. Sunlight crests the horizon. Birds rise in song, greeting the day, seeking out their mates. Except the sound is coming from up in the heavens. It's galactic audio. Emanating at random intervals and just a little harsh—abruptly coming and going, yet strangely inviting somehow. Friendly. But also musical, no question. Today the phenomenon is generally referred to as "dawn chorus."

In MacLean Hall that day, Roger Gallet explained that the existence of chorus was first recognized sometime during the 1930s. But not widely. Only a few speculative papers had been scratched together. Nobody really knew for certain what might be causing these sounds to occur. "He hardly had any theory of that at all," was how Gurnett recollected Gallet's progress on the topic. "It was just an unknown."

Hearing chorus flipped a neural switch deep inside the brain of the young Iowa student. In an instant had his fascination been ignited. "Yes, I will say that is definitely the case," declared Don Gurnett. "I have an inherent interest in a lot of weird things!"

Everything Gallet played back that day had been recorded on the ground. In a configuration not unlike what those Germans had rigged up, Gallet and his colleagues ran long strings of antenna wire back to amplifiers and recording equipment, and basically waited for noise of any kind to announce itself. What they might get depended on luck and lightning and accidents of physics. But what they'd already gotten had opened the eyes of at least one person in the audience that day. "I'd never heard of *anything* like that before, and I found it fascinating," gushed Don Gurnett. The mystery of it all—the presence of naturally generated sounds that nobody understood the origin of—really grabbed him. "Well, people would guess they were somehow from space, but people didn't know at that time."

Clues did exist. A 1957 paper by one of Gallet's associates recognized, after years of worldwide data gathering, a correlation between chorus and increased activity across Earth's magnetic field. It led to a working hypothesis that "clouds of positively-charged particles" were slapping the earth's outer atmospheric layers, similar to waves crashing on the beach, then falling into alignment with the planet's magnetic field. And along the way these noises were generated, happening to mimic flocks of birds singing in the morning. As Gurnett learned more about it, "They kind of knew from laboratory plasma experiments that if you shoot an electron beam into a plasma, you'll make all kinds of noise." But what exactly was the mechanism of that?

So as Gurnett summarized the transformative MacLean Hall event: "This guy comes here, plays some weird sounds that they detected on the ground, that *came from space*, and nobody knew what they were! And those sounds just interested me. Yeah." Mystery sounds from nature. And to boot, the kind of nature that wasn't animal or plant or anything living. Well, maybe that's not profound enough? Gurnett offered context: "If you sit in a room, a very quiet room, you don't expect the molecules up in a corner of the room to suddenly start whistling at you."

After the seminar ended, Gurnett couldn't clear the sonic mystery from his head. He got to talking with one of his physics professors about it, who suggested, "You know, I think we should try to build a radio." At that time, Iowa physics had a new research satellite in the works, to be called Injun 3. Under construction right there in MacLean. It'd carry aloft cosmic-ray detectors plus equipment to study auroras. Why not add in something to listen for these low-frequency spectaculars? Room could be made for it.

Creating specialized equipment from nothing, rugged enough to fly on a satellite, is not exactly Tinkertoy stuff. The Physics Department ended up hiring the Raytheon Company to fabricate a custom spectrum analyzer tuned to six chosen frequencies. Iowa wanted more but lacked the weight margin. Atop Injun's spherical metal framework, the Iowa physicists mounted a loop antenna because the same type worked well for ground recordings. Once assembled, every part of the spacecraft then had to be evaluated for potential showstoppers. As only one example, the satellite's own radio transmissions back to the ground could likely be picked up by Gurnett's antenna also, creating a feedback loop—again, the same thing Ray Charles would get if too close to a speaker while belting "Hit the Road, Jack" into his mic. Gurnett minimized the feedback problem, in part, by wedging kitchen tinfoil between Injun's antenna and chassis—a solution that typified these early DIY days of space exploration.

With everything more or less in place, Don Gurnett now needed to test his creation as any budding engineer would. And doing so, on sensitive radioelectronic instruments, meant finding a low noise environment. One far away from the droning hum of overhead power lines that would easily contaminate his signals. Gurnett headed for the farm.

It's where he'd grown up—Fairfax, Iowa, a location not-so-far northwest of Iowa City and not-so-far southwest of Cedar Rapids. But far enough away, and primitive enough in trimmings, to serve nicely as a test bed. "Pretty old, pretty big, well-built farmhouse," said Gurnett. As close to the space environment as he could get on a budget. His parental pitch went something like this: "Hey Mom and Dad, I'd like to turn off the power to the whole farm for a while. Is that okay?"

"Yeah," Gurnett recounted, "my dad was very cooperative about stuff like that. No problem!"

On the concrete floor of the 1900s-era farmhouse basement, right near its oil-burning furnace, Gurnett prepped the space-age satellite's radio system and then headed back upstairs and outside. "There's no problem with the radio signal getting in the house or anything like that," he reassured. The farm did have power but overall lacked power lines; incoming service fed a single pole out in the yard. Gurnett moved a switch and the place went dark. Returning to the basement, he flipped on the radio gear, which was all battery powered, including a homemade receiver to process whatever his antenna might snatch from the sky.

After half an hour of listening, he'd heard nothing.

The next night, another half hour. Still nothing.

But on the third night, "whistlers, very clear," came the report. "And that's what made me confident about my design."

Gurnett now knew his experimental rig would capture that elusive natural noise. But Van Allen still wanted more in the way of testing. So once the entire spacecraft came together, a whole group rolled back to the farm and parked it *way* out in the field and away from any electrical interference. It checked out just fine, and Injun 3 next went to the Des Moines airport. The U.S. Navy picked it up with a Lockheed Constellation and flew the thing to Lompoc, California. It went atop a rocket at the air force's Western Test Range and then into space.

Back in Iowa City, the Physics Department had rigged new antennas on the roof of MacLean Hall with cabling strung down to a simple receiving station. "And the first thing we heard was all kinds of whistlers and chorus and— you know, coming from *my* instrument on the spacecraft! That to me was just something I'll remember to my dying day." As the mission evolved, people would spontaneously drop into MacLean Hall just to experience the crazy sounds. They kept a logbook of listening sessions. One entry read, "Sounds like the gunfight at the O.K. Corral."

That spring, Gurnett wrapped his bachelor's in electrical engineering and immediately rolled into a master's program to continue studying natural noise. By then he understood it more. Natural noise depended on the presence of ionized gases flowing in space—what's also called plasma. Any analog sound—such as dawn chorus—is a wave signal, so the object of his affection picked up the title "plasma waves." In a sense, the concept as a whole would become Gurnett's original brand.

What a departure from his original ambitions. Donald A. Gurnett, future professor of physics at the University of Iowa and internationally recognized

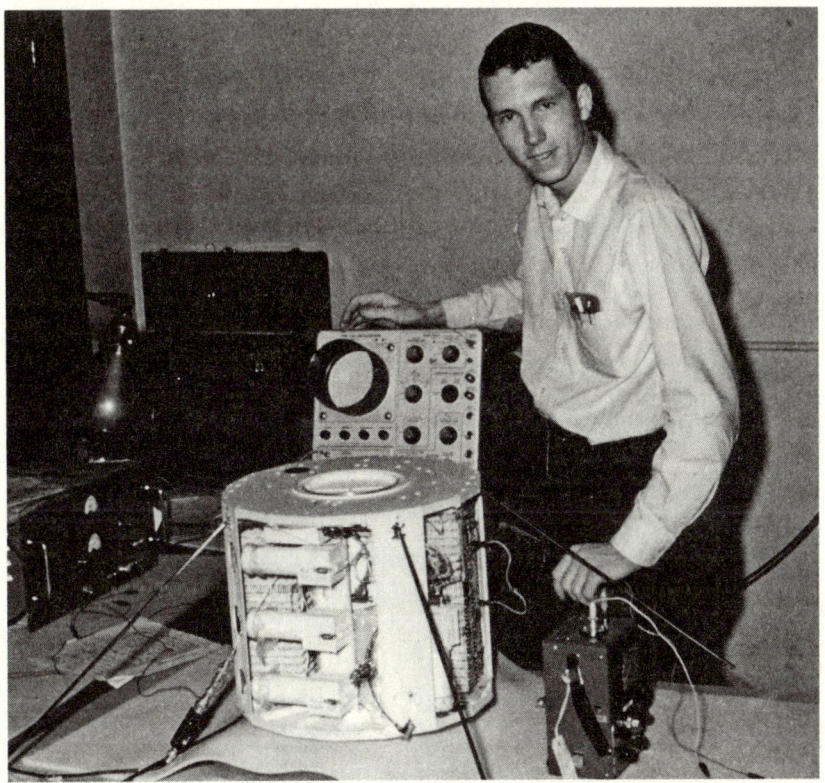

17. In his element, an upbeat Don Gurnett poses with Injun 1 during tests in 1961.
Courtesy the University of Iowa Department of Physics and Astronomy.

pioneer of plasma wave research, had grown up a model airplane builder.
He credits his father for kindling the interest. As a teen Gurnett worked at
Chandler's, the local Cedar Rapids hobby shop, right there on Second Avenue by the railroad tracks. All the while he tinkered away in the basement
on free nights and weekends to create his own models and participate in the
local club. "They used to meet to fly model airplanes out on the runway of the
Cedar Rapids airport. That seems unbelievable now."

With his parents, young Don traveled the country for model airplane
competitions and even brought home top honors on occasion. "Those model
contests used to be run by the navy, and they were sponsored by Pan American World Airways. And they used to give really pretty good-sized prizes for
winning a model event, like a $200 savings bond." He laughed. "You know,

that's worth a couple thousand bucks now!" It led him to think about life after high school and maybe aeronautical engineering. "That was gonna be my life!" he predicted.

Cedar Rapids was home to electronics giant Collins Radio, and Gurnett knew a fellow builder there who'd been putting radios in his large models so they could be controlled from the ground. That led Gurnett to try the same thing himself, revealing a previously undiscovered attraction to electronics. His inner fascination bloomed around the same time Explorer 1 launched in early 1958 with Iowa's cosmic-ray instrument aboard.

An Iowa freshman, Gurnett was amazed that an epicenter of the space age happened to be where he attended school. That April he'd marched over to MacLean Hall and Van Allen's office, hunting for a job. Gurnett made it as far as a secretary. "She had me fill out a form, what my interests were, and I said, 'Radio control electronics.'" He handed back the form and left the building and heard nothing back. "I didn't know anything about Van Allen at all," he admitted, beyond what showed up in the headlines.

Despite the nagging silence, Gurnett couldn't help but think positively. He'd learned a little about the radio system aboard that headline-making Explorer 1. And as cool as it was, it could've been more advanced. "What they flew in Explorer 1 was just crude," he alleged. "Very difficult to analyze." Analog signals coming down had to endure a soul-crushing manual process of "hand reduction" before any usable results became available. Why in the world couldn't the signals be formatted some other way, and be useful right off the bat, without all this processing? Gurnett had been applying simplified approaches to controlling his model airplanes and hoped beyond hope that the Iowa Physics Department might want him aboard to help streamline their own data workflow. Side reason: "I needed money. And they didn't have an aeronautical engineering program at Iowa. And I'd already gotten so interested in electronics!"

At the time, Gurnett lived in a rented Iowa City room on South Johnson Street. A solid fifteen minutes' walk to MacLean. An envelope finally showed up for him in April. From it Gurnett extracted a handwritten letter composed by Van Allen himself. *And?* "He wanted me to join his cosmic-ray group to do electronics design."

What an evolution of purpose! Decades later, assessing a home office filled with old electronics and award-winning models, Gurnett made a fulfilling mental connection. "My work on these radio-controlled airplanes," he smiled. "Almost like somebody was priming me to go work in space physics and space engineering."

After finishing a master's in physics in 1963, Gurnett—at Van Allen's suggestion—began PhD work at Stanford, again in physics. Gurnett figured on staying. "I'd work in the aerospace industry out there, somehow." But he ended up returning to the Midwest. "Look, when Van Allen offers you assistant professorship?" he blurted. "I mean, I'm going to take it. Not only that, I knew the whole engineering operation here at Iowa. You know, the machine shop, the electronics. I mean, I worked with all these engineers and stuff, so it was pretty natural for me to come back here."

The doctorate was complete by 1965. That autumn, Gurnett started teaching "university physics"—a two-semester ordeal of doom for freshman and sophomore undergrads. Some college professors will argue that taking a class is nothing compared to instructing it, and Gurnett had to juggle his new teaching duties with ongoing space missions. See, completely separate from anything NASA, Iowa physics ran its very own space program courtesy of a longtime collaborative relationship with the U.S. Naval Research Laboratory. It was modest yet bountiful. Altogether there were teeny missions and premier ones, and come the late 1960s, Don Gurnett wanted aboard *Pioneer 10* and *11*. Van Allen had made the cut with a cosmic-ray detector. But Gurnett? Despite a growing résumé of flights and discoveries, he had no such luck with his radio and plasma wave experiment.

"NASA did something I think was very poor," came his opinion of the Pioneer selection process. "They would invite you *and* your competitor to give a presentation in front of the selection group." But when finished, each was expected to critique the other guy's proposal, right there in front of everyone including the competitor. Just pick it apart right down to the crumbs. "It was awful," alleged Gurnett. "It led to lifetime animosities between certain people in various fields of research."

His Pioneer opponent happened to be colleague and fellow plasma researcher Fred Scarf from the Space Sciences Department of TRW Systems. A well-respected and well-liked colleague. "It was brutal criticizing each other. He—he wanted short electric antennas, and I wanted rather long antennas." Gurnett explained that shorter antennas were mechanically simpler—no need to fold for launch, then deploy with some convoluted system that added weight. But longer, thirty-foot ones, in Gurnett's opinion, offered more sensitivity.

There was no way he and Scarf were going to agree on such a fundamental difference in approach! It came out in full force during the critiques. And neither one got aboard a *Pioneer.*

"So that was not a good start for our relationship."

When an invitation went out for Grand Tour experiments, "I joined with Fred Scarf," Gurnett explained, in neutral fashion. According to him, the collaboration really came down to their scientific pursuits and the quest for legitimacy. "Plasma waves were regarded as just a curiosity. Strange sounds that weren't important," he claimed of the prevailing attitude at the time.

JPL's original Grand Tour pitch championed four unprecedented flights: a dual launch going Jupiter–Saturn–Pluto and another dualie going Jupiter–Uranus–Neptune. All the regulars—imaging, spectroscopy, cosmic radiation— would want aboard. Why give precious instrument space to a new-in-town wannabe like plasma waves? Who'd even care? At press conferences, NASA always handed out printed eight-by-ten photos from the spacecraft cameras. Would they soon have to start handing out reels of audio tape?

"Fred and I both struggled to try to get them to put a plasma wave instrument on," Gurnett continued. In 1971 the duo convened half a dozen others to address what kind of equipment made sense to have on board. As might've been predicted, the contentious booby trap of antenna style detonated almost immediately. Scarf wanted his shortie-guy antennas. "The argument that he made was not really based on the physics merits," Gurnett insisted. "Rather, on the basis that this type of short antenna was much lighter and more easily accommodated on the spacecraft."

Scarf wasn't alone with his preference. Would that really help them get on?

To the group, Gurnett brought up a sensitivity advantage of using longer antennas. Naysayers argued that Jupiter, being the hugest radio source anybody knew of, made sensitivity a moot issue and that only the strongest waves would be important out there.

Don Gurnett had to make a tough decision. He recognized the advantage of size reduction. "It was apparent that there was going to be considerable competition to get instruments on this spacecraft, with many potential conflicts." So he caved and went in on a proposal using short antennas with Fred Scarf as the principal investigator. It was kind of a half-court shot.

Thirty-one year-old Gurnett felt excitement for the mission. But then received a mailing from Scarf, just weeks into the new year, addressed to everyone on the plasma wave team:

> As you know from the newspapers, Grand Tour is now dead, but there is a chance that a Mariner-Jupiter-Saturn (MJS) mission with two launches in 1977 may survive. We should know more about this within the next few months.

Indeed, NASA had pumped the brakes on a four-ship Grand Tour. It was too ambitious. Too expensive. Congress would never fund such blue-sky fantasy. In response, the Lab hastily churned through options for a drastically scaled-back trip, attempting to erect a shield against the Grim Reaper's scythe. "MJS-77," as the downgrade was uninspiringly termed, would visit two planets only and use a proven spacecraft design to do so.

Scarf followed this bad news with a copy of his letter to Ed Smith, a JPL-based researcher of space plasma. Over four taut pages, Scarf (on behalf of a larger "Outer Planets Grand Tour Plasma Wave Team") tried to rationalize the accommodation of plasma waves on MJS. What the researchers were proposing "imposes no special requirements on the spacecraft trajectory, data system, storage, etc.," as Scarf put it. And the instrument would be capable of detecting lightning at Jupiter. Great goonies! Didn't anyone want to know if Jupiter had lightning?

"In addition, these measurements will provide invaluable general information that can be applied to other areas of plasma physics and astrophysics." Scarf asked for only 10–15 percent of the available telemetry during cruise to the planets. He didn't even care if it went aboard a stabilized spacecraft instead of a spinner.

NASA's formal announcement of opportunity to propose MJS instruments hit the streets on April 14. Pay attention! If you want on board, have your paperwork in to headquarters, attention Mike Mitz, by 4:30 p.m. on August 31 or else don't even bother. Use a descriptive cover page. Limit the abstract to 250 words. Follow it with a one-page summary of your physical hardware. List the scientific objectives and anticipated results. Budget out a cost plan involving one breadboard unit, a prototype, and three flight units. Dammit, think: How are you going to calibrate these things? How much computer time will you want? What is every single expense you'll incur greater than $50,000?

Oh, and break everything down by fiscal year, would you? Don't forget about completing appendix J, the "Cost Summary Form."

Your delivered, cover-to-cover proposal can be no fatter than twenty single-spaced, typed pages. Obey the warning sentence on page I-2: "Photographic reduction of conventional typing is not approved." So no cheating by shrinking down the type size, because they're on to that! Want them to acknowledge receipt of the proposal? Then tuck in a self-addressed postcard, and they'll mail it back. Expect the first meeting of the experiment team to happen in December.

If you get on.

As team leader, Scarf oversaw the proposal's construction. He collated everything and sent it up the chain and wished for the best. Don Gurnett recalled his own disappointment upon hearing that December's bad news. "We didn't get selected," he mourned, unhappily. Ninety other scientists were, for a total of eleven experiments.

Precious weeks ticked away, during which JPL fleshed-out the MJS mission plans and hardware requirements. Drafters sketched out the lines of the ship: electronics compartments, radio antenna, boom arms. Excitement built on the mission team and in the press. Being left out put a crease in Don Gurnett's heart. "It looked pretty hopeless, actually."

Equally unhappy was Gurnett's boss, James Van Allen. From behind his ever-present tobacco pipe, Van Allen voiced a polite level of frustration about Scarf's pecking order. Okay, fine, they were all losers because the proposal had been rejected, and nobody would be sending a plasma wave experiment past Jupiter. But even if they went, how come Scarf would've been king pooch—the principal investigator? As Gurnett pointed out, "The instrument was being built at Iowa, and we had all of the responsibility to produce the instrument." So why not Donald A. Gurnett as PI?

For the immediate future, those eleven MJS experiments would inhabit a kind of gray zone because the requirements for each had to be verified against what the real spacecraft could ultimately provide. During this interval, Van Allen prodded Gurnett. "If you want to get something done with the government, send them a specific proposal. They are obligated to respond." Send it right down the throat of that Mike Mitz guy.

Today Gurnett laughs about the supposed obligatory response. "I don't know if that is legally correct, but it does give some insight to how Van Allen dealt with issues of this type!" And for the kicker, Van Allen wanted Gurnett, in the new proposal, to add some torque to the plea. "He actually urged me to negotiate with Mitz over my becoming PI—potentially not agreeing to build the instrument unless I was named the PI." Van Allen told him the original pitch was a dead horse anyway, so why not swing for the fences?

Gurnett felt unsure. "This was obviously a difficult situation. I discussed the options with Van Allen. And not being certain I could win such a confrontation with NASA's decision, stated that I definitely wanted to be involved in the first plasma wave measurements" at the outer planets. "So I called Fred Scarf and discussed the situation with him. We were on pretty good work relations." Yeah, there was that weirdness when trying to get on the *Pioneers*, but things had been patched up during the original Grand Tour proposal.

"My priority was basically *science*," Gurnett affirmed of his intent. He wanted to fly a plasma instrument to Jupiter badly enough that he was willing to sacrifice the prestige of being principal investigator. "I proposed that we have a mutual agreement in which we share equally in the science." After each planet, their initial observations would be submitted to some prestigious journal. And they would trade off whose name came first as they went along.

Scarf embraced the plan. Gurnett offered more details: "Also, he agreed that my students could play key roles in any such papers that came about as the result of our collaboration. Nothing was said about interstellar space." After formalizing this relationship in a letter of understanding, the men now needed a minimally invasive way to get their hardware aboard the MJS ships. Not necessarily a trick—just a clever angle. Something so easy that Mike Mitz couldn't say no.

Measuring plasma waves means using antennas, and the guy with a radio astronomy experiment on MJS was slated to have two honkin' thirty-footers all to himself. Formed from an exotic mix of beryllium and copper, they'd be a half-inch thick apiece. Gurnett and Scarf approached the guy about modifying his radio receiver. Maybe he could extend its frequency range downward enough to catch their waves? According to Gurnett, the guy didn't seem real interested in plasma wave research. But he tried to work something out anyway. Said Gurnett, "I think he just wanted to see it happen because the antennas were there, and it just made sense."

The air filled with question marks. Could Scarf and Gurnett tie into those antennas without degrading the performance of the radio astronomy hardware? Could piggybacking lead to some unforeseen in-flight failure and bring the house down? The radio astronomy instrument was already approved and paid for, and here they were trying to monkey with it.

That's when Don Gurnett ginned up his fabulous angle—recycling. In November 1971, a modest satellite called Explorer 45 had launched into Earth orbit carrying a plasma wave instrument of his. It was tiny and low-powered and, best of all, proven. Why couldn't he and Scarf reuse the design?

With no time to lose, Gurnett dispatched a direct appeal to Mike Mitz, program scientist for the Outer Planets Mission at NASA HQ:

The University of Iowa
16 February 1973

Dear Dr. Mitz:

We all feel that it is very important that plasma wave measurements be included on the MJS missions.

Since an electric antenna is already available on MJS (for the radio astronomy experiment), it is only necessary to provide the electronics for the plasma wave measurements. In order to minimize the cost and provide an instrument with flight proven reliability, I have proposed that we fly an instrument essentially identical to the University of Iowa plasma wave experiment flown on the S3-A satellite.

The estimated cost to provide this experiment is $479K. Since this cost is based on instrumentation which we have already produced and since we have had considerable experience with similar instrumentation on other projects, I am confident that this is an accurate and realistic estimate.

I sincerely hope that some arrangement, either with this or similar instrumentation, can be worked out to provide plasma wave measurements on MJS.

Mitz took a month to respond, sending only five brief sentences. One read: "We have no plans to reopen or increase the scope of our selection." He dangled one measly carrot by admitting how the design phase was still underway. If things changed, he'd be in touch.

And in late July 1974, they did. Gurnett himself learned about it after trailering his sailplane from Iowa City all the way down to Hobbs, New

Mexico, for a soaring competition. The news reached him sometime during the twelve-day event, and he mentioned it to a reporter on scene for the local *Daily News-Sun*. (The Foodway store in Hobbs had whole chicken fryers on sale that week for 39 cents a pound, FYI.)

Scarf and Gurnett were in. They'd lucked out at someone else's expense; an already-selected French ultraviolet photometer would have to come off. "That experiment did not get confirmed because we had the *Pioneer* flyby of Jupiter, which found that the radiation was a thousand times more intense than expected." So came the explanation from Ed Stone, JPL's project scientist for the entire MJS effort. Such high radiation levels meant the MJS ships would have to fly much farther away from Jupiter than originally planned, thus precluding useful results from an ultraviolet photometer.

Iowa's surrogate hardware would mount directly atop the electronics box for the radio astronomy experiment. And with this sort of parasitic arrangement, took on the name *plasma wave subsystem*. An April 1975 formal agreement letter from NASA spelled out the official hierarchy:

Dear Dr. Scarf:

I am pleased to confirm the selection of your Plasma Wave Investigation for the Mariner Jupiter/Saturn 1977 mission with you as the Principal Investigator and Dr. Donald A. Gurnett, University of Iowa, as your only Co-Investigator.

So what if Gurnett wasn't the PI? They had a green light—not to mention the greenbacks—to try and snatch the as-yet-unheard sounds of Jovian dawn chorus from the ether. And the effort to build the equipment to capture those sounds, and to calibrate it, and to attend all the meetings and project reviews and encounters to support it, and to publish all the findings? That effort commenced the instant Don Gurnett saw their names in attachment V of the paperwork.

11

A World with No Corners

It's much easier to get in the paper talking about the Shuttle
than it is about some unmanned spacecraft.

—Brian Duff, former NASA public affairs officer

NASA WEIGHS DEFERRING 1982 MISSION TO JUPITER
By Thomas O'Toole, *Washington Post*, September 4, 1979

The National Aeronautics and Space Administration is considering
a plan to delay its 1982 Galileo mission to Jupiter and split it into two
flights that would add more than $60 million to its cost.

The two-year delay, if it comes, is the result of delays in developing
improved versions of the space shuttle engine called the "109 percent
engine" that burns at higher temperatures, delivers more thrust, and
can carry more weight into orbit.

On Valentine's Day 1980, the Winter Olympics had just opened in Lake
Placid, New York. JPL director Bruce Murray was not attending and had no
idea of the impending "Miracle on Ice." Instead, Murray stood in a Lab meeting room before visiting members of the House Subcommittee on Space Science and Applications to essentially make the case for his job.

"Mr. Chairman, members of the Committee," he began, "I would like to
present a personal overview of how I see the planetary program at this point.
In a sentence, it is JPL's feeling that the planetary program is in deep trouble."

The referred-to chairman, Representative Don Fuqua, applied a patient face
and listened as Murray expanded. They were coming off some great years—what
with Pioneer, twin Viking ships to Mars, and now the *Voyagers* at Jupiter. Murray termed them "an awesome demonstration of excellence and commitment."

He now summoned everyone's attention to a seriously overcomplicated
graph he'd prepared but unfortunately failed to edit for clarity. If stared at
long enough, it showed an upcoming and pronounced gap in planetary explo-

ration. He termed that "a blackout" and "the disaster," and it was only there because of the shuttle's budgetary chokehold on most every other NASA effort.

Firing from various angles, Murray suggested that JPL efforts benefited all of humankind. "We have led this process of peaceful exploration, the one that is almost universally without negative side effects, and we are appreciated throughout the world for that." He insisted the Lab spent its money as carefully as possible. And facing a true blackout of missions and science and discovery, JPL now needed a top-down commitment to the future of planetary exploration. "We have been living off the capital of the 1960s and 1970s. And unless we make the decisions to reinvest, the game is really going to be over."

Back in 1972, and working through his first decade in politics, Don Fuqua had begun chairing a now-defunct subcommittee on human spaceflight. It'd evolved into the current one on space science and applications, which sat under the larger Committee on Science and Technology.

"When I became chairman of the full Committee in 1979," Fuqua later recalled, "I made a very conscientious effort to reach out to the science community and say, 'Listen, we've got the shuttle program. But also, I'm very interested in science. Now, we don't have an unlimited amount of money. But let's see if we can't come up with some of the best science programs and I'll fight for you! I will lead a fight.'" So with this particular representative—a lifelong Floridian and ex-dairy farmer and self-proclaimed space enthusiast—the Jet Propulsion Laboratory had a known advocate. JPL needed this guy, even if he'd also wanted shuttles.

And for a man who'd spent his career in politics, Don Fuqua lamented how political the space shuttle had become. "I thought it made sense; I'll tell you why," he'd later say about backing the controversial program. "Once we landed on the moon, there was no more use for Apollo. But what do we do?" He loved the idea of an *ongoing* program. Of no longer throwing away launch hardware. "We were going to build seven, eight, or ten of them and fly them. And it would be so routine. It would be like flying an airplane." Alongside a Republican committee member, Democrat Fuqua had attended the shuttle's first landing test in August 1977. "We both had tears in our eyes and we hugged each other," he recalled of the safe landing. "It worked!" Imagine that—a literal bipartisan hug.

Murray handed the floor over to JPLer Henry Norris, who talked up a proposed international collaboration to explore the sun with two separate spacecraft: one American, one European. Norris's project, which went by the fun-to-say name Solar Polar, would also rely on a shuttle for its initial boost to Earth orbit. Everybody working Solar Polar was holding their breath over whether the shuttle could lift its advertised sixty-three-thousand-ish pounds and carry both spacecraft in one go. If not, plan B was dual launches. It prompted some awkward discussion in the room over whether the Europeans might be retroactively asked to pay for their own launch.

"Next we will look at Galileo," said Henry Norris, and introduced John Casani.

"Good afternoon, Chairman Fuqua, members of the Committee," began Galileo's project manager. Casani had easier-to-understand slides than that of his boss, including what hoped to be a positive twist on the recent shuttle delays. Namely, what the *Washington Post* had mentioned in its news article— divorce. Splitting the spacecraft and launching via two shuttle missions.

What—chop the ship in two? How in the hell could such a thing be split? Were they maybe doing away with the whole dual-spin approach and going instead with two orbiters, where one did the spinning while the second one didn't?

Looking back decades later on these times, and sitting just a few miles from where the Valentine's Day meeting had gone down, John Casani accepted another glass of red wine from the waitress with whom he'd been only lightly flirting. Casani's hairline had changed but not his command of specifics.

"Different launch years required different launch *energies*. When the launch energy requirement exceeded what the shuttle could deliver, we had to find a solution," he began. "One solution was to take *the probe* off the spacecraft." Such a reduction would allow the shuttle to achieve the required launch energy.

Whoa, whoa, take the probe off? So what would happen to it? Would some luckless bastard have to call up every probe experimenter and tell them, after years of development, "Um, sorry, it's not going"? Not at all. It'd fly on a second shuttle. But by itself, the probe was helpless. It couldn't just rattle around loose in the cargo bay like the marble in a spray-paint can. And it surely couldn't get to Jupiter all by itself, either. It'd have to be mounted on something—a little carrier ship, with a radio and maneuvering thrusters and other bare minimums.

To be clear: splitting the mission meant creating a brand-new probe carrier spacecraft. Able to navigate all the way to the outer solar system. Able to radio back every second of probe data during its suicidal descent. At least it would fly directly from Earth. "Lighter in terms of weight and requires less from the launch vehicle," Casani explained between sips.

The orbiter would still angle past Mars and pick up a gravity assist. But not a big enough one compared to their original plan. "Mars has moved considerably, relative to alignment with Earth and Jupiter, making the gravity assist much less effective than for 1982." That meant a supplemental "flyby module" kick stage on *Galileo*'s back end. It needed to be configured and tested and fitted to the spacecraft, and would not cost zero dollars. A contract for the probe carrier went to Hughes.

"This was a major change to the program. But essentially no change to the probe," Casani summarized, trying to sound positive and reinforce the upside. He daubed his mouth with a napkin and waved at some women departing a nearby table. "Bye, ladies! Bye!!"

Going first would be the orbiter, in February 1984, as an intact whole. Probe and carrier would leave about a month later. Luckily though, all these delays and changes were going to have a minimal impact on the overall science objectives. That's what Casani told the representatives in the conference room on that February day in 1980: "Some are enhanced somewhat and some are eroded somewhat. But, on balance, I would say that the full science value has been at least preserved."

In speaking so high level, Casani deliberately sidestepped the agony of discarding JPL's otherwise complete mission design. He didn't bring up any of it: the nuanced arcs of the journey to Jupiter. The exact moment of probe release. The trajectories of every Jovian orbit. The specific moons *Galileo* would fly closest to. The exact usage of the onboard science instruments. The exact angles by which sunlight would illuminate photography targets. The durations of closest encounter. The intervals of committing science data to *Galileo*'s onboard tape recorder and the timing of playback for transmission home. All that meticulous planning was out the window. But the subcommittee had been spared these details because who cared about some guy in an office planning trajectories? He could just plan them again.

"What was the cost of the delay?" called out Representative John Rousselot (R-CA). This congressman wasn't on the subcommittee—or even on the parent committee. But he'd tagged along to the proceedings because he was local.

Answered Casani, "The cost of the delay is in the range of $150 to $200 million."

"You will explain to us how it was worth the delay?"

"I will do my best."

Don Fuqua jumped in on defense. "It wasn't their fault," he soothed.

"I know that," sneered Rousselot. "I thought I would bring it up."

There's an old joke that everyone is in sales, no matter what they do. And despite avoiding the candy-selling business, part of John Casani's job was selling space missions. "It is still a bargain," he countered to Rousselot, carefully navigating the room's tension. Quickly he changed gears with a fun chart of updated hardware. "So the new work that we are concerned with is the probe carrier, the Mars flyby module, and the interfaces that have to be accommodated."

He didn't have it handy at the time, but Casani's delay cost broke down like this: First came an estimated $56.5 million to create a new, didn't-know-we-needed-that probe carrier. Plus $105.9 million for major incidentals such as modifying *Galileo*'s thermal protection and redesigning the mission. Add inflation. Then factor *out* a savings of $21 million on account of a shorter program overall. The final number at the bottom represented an increase of $177.8 million—smack in the middle of Casani's projected range.

To the group, he now reinforced how dependent their revised mission still was on the space shuttle's lifting capabilities. Another chart went up—Casani's last—translating this capability into the *Galileo* orbiter's resultant size. And mission. While explaining this, Casani spoke of "burnout mass"—essentially, the weight of the orbiter minus any propellant it carried. Even without the probe, they had to add some 1,300 pounds of propellant to account for an unavoidably inefficient flight profile. This drove up the overall weight of the flown orbiter. Not as much as with the probe back in . . . but a not-insignificant amount, either.

Don Fuqua threw out a question. "Are you going to have any problem with having too much weight?"

Wouldn't it be great to answer with a simple yes/no? But Casani's reply had to accommodate the fluctuating estimates of shuttle capacity. Numbers seemingly changed with the wind. (Indeed, the final test of a shuttle's solid rocket motor had occurred only the day before. And initial testing of that "109 percent" engine wouldn't happen for another month.)

Galileo planners were counting on roughly sixty-four thousand total pounds of shuttle capacity. But nobody yet knew if JPL would get it. Casani kept talking. "Now, should that performance not materialize, there are a couple options available to us." For starters, they could leave out some of that additional propellant he'd just spent the last couple of minutes rationalizing. The ship would still reach Jupiter but fly a shorter mission and fewer orbits.

"If the shuttle comes in, say, at 61,000 pounds," elaborated Casani, tracing a hand downward along the graph, "then we have to take fuel out, and we would move down this curve here." At that decreased capacity, they were talking fewer than five orbits. He paused near the figure of 60,500 pounds. "And at this point, as you see, we would only have enough fuel left to do a single orbit."

Casani faced the group. "That would be judged, I am quite certain, not a worthwhile mission." They had a backup trajectory, he acknowledged, requiring much less shuttle capacity. It'd add two years of flight time and $50 million to the cost.

"The Appropriations Committee has been talking about using a Centaur," piped up Chairman Fuqua. He wanted to know its potential impact. Casani told him that with Centaur instead, they could do the whole mission with a measly fifty-five thousand pounds. Bruce Murray jumped in—contending that even with Centaur availability, the shuttle should still be engineered to its full potential.

"I think it would be a real disservice at this point in time to switch to Centaur," preached Murray. They'd put all this effort into splitting the mission. Just leave it the way it is. Spreading the mission over two launches is kind of like spreading the risk, don't you think? His comments implied a distinct lack of excitement for planning the mission yet a third time: "This project cannot stand the disruption of another major change in launch vehicle."

Murray took another challenge from John Rousselot—this time about the rationale for undertaking such explorations in the first place. Murray went philosophical, talking up how much the general public loved pictures. "Here at the highest level of the government," he then screeched, "this poor little project is being scrutinized with a magnifying glass!" Murray plunged into another colorful justification, focusing on a somewhat nebulous theme of "the right thing to do."

"And Galileo *is* the right thing to do, as far as Jupiter is concerned," insisted JPL's director.

The meeting ended and the group dispersed. The House subcommittee faced busy days to come, with a trip the next morning to Rockwell in El Segundo—to hear of that contractor's progress in building America's first space shuttle.

Tense meetings had become a major aspect of John Casani's work life. Before heading home, another aspect would sometimes begin to play out. Often it started with a quick phone call to Lynn, announcing his departure. She'd report a recent act of insubordination on the part of their children.

"When you get home from work," she'd say, "I'll tell you which one needs some corrective action. And you'll exercise it." Having arrived home, Casani would be greeted with her saying, "Hi, dear." She'd then point accusingly at one of their preteen boys. "*Him.*"

Typically, Casani never asked why or what happened. Didn't matter. Wordlessly he'd march the defiant offender aside and tell him, "Face the wall." Followed by, "Assume the position." Every kid in the house knew the drill: Put your hands up high on the wall, and stick out your butt. One open-handed smack to the tuckus, and that was it.

"I tell people about it now, they're absolutely horrified," Casani informed his restaurant guest. Out of tune with what's right. Inhumane.

"Well, it was pretty effective!"

Galileo's divorce ultimately failed. Cost and technical issues had been plaguing the special three-stage air force booster; NASA's outgoing administrator really wanted to shitcan it. Besides, high-ranking advocates of the whole Centaur-in-a-shuttle arrangement had been championing that thing for heavy military satellites. Where there's military, there's money, and if Centaur became a reality, the air force itself looked to be an even larger shuttle customer.

Congress approved it on January 15, 1981. Now, as envisioned by Bruce Murray, "the projected Centaur-in-Shuttle combination would be so powerful, that an all-up *Galileo* spacecraft could be launched directly to Jupiter with a single Shuttle."

Um, probably there's going to be some bad news, right? "The change would add another year's delay to Galileo," Murray conceded. Launch day now pushed to April 1985. With sadness, Bruce Murray noted how Casani's project team—already in place for more than three years—was still as far from a launch as when they began. This new direction of recombining the orbiter and probe back into a unified machine, not to mention a transfer over to Centaur, also jacked up JPL's own costs even further.

So that custom probe carrier spacecraft? Yeah, stop with that and get back to work on the single-shot mission. All you trajectory people? *No* close encounter with Mars on the way out. Stop working on that. And hey, mission designers! All those orbital calculations and science instrument usage plans and data-collection schedules and everything else you've been slaving away on? Guess what? Start over on a new mission, again, from the beginning, for the third time.

Reconfiguring proceeded despite fresh threats. Bruce Murray learned that the new Ronald Reagan administration's budget director wanted to cut $629 million from NASA's budget, and—no question—it was going to hurt *somebody*. Galileo made the short list. One week had elapsed since the Centaur switch. No shuttle funds were to be impacted by the impending cuts. Murray knew the reason: "national security."

Transient thoughts of frustration and despair echoed up and down the sometimes steep roads and walkways of the Jet Propulsion Laboratory. And John Casani picked up on these feelings. His Galileo team needed positivity. To know that things would come out okay. He took to the printed word as JPL began publishing the *Galileo Messenger*, an internal quarterly newsletter. In a bit of homage, the name honored Mr. Galilei's landmark 1610 publication *The Starry Messenger*, in which the discovery of Jovian moons was first announced.

On average the *Galileo Messenger* ran a breezy four to six pages, addressing the latest news, profiling a member of the team, and highlighting other mission aspects such as one of its science investigations. But no matter what ended up in the newsletter, it always expended a few column inches on a page 1 address titled "From the Project Manager."

For its inaugural issue, Casani authenticated the rumors about *Galileo*'s new Centaur-based delay. He mentioned the effort to hold down costs. And tendered a confident message of hope:

> I would also like to remind all of you, especially those of you who may be depressed because you have now seen the launch date slip twice, that Galileo has received extraordinary support during the recent budgetary discussions with the new administration.

In encouraging yet somber prose, Casani noted how Galileo currently stood alone as NASA's only mission of planetary exploration. Yeah, it wouldn't leave the ground for a very long time but would emerge a survivor all the same. JPL's first issue of the *Galileo Messenger* came out on April 10, 1981—only two days before America's space shuttle, our Father who art in heaven, finally flew for the very first time. It was years over schedule, hundreds of millions over budget, and hyped to all get-out. But up it went to circle Earth. And hallowed be thy name, it worked as intended.

April gave way to July. On page 1 of the next *Messenger*, Casani conveyed to each reader the raw heat of budgetary demands. "In a very real sense Galileo has made a fixed price commitment to NASA Headquarters and the new administration," he wrote. "I ask each of you to join with me in increasing the cost consciousness of the Project at all levels."

The Lab's collective mood undulated with the realization that Galileo might be everyone's final hurrah for the near future. "We have grown accustomed during the last decade to a new unmanned planetary spectacular every year or two," complained engineer Gentry Lee in that same issue. "And now it will be at least five years." That "disaster" of a gap prophesized by Bruce Murray had become a real thing.

It happened suddenly: President Reagan's budget director killed Centaur at the end of November 1981. He did this by decree and, while under such momentum, axed the whole of Galileo. Also slated to die was funding for *Voyager 2*'s eventual Uranus encounter. For the moment these cuts existed only on paper, courtesy David Stockman, director of the Office of Management and Budget (OMB). (To clarify, the OMB dwells within the executive branch and administers the entirety of the country's budget.) "I was as much of a space buff as the next person," defended Stockman. But a few spacey specifics apparently had to give in order to arrest the country's budgetary expansion. "NASA was hardly suffering," he alleged. "Even with the cut, its 1982 budget would be 11 percent higher than 1981."

What a blowout. *No* new planetary exploration startups were set to be approved for the whole of the following year. Bruce Murray's gap had widened into a monstrous crevasse. He'd spend the Christmas season trying to fill it in.

December 10 saw President Reagan's science adviser, George Keyworth II, testify before the House Committee on Science and Technology. In his position as committee chair, Don Fuqua beckoned colleague Representative Ronnie Flippo (D-AL) for introductions.

"It is a pleasure to have you here, Dr. Keyworth," began Flippo, who then gracefully deployed his ambush. "In the *Washington Post* you were reported to have recommended against the continuation of NASA's planetary program. Is this true?"

While acknowledging "that rather crisp statement attributed to me," Keyworth failed to directly answer. Instead came a suspiciously prepared-sounding dodge: "We have produced a new era in space science with the new capability of the Space Shuttle." He demurred on explaining what the era or capabilities were or how the space shuttle could be considered an element of scientific research. Good thing for him Bruce Murray wasn't attending.

Digging for subjectivity, Flippo sent another cannonball arcing. "How do you feel about the Galileo project? Have you supported it?"

Again, Olympic-level waffling from the presidential science adviser. "I think it is a good mission," he stated. And then, finally, a hint of things to come: "The question of whether we can afford it is under consideration now."

"Have you specifically supported it?" charged Flippo.

But Keyworth didn't budge, insisting that he couldn't comment "on a specific budgetary issue under consideration at this time. Excuse me." The courtroom-style exchange trailed off into competing disagreements over the relative merits of military-based versus civilian research and development. And concerning the relative merit of planetary exploration, that was the end of it.

At least, that was the end of it as far as testimony that day. In a newspaper interview appearing the same month, Keyworth recommended a cold halt to all new planetary space missions for at least the next decade. A piqued John Casani would confront him the following February at the National Space Club in Washington DC and more or less demand, Are you trying to cancel Galileo?

Keyworth clarified. No, he protested, that wasn't the intent at all. Really, he'd been referring to new start-ups of additional huge projects *like* Galileo—or Voyager, for that matter. These $500 million to $1 billion projects, claimed Keyworth, are just too massive for taxpayers to swallow these days. Give it a rest for the next decade. In the meantime, think more about budget missions—$200 million a pop, or maybe $300 million if you've got something really clever.

George Keyworth, as popular as he wasn't with JPL types, had nonetheless foretold their future.

Despite an eventual restoration of funding to Voyager—and Galileo—JPL had lost its battle to keep Centaur. The world celebrated New Year's Day 1982, but it came off as bittersweet in Pasadena because the month had arrived for *Galileo*'s original launch. Instead of sending their dream ship aloft with excitement and cheers and promise and fist pumping, Galileans once again would be redoing the spacecraft design, interfaces, mission planning, and everything else. Back into the mix went the underpowered air force booster—in a downsized, two-stage configuration no less. To it would be added a supplemental JPL kick stage fancifully called the injection module. Launch dates pushed back by only a month, from April to May 1985.

But the trip now entailed extraneous maneuvering to reach Jupiter with the low-energy booster. From a shuttle they'd launch toward the sun and, after a year, blast the orbiter's maneuvering rockets big time. Then wait another year before picking up a gravity assist—back near Earth this time, only 140 or so miles above terra firma. One big two-year loop just to collect more energy! Following the earth flyby, *Galileo* would coast along for a couple more years until reaching Jupiter in late 1989.

They still had a mission. But this latest change forced more redesigns—partly due to that huge maneuvering burn. It'd practically empty the current tank sizes, whittling down their safety margins during the flybys of the Jovian moons.

Casani glanced over his low-tech notepad. Modifying the spacecraft, once again, for a new configuration, plus reworking the thermal shielding, plus enduring a new mission analysis, plus developing the new JPL-designed injection module, added up to $45.6 million. Day-to-day operating costs went up also, by some $122 million, due to the strung-out mission time. The total project budget now cleared $829 million.

It really seemed to be a go this time. By mid-May 1982, most *Galileo* hardware had escaped the purgatory of design for the greener pastures of fabrication. The probe began integration tests. Managers looked forward to orbiter part and subsystem deliveries later in the year. The team had begun moving off paper to reality.

JPL wasted six months rejiggering things for the air force booster. The project team discovered this on July 20, when Casani brought to order an all-hands meeting inside the Lab's famed Theodore von Kármán Auditorium.

Located on the west side of the campus, it's handily adjacent to the parking lots, in Building 186. There, Casani informed everyone that large aspects of the mission were changing yet again. A bunch of stop-work orders had just been sent out to every subcontractor on the special injection module. It was coming out because two days prior, Reagan had signed an appropriations bill restoring money for the Centaur stage.

Not knowing for certain, Casani suspected that the about-face happened because commercial satellite customers were flocking to Europe's Ariane space booster. With Shuttle/Centaur back in the mix, America could maintain a competitive edge. Even absent Ariane, the U.S. Air Force was going to need a larger stage anyway. Why not go with America's proven workhorse—the Centaur?

Obviously, switching back involved headaches and compromise. Casani remained adamant about minimizing hardware changes. So they planned, for example, on devising a custom interface plate to fit between *Galileo* and the Centaur. One side would mimic the injection module and mate to *Galileo*'s current back end. The plate's reverse side would imitate *Galileo*'s previous back end, when fitting to Centaur the first time around. Launch day slid even further down the calendar, to May 1986. Cost impact?

Replace air force stage with Centaur: $131.4 million.
Money saved from cancelling air force stage: −$32.8 million
Balance forward: $98.6 million
Decrease in overall costs due to a shorter mission: −$94.8 million
Total project cost change: $3.8 million

A bunch of already-done work for the original mission, subsequently redone for the split but then undone, now had to be un-redone for the new-old single-ship version. As only one example, check out the relay radio scheme. See, the probe's radio lacked enough power for transmitting data all the way to Earth. So as the probe entered Jupiter's atmosphere, it would instead transmit to the orbiter. This required a three-foot-diameter receiving dish to swing out from the orbiter's body. Splashing into the dish on dual redundant frequencies, the strength of the as-received signal would vary—along with the radio frequency itself—as the probe sank increasingly deeper into a fundamentally mysterious environment. The orbiter's relay radio system would have to find this signal, from more than 124,000 miles away, broadcasting on a scarily narrow thirteen-degree beam width. It had to do this in a completely automatic manner, within fifty seconds of the probe initiating broadcast, and remain locked for the duration. Priceless data would then travel through internal wiring to the orbiter's massive high-gain antenna, which always faced Earth, and depart for home. It had to work the first time, and the only time, for sixty minutes of lifetime.

JPL drafters had already un-drawn the relay system from the probe carrier and redrawn it back onto the orbiter. But hold up. The receiver antenna would have to be relocated on account of the orbiter's changed orientation at probe release. It'd now be facing differently as the probe began descending. Equipment on the orbiter's spinning upper half could potentially obstruct the relay antenna's field of view and disrupt reception of critical probe data. The geometry was all catawampus.

These days, managers can track issues using elaborate software capable of generating colorful diagrams showing who must do what and when. Milestones and dependencies and email groups and web-based collaborative documents can all be configured with granular editing permissions. But John Casani managed the entire Galileo project using three-by-five index cards that lived in his shirt pocket. He got the idea from fellow JPLer Bud Schurmeier—Voyager's project manager until Casani took over.

"He was constantly writing notes to remind himself of what he had to do. And I sort of took up that practice," Casani explained. "Maybe you'd be workin' on two or three or four cards at one time. And you fill a card up, you throw it away, and work on the next." He stressed the importance of adding issues to a card immediately upon hearing about them. "Sometimes, goin' on the card just meant that I was gonna talk to somebody else that worked for me and say, 'Look, here's an issue. I want you to handle it.'"

How much space an issue consumed depended on its relative importance. "If it was a big issue or something very different, I'd start a new card on it. But usually it was just the sequence of things that needed to be done, and the order that they came up," he said. "They went home with me usually. They stayed in my pocket, and then got transferred to the shirt the next day!" If a card went missing, he'd lose whatever was on it.

For everyday meetings, Casani also maintained an expanding inventory of notebooks into which he scribbled relevant points, in real time, as the proceedings unfolded. "I had a date on each page, and who I was talkin' with, and what we were talkin' *about*, and, you know, if we made an agreement. I was always careful to document what I thought was an agreement," he emphasized. "If I'm havin' a conversation with you, and maybe there's a little bit of issue? We're not seein' eye to eye, and finally I said, you know, 'Okay, well, what can we agree on?' You know, and I would write the agreement down. And if there was *nothing* we could agree on, I'd say, 'Well, what are we disagreeing about?' And I would write *that* down. You know. And so those were the key things that I was focused on."

As autumn approached, Casani offered supportive words of encouragement to the troops by way of the *Galileo Messenger*. First, he praised the injection module and mission design teams, calling attention to how their work, interrupted as it was, nevertheless "demonstrated our resourcefulness, our ingenuity, and our adaptivity to rapidly changing circumstances. We must reach for those same qualities again as we begin the process of adapting to the '86 launch.

"We must always remember that our ultimate objective is to deliver a fully functional Galileo spacecraft—Orbiter and Probe—to Jupiter. With the reinstatement of the Centaur we can achieve that goal in 1988."

Years down the road—after *Challenger* and long after *Galileo* finally met Jupiter—Casani ran into a guy from NASA Headquarters who tendered a confession.

"You know, Casani," he started, "you know how we used to refer to you in the launch vehicle area at headquarters?"

"No, what?"

The HQ guy drew himself up. "Casani lives in a world with no corners."

Galileo's ex–project manager had to request an explanation, and he communicated what that was at the table of the Italian restaurant. Those people at headquarters? "What they were talkin' about, every one of these reprogramming activities, *they* thought was a showstopper. That we were in a corner, and there was no way out, you know?" Name any pitfall: shuttle delays. Radiation. Splitting. A ricochet at Mars. Recombining for Centaur. Losing Centaur for the air force booster. Resizing tanks. Getting Centaur back. No matter what happened, John Casani persevered and always, *always* found a way.

He smiled broadly, and midday sunlight filled the creases of his face.

"They said, 'This guy lives in a world with no corners.' And I took that as the greatest compliment from those people."

12

Very Rad Hard

Clyde King was the boss but he was first a friend. I am not sure if
anyone knew how to be my boss. But Clyde knew how to be a friend
first, and the boss thing just came along.

—JPL cabling technician Mary Reaves

"For Fred Scarf, there was no greater privilege, no higher form of existence,
than to be a principal investigator on a NASA spacecraft. Fred spent hours every
day on the telephone with other PIs working through the interpersonal, organizational, and engineering complexities of his missions." So came a behind-
the-scenes account from Charles Kennel, a member of Scarf's plasma wave
team on Voyager. "Getting a dozen or so specialized experiments to squeeze
together in the same tiny volume was never going to be easy. And Fred spent
those hours wheedling, cajoling, and flattering, in search of the give-and-take
that enables experiments on board to meet their minimum requirements."

And then as far as Don Gurnett was concerned, nothing quite compared
to scanning the ether with a tool never before used this deep in space. "When
we flew *Voyager* to Jupiter, we heard whistlers. And that is the first discovery
of lightning at Jupiter," he signified with understandable pride. A *National
Geographic* cover story describing *Voyager 1*'s exploits listed it as one of the mission's top highlights to date. "Crackling continually with lightning," squealed
the article in reference to Jupiter's atmosphere. "Many scientists are now convinced that this lightning must be triggering the formation of organic molecules, the chemical foundations of life, in the cauldron of Jupiter's clouds."

Nearly two years beyond that initial discovery, with *Voyager 1* in the vicinity of Saturn, its plasma wave subsystem captured even more juicy material
for the headlines. To experience what happened, let's crash JPL's von Kármán
Auditorium—swarming with throngs of reporters for the daily news conferences. At one in November 1980, Don Gurnett described strong radio waves in
the vicinity of Saturn's moon Titan. They'd found twenty kilowatts' worth—

enough for a legit AM radio station. With honesty, Gurnett disclosed that his group didn't yet understand how such radiation was being generated. These waves were exposing new mysteries in real time.

But what stopped everyone cold were the sounds. As both *Voyagers* negotiated Saturn's greater metropolitan neighborhood, their plasma wave receivers captured audio of whistlers and dawn chorus shrieking through the local environment like the soundtrack to *Friday the 13th*. JPL employee Anita Sohus happened to be talking with Voyager scientist Ed Stone when Fred Scarf "came roaring into Ed's office with his little cassette player," she recalled. "With a big grin on his face. *So* excited because he had the sound of the spacecraft being bombarded with particles as it went through the ring plane of Saturn. Ed just lit up like a Christmas tree." During one news conference overflowing with press and other attendees, Scarf played sounds for a much larger audience. Booming responsive applause was quite the rarity for an instrument that didn't even take pictures.

Only a day later, after *Voyager 2*'s scan platform stuck, the plasma wave team offered evidence of a culprit. To a crowded morning press briefing, Scarf dropped his newest track—a hailstorm cacophony of recorded noise. "It sounds very much like material impacting the spacecraft," he suggested. "I think the most likely explanation must be impacts of small dust grains." Bad news, certainly, but the team had been pleased to contribute to an understanding of the problem.

A *Voyager* sent data as digital bits, which for experimenters and engineers alike are almost a form of currency because bits are essential yet scarce. All kinds of tense meetings saw negotiations and tugs-of-war over exactly how many bits of homebound data each team would get for its instrument or system. People made deals—trading Jupiter approach bits for flyby bits, as only one example.

Long before launch, Don Gurnett had happened upon an unexploited method of sending "free" bits. It incorporated *Voyager*'s onboard camera, or imager, which by orders of magnitude operated at the highest bit rate of any instrument. Whenever the imager idled between pictures, while being repositioned on its boom arm, it output no bits to the ship's radio transmitter. So the homebound signal contained *unused bits*. They were like empty seats on a roller-coaster ride. Recognizing this, Gurnett formulated the idea of disguising plasma wave data—having it masquerade as something it wasn't. "If we could just make that look exactly like the imager," explained Gurnett of his scheme, "then we could just have a switch between the imager and our instrument."

It worked, resulting in the transmission of bonus plasma wave data that *didn't* have to be negotiated or bargained for. Netting them actual sounds from the cosmos. Laughed Don Gurnett, "Like you had earphones on the spacecraft, listening to the signal detected by our antenna!"

The year 1982 dispensed all sorts of activity. Commodore International unveiled its C64 computer at a January trade show in Las Vegas. Less than a week later, Air Florida's iced-up Flight 90 crashed right after takeoff into the Potomac River off Washington DC. By March, Argentina had unwisely decided to invade the Falkland Islands. And then in June, an apparently restless male adult climbed through an unlocked window at Buckingham Palace and wandered around long enough to grow bored and leave without anyone ever approaching him. What potential to be a strange year indeed.

Perturbations coursed through the Jet Propulsion Laboratory as well. Its top employee, Director Bruce Murray, abandoned his post at the end of June and left behind a ragged-edged vacancy. "What I enjoy more than anything else in the world is teaching and working with graduate students doing research," he said years after leaving. "I went into JPL full of hope that I could reverse the trend of backing away from space exploration. Which started in '72, and by the time I got there in '76, was in full steam. And found I couldn't." Murray retreated to his beloved Caltech and its faculty and never-ending rivulets of grad students.

Galileo project testing continued. On July 17 General Electric (GE) and Hughes convened at White Sands Missile Range in New Mexico. Although Hughes was supplying the probe itself, GE had been contracted for its two-piece heat shield plus a parachute system plus an overall method of separating everything at the magic moment. Safe to say, the two companies were getting cozy.

Underneath a massive, four-hundred-foot-long balloon, their test model of the Jupiter probe ascended to a hundred thousand feet—whereupon three explosive nuts let go from a supporting gondola. The probe free-fell halfway down, kicked out a pilot chute, shucked its aft cover, then almost immediately deployed the main chute. A lower heat shield fell away. Preset commands destroyed the balloon. So far, so good. Things started cratering into the desert floor: aft cover, lower shield. The probe kissed the dirt at thirty-eight miles an hour, and crews rolled out to sweep up the mess. Then commenced a series of follow-up analyses to determine whether everything worked as expected. The gondola had film cameras rolling. Could someone please get that in to be developed?

Nobody formally took charge of JPL for the whole summer. But come October 15, brand-spanking-new director Lew Allen watched with a group as technicians performed deployment tests of *Galileo*'s flight antenna—manufactured by Harris Corporation in Melbourne, Florida.

Lew Allen had spent his career in government. Years prior he'd been director of a quite different entity—the generally distrusted National Security Agency. During that time he gamely attempted to sidestep any agency involvement with the Watergate investigation. From there he'd become chief of staff of the air force, retiring early in 1982 as a four-star general. That should've freed his schedule to experience an incredible summer of cinematic entertainment as *Blade Runner, The Thing, Mad Max 2: The Road Warrior, Tron*, and *Star Trek II: The Wrath of Khan* hit theaters. During this same period, the child inside him could also have taken in *The Secret of NIMH*.

But Lew Allen had little time for movies because he now worked at such a cool place. "The missions at JPL are of course very exciting, and the people at JPL are highly professional. Which meant I had absolutely no difficulty in beginning productive interactions with the people here." Allen's existing deputy director also happened to be a retired air force general and had been at the Lab for over ten years already. It helped ease the new director's awkward transition into a politically uncomfortable environment. "He provided a very nice bridge," Allen praised, "in that he clearly understood very well the culture from which I came. And he understood the JPL system very well."

Allen dismissed any fish-out-of-water notions with regard to a military man running what amounted to a scientific campus. "Cost, scheduling, and engineering problems are the sort of issues that always come up. So the background was not really inapplicable," he waved aside. "These were things I was certainly familiar with."

Another certainty: Lew Allen had not come to JPL ready to implement some long-term vision of how the Lab would evolve. "No, I'm sorry to say that I did not!" he laughed, recovering. "No, the issue really was to deal with the circumstances that existed at the time." He spoke of decreased budgets, layoffs, and JPL's overall mood as "bearish." But in his eyes, the Lab's mission and ellipses of expertise had actually been growing—which, among other things, brought into focus the deterioration of its facilities. Some buildings, noted Allen, had been around since the earliest days of the 1930s.

The new director looked on as *Galileo*'s massive antenna, clamped upright to a large horizontal pipe, slowly unfurled its eighteen composite ribs as would a mechanized, high-tech flower struggling to bloom. It was gorgeous. Cabling spiraled around the support pipe and trailed away to power and control sources. Never before had JPL tried a collapsible antenna; the Voyager ones were cast. Technicians poured liquid epoxy into a mold and out came a single inelastic piece. Just like baking an antenna-dish pie!

Higher data rates on Galileo—supporting the promise of real-time image transmission—mandated a larger dish. But one that had to squish inside the shuttle's payload bay regardless. The tip of each rib unlatched from a central hub. Near its base, dual electric motors spun a half-inch-diameter threaded rod, on which a ball nut slowly ascended. Attached to this nut like bicycle spokes, eighteen spring-loaded pushrods drove the ribs outward on pivot points at their base ends. The ribs would halt at mechanical stops, after which a whole separate system tensioned mesh against ribs. The Soviet Union had been doing a similar thing for years on their Venus and Mars flyers because constraints such as "must fit inside the launch vehicle" cross all international borders.

Nobody knew it at the time, but the horizontal way in which Harris had transported its antenna from Florida to California signified one milestone leading toward catastrophic in-flight failure. It would never work properly. No matter how many deployment tests JPL or anyone else performed, the thing was doomed. More tests would only wear on the mechanicals and further hamper their correct operation. Lew Allen had no idea. His clueless gaze crisscrossed the lightweight antenna assembly, sizing up its detailed engineering cues. What a beauty!

Years later he would offer perspective regarding the situation that greeted him upon arrival at the JPL gates. "When I came in, the relations with NASA had been kind of strained," he said, trying to infuse context to the state of affairs. "My predecessor, Bruce Murray, had left several months earlier. He had left in kind of a huff, because he felt that NASA was not supporting the planetary program with sufficient vigor. He was very disappointed, and he decided he would leave to do other things."

Part of Murray's concern had to do with the Reagan administration's efforts to slash costs—plenty of which targeted robotic space exploration. "We're sitting here watching the coffin being nailed shut," Murray had squawked to a reporter one year before fleeing. "I wasn't appointed Director to preside over the dissolution of the U.S. space exploration program.... I'm not going to be squeezed down to nothing."

Allen understood Murray's point of view—categorizing Reagan's transition team as "slightly to the right of Attila the Hun" and as planning a high-flying, from-the-top-rope diving elbow drop to the budget as soon as the bell rang. "So when I came in, there was a sense that the planetary exploration ambitions of JPL were not receiving appropriate support." Allen faced an uphill slog. He described the Lab as attempting to diversify a percentage of its activities and disputed the notion that he'd come aboard specifically to increase how much work JPL did for the Defense Department. "Of course I was interested in that, and did follow it with care. But it was not particularly a priority of mine."

With *Galileo*'s antenna ribs mostly unfurled, the nearly sixteen-foot-diameter assembly tilted forward on its mounting pipe, and Allen could now make out what hung between the ribs—a fine mesh created by knitting extremely thin, gold-plated wire into a metallic fabric that offered all the properties of a solid dish. The machine used to create this fabric originally made pantyhose. On the dish's reverse side, a series of cords and tie wires maintained the wire mesh in a precisely defined shape.

Lew Allen huffed it all in. As far as he was concerned, the White House's proposed cuts weren't particularly drastic—or even illogical—and certainly weren't the gloom-and-doom scenarios prognosticated by Bruce Murray. "It wasn't a high priority with them," Allen suggested of the Reagan administration's overall attitude toward planetary exploration. "But they still were perfectly willing for NASA to set up a reasonable base of priorities. And NASA itself had a plan that involved somewhat lesser expenditures than had been envisioned earlier. But was still a sensible plan that they seemed quite willing to pursue in an orderly way." Allen preached quality working relations with NASA HQ, "being careful that we were seen as a *supporter* of NASA. As a laboratory that was prepared to establish and implement programs that NASA wanted."

People work in different ways. JPL's new director contrasted his own style with that of his predecessor. "Bruce Murray was, of course, a renowned scientist with a long academic background. And he operated in a manner that one would sort of expect from that," Allen suggested. "A lot of interest. A lot of enthusiasm in particular things. But not as orderly a manner as one might have thought was appropriate." Allen planned instead to focus on a smooth-running organization: eliminating managerial gray areas. Promoting clarity. Rebuilding a sense of teamwork. Reinforcing the Lab's focus on planetary exploration while supplementing with outside jobs as needed. And always excelling at everything they did.

For guidance, Lew Allen planned to rely on the work of an outside Solar System Exploration Committee—formed by NASA Headquarters with the idea of creating a logical road map for upcoming efforts. The committee's recommendations were just hitting the street at the time of Allen's arrival. Retain Galileo momentum, they said. Begin planning for a Venus orbiter to map its surface. Begin another project to photograph Mars in great detail. And then start conceiving a behemoth of a mission to explore Saturn and its own system of moons.

"It represented a fairly sound plan," commented Allen of the high-level recommendations. "It was endorsed by NASA and provided a priority list of how to go about these things. And obviously, with normal concerns about funds and location." Allen further regarded the committee's work as helping to improve the relationship between JPL and its single largest customer—NASA. This committee, it seemed to be proposing good things for everyone.

Lab technicians did much more than test whether *Galileo*'s folding antenna opened to its maximum diameter. Flawless deployment was certainly important but represented only one single checkbox on a list of many. As the days and weeks ticked by, electronic signals coursed through the dish to benchmark performance. It got shaken to approximate the stresses of launch. All seventy-six pounds of it were stuffed into a vacuum chamber and exposed to a merciless space-like environment. They test-opened it there also, in vacuum conditions, and the test went successfully. Alas, that procedure did nothing to expose what would later bring the mission to a near standstill.

Elsewhere, the elements with which to create a flightworthy spacecraft were slowly accreting. So much to sort out. Take Jupiter's highly radioactive environment, for example. *Galileo* would bathe in it for months. How best to deal with the radiation and create a ship able to perform while immersed in it?

18. Ever the entertainer, John Casani (*left*) donned period gear to throw a team party in honor of Galileo Galilei's birthday. Lew Allen (*right*) kindly showed up. The cake's inscription "421st Birthday" corresponds to the year 1985. Courtesy the Casani Family.

"We *thought* we knew quite a bit about 'rad hard' electronics," grimaced John Casani. He used the I'm-an-insider abbreviation for *radiation hardened*. "And we were busily going along designing the computers and everything else based on a family of integrated circuits that was rad hard, very rad hard. And very well established, very mature, and there was like six or seven individual integrated circuits that were in this family." But some already-flying spacecraft were discovered by engineers to have developed "funny things in their data that they didn't understand," as Casani put it. The funny things turned out to be cosmic rays. Memory cells on computer chips were getting struck by rays and flip-flopping zeros to ones. Or the other way back.

"This is a serious problem!" emphasized Casani. "The circuits that we were planning to use on *Galileo*, and we had designed the computers around, and *already* had built circuit boards and everything else, weren't gonna work! And so we were scrambling around, trying to figure out what to do."

While Casani worked the delicate situation, the mission's revised flight path started coming together. To plan it, the Lab's department of trajectory experts used numerous in-house computer programs. One in particular could plot both mission phases: *Galileo*'s initial one-way outbound trip and the elaborate course through Jupiter's neighborhood of moons. They had the former nailed down. One of the trajectory guys asked his supervisor if they should delete that section of the program because nobody needed it anymore. *Go ahead?* The supervisor thought about it and said, No, let's keep that in case something happens.

Autumn dawned. An entire week that November went toward designing the ship's fault-protection system, as well as planning such activities as design verifications and system testing. It was metawork—work about work. In December an updated probe model came through JPL's gates, destined for a battery of tests that might run for months. All the remaining spacecraft bits and pieces, currently strewn about on assembly tables at a prodigious number of locations (including many outside contractors), were due to begin converging in February of the following year.

And always, riding atop JPL's existing base of experience, were shuttle complications one after another: brand-new shuttle-to-Centaur interfaces. Support cradles. Beefed-up safety requirements. And, god, the contingencies. What would happen if the shuttle had to abort during ascent? Come back with Centaur and *Galileo* still camped in the payload bay, fully fueled and radioactive? Could the landing gear take that weight? No other planetary mission ever had to consider these scenarios.

Life in the shadow of the shuttle.

By the time February rolled around, progress wasn't exactly smooth as glass. Multiple unresolved issues plagued the actual flight version of the probe. Its brain, the data and command processor, didn't work properly. They had a revamped brain going, but it wasn't yet together. The probe team had six months to iron everything out before the flight unit had to be at JPL for initial mating tests with the real orbiter. Those tests were slated to run for a month, after which the probe would return to Hughes for installation of its flight-qualified brain. To remain on schedule, the finished probe would then need to be back at JPL by January 1984. Casani scribbled an update on his index card and moved to the next probe item—batteries. Their lithium cells needed increased resistance to high-temperature corrosion. New ones would have to be manufactured.

Everyone trudged forward, the giant machine of the project slowly turning, generating incremental progress. By March 1983, a mockup of *Galileo*'s eight-sided primary bus had taken up residence in a crowded south-central area of JPL's campus known as Building 179, Spacecraft Assembly. Still in existence today, its two hangar-like bays represent an ultimate workshop for servicing high technology and form the locus of activity prior to heading for the Cape.

Either bay makes the average surgical ward look like a Missouri gas station bathroom. Filtered, almost perfectly pure air flows through the bays at a constant 72 degrees Fahrenheit and 50 percent humidity. The cleanest of clean rooms. Inside, trained experts fawn over exotic hardware with near-religious reverence and crazy-clean tools. Access is privileged. It's tough getting into Building 179 with a part or instrument or whatever the heck thing it is because the hardware in question must first endure severe end-to-end testing. It must justify its presence and prove its worth. Some people labor for years before they're allowed in. On paper, *Galileo* consisted of more than twelve major subassemblies. Here in Building 179 is where everyone's parts would unite to form one true spaceship.

The state of *Galileo's* bus occupied high bay number 2. A large sign reading SPUN balanced atop the core's upper half. Only a couple feet away, beneath a giant DESPUN placard, sat the lower half. West Germany's retro-propulsion unit would fit essentially between them and wasn't expected until later that year. Nearby, a large metal supportive framework labeled SCAN PLATFORM waited calmly until the day it would receive a boom arm for the camera.

Just a few feet away from the spun section perched about a dozen poster-size rectangular circuit boards. All sat vertically on their short edges. Together they formed a near-complete circle, fanning around in one three-hundred-degree arc. Cabling erupted from each board top, merging into bundles that spilled down the front and ran across the floor. Meet *Galileo's* brain, the command and data subsystem—which at that moment was sort of maybe partially up and running.

Test equipment, from oscilloscopes on carts to a thick TV on a stool, littered the floor space. Assemblers and technicians sported white coveralls that guarded against static discharge more than germs. Their outfits lacked the full face masks utilized when true sterility mattered for something akin to a Viking lander, two of which had put down on Mars only seven years beforehand. The assembly of *Galileo* needed to occur under *clean* conditions, okay? One stray bit of debris could find its way into a critical part and hinder the ship while in flight. But nothing about the mission called for *sterility* per se.

One day in the not-too-distant future, *Galileo* would depart California, launch from the shuttle, and make history. Such notoriety! Such glamour!

But in gloomy comparison to this ritzy-glitzy environment, consider what is perhaps the most luckless piece of hardware associated with the entire mission. Created exclusively for a sad, serf-like life of service through pain and hardship. A defenseless thing. Uncelebrated. Total lab rat. Destined only for injury and eventual storage-room abandonment.

It went by the name of development test model. Nothing less than a full-scale analogue, the test model represented a pretend *Galileo*, an empty structural twin, born into bondage. Devoid of any working electronics. Grounded by design. It would never get to ride in the shuttle. It would never experience those special feelings of importance when photographers took its picture for tomorrow's front pages. Oh sure, parts of it might spend time in one of the high bays. The test model is what they'd bolted the model probe to, checking whether things fit together as intended. But who'd take pictures of test hardware when the real deal was already coming together right there in the same room? Maybe the sort of people who'd go see a Neil Diamond impersonator when the real Neil was still alive and touring?

Galileo's design looked great on paper, but the team had to be sure. By force with ungentle hands would the test model be used to verify calculated design parameters. If anything unexpectedly showed itself—torsion or bending moments or stress cracks or buckling or interference or temperature intolerance or whatever—it absolutely had to be discovered in advance. *Before* making that supreme last round of parts for the actual flight article. Discover what's wrong in advance. Build your best guess, as if it *were* the real deal, and then exercise the thing through its full physical range of motion. Shake it like a daiquiri to simulate launch. Torque it as would happen when the boom arms deploy. Contort. Look at everything possible. Yes, even test with Super*Zip.

With a mocked-up flight chassis in hand, engineers and techs began test-fitting the nearly twenty-five thousand feet of wiring responsible for moving signals around. Think of these wires as the ship's blood vessels. Branching bundles would plug into some seven hundred connectors peppered kind of everywhere, and had already been in fabrication for about a year. This work fell under the purview of a highly experienced team of about fifteen people who did almost nothing but specialized wiring. Galileo had seen them pushing through sixty-hour weeks, in split shifts, with the goal of delivering the main bus assemblies by March 7. Once those were laced through the chassis, the team would examine how every last wire fit—identifying what had to change as well as *when* to remake it—on top of readying other bundles for *Galileo*'s despun bus. Oh yeah, plus the ones for the science instruments too. If the pace held, they could probably deliver the latter by summer.

This work suits methodical, detail-oriented people. It's not for anyone in a big fat hurry or emergency-room types who get off on urgent, hyper-acute fire drills with blood on the floor and alarm bells clanging. *Galileo*'s angel-hair wires had to be individually soldered or crimped to electrical contacts, dressed into bundles, then held together with fabric ties. Every single solder joint on every single wire had to be meticulously examined. Everything done by hand. Everything photographically documented. On occasion the work required tweezers. Prepped bundles went off for a step called potting in which the electrical contacts were seated into connector frames and then filled with a molding compound. Some connectors had as many as fifty pins on them, with each pin connected to an individual wire leading in turn to a different part of the spacecraft. A single complex connector with many pins could branch to eight or ten different places. Any exposed cabling got additional fiberglass sleeving for protection against micrometeoroids.

"About four months to cable the despun section, and about six months to do the spun section." That was the prediction of Clyde King, forty-seven years old at the time and a veteran JPL cabler going back to the Pioneer days. The Texas native and father of four wouldn't retire until the end of 1998, spending over thirty years at the Lab.

"He really did not talk much about work at home," recalled Clyde's eldest son, Seaton. Regardless, Clyde left an impression on his coworkers. As Seaton continued, "I do remember going to the annual JPL Open Houses. We would go from building to building and presenters would be giving their speeches." Oftentimes, one of them would spot Clyde among the onlookers and holler, "Clyde! Did I leave anything out?" Or, "Clyde! Was that accurate?" Seaton's brother, James, verified the story. Both loved the shout-outs their dad received. "I remember as a young man, how proud I was!" beamed Seaton King.

His most poignant memory of their dad's work, however, involved the 1977 launch of *Voyager 2*. The family visited JPL early on launch day to watch the proceedings "and then left on vacation. We did an annual camping trip to South Carlsbad State Beach," explained Seaton. It sits just over a hundred miles south of Pasadena. Right on the coast. The King family loved being out-doors and looked forward to a nice getaway. But this particular vacation got cut short for one particular individual. As Seaton picked up the story, "The park ranger left a note on our site asking dad to call his boss from JPL." That couldn't be good.

Although *Voyager 2* had embarked for the outer solar system, it was floun-dering. The spacecraft employed two boom arms; power hung on one with science instruments on the other. Both arms folded down to fit in the nose of the launch rocket. After shedding that nose in space, everything was sup-posed to unfold like a Transformer toy. Alas, explained John Casani, "one of those two booms did not fully extend to the correct position. It went *almost* as far as it needed to be." With *Voyager 1* supposed to blast off in ten days, JPL faced a legitimate crisis on one ship with a ticking clock on the other.

Some of *Voyager*'s cabling ran down the instrument boom to the fold-down hinge. There, it formed an exaggerated camel hump so as to clear the mov-ing parts and leave room for the arm to lock firmly in place. But the excess cabling could also bump into the giant antenna dish sitting just above and affect boom arm deployment. They needed Clyde. *Where the hell is he?* Some-body found out and then sent a park ranger.

"Dad was livid," winced Seaton King. "I remember him telling us how he warned them this would happen 18 months prior to launch, and they ignored him. Now he is on vacation and they need him to go to Florida to fix the prob-lem. *In Florida!*" The kids could only stand there in swimsuits and watch as a car picked their dad up at the campground and drove him straight to the airport for a plane to the Cape. That right there was the end of Clyde King's vacation, by golly, because after modifying *Voyager 1* to ensure the same thing wouldn't happen, he then spent three weeks helping unfold *Voyager 2*.

Laughed Seaton, "I was proud, awed, and impressed with his vocabulary when he found out he had to leave his family and fly to Florida."

Now his old man had moved on to this new Jupiter orbiter. Separate from the spacecraft's internal wiring, a different set of much thicker cable bundles now traversed the floor of the high bay. Starting from *Galileo*'s partial hard-ware assemblies, they meandered like Amazonian anacondas into an adjoining room and connected to a ring of thick metal cabinets filled with test equip-ment. A line of control consoles packed the middle, resting atop a grid of heavy white floor tiles that lifted out and permitted access to a netherworld of race-ways underneath. Once Clyde and his crew finished the internal spacecraft cabling, these consoles would be used to power up the creation and execute batteries of tests. The schedule mandated this phase to finish late next year—completing the first round of integration testing and pronouncing the electronics ready to move into the flight chassis.

In late May, Galileo's development test model weathered a new round of hell. Lacking the dignity of an antenna dish or extendable boom arms, the it's-not-even-a-real-spacecraft was dangled from a crane and then tickled with stick-on mechanical torture devices that revealed natural frequencies and other structural parameters. This necessary cruelty went on for three days. The model came down on May 25 and underwent reconfiguration for its next ordeal. The antenna, arms, and mock probe all went on. Weighted boxes went aboard to simulate electronics and science instruments. Easy-to-handle alcohol and Freon went in the fuel tanks to simulate real-world liquid fuels. Underneath sat an adapter simulating attachment to a Centaur.

Afterward, instead of riding up in the air again, the whole cursed thing went atop a massive block of concrete sunk fourteen feet into the ground. Hundreds of gauges and sensors hung all over the model like Australian sand flies. It'd now been prepped for a fresh round of tests to investigate launch stresses. The schedule said this would last a good month and a half. Q: Why bolt the model to the concrete? A: To isolate it from extraneous vibrations, no matter how small. Such as those from passing trucks, which theoretically wouldn't be around in the skies above Cape Canaveral.

Once all that finished, the luckless model would be transferred into a special chamber and yelled at by enormous loudspeakers at an average level of 147 decibels. Which is pretty loud. In 2017 crowd noise at the University of Kansas's Allen Fieldhouse hit 130.4 decibels during a basketball game between the Jayhawks and the West Virginia Mountaineers. It set a world record for indoor crowd noise. It's close to what a jet engine produces and is above the pain threshold for humans, yet failed to reach what the test model endured. That 147 decibels are greater than a level blurring your vision and reliably inducing nausea after only a few minutes' exposure. In the case of Galileo's test model, this massive amount of sound would replicate conditions inside

19. In a Building 103 workroom, Clyde King reviews cable harnesses test-fitted onto a mock-up of the *Galileo* orbiter's despun chassis. Colleague Mary Reaves indicated that cabling interconnects would sometimes end up in undesirable locations and force the remaking of bundles. Courtesy NASA/JPL-Caltech.

the shuttle's payload bay precipitated by the acoustics of launch. The nightmare would run for about three days. Afterward they'd squeeze in two days of testing explosive bolts. And then beginning in September, the entire model was slated to suffer another ten weeks of misery as hydraulic rams pulled at it like vultures on roadkill so as to imitate structural loads.

Progress abounded. By mid-June 1983, with cabling well underway, flight-ready elements of *Galileo*'s power subsystem had been joined to a mock-up computer and other electronics. The radio went in. A mechanism to unfold the boom arms and snap 'em into place also went in. Three real science instruments, including the dust detector from Europe, had even been added. If a need arose for some piece of flight hardware not yet available, engineering models or prototypes were used to keep moving. Versions of five additional science instruments, built strictly for testing and not flight worthy, arrived at JPL for fit checks with the actual spacecraft. They joined four on deck to go in; one was from Don Gurnett and his team. Other elements of the flight hardware slowly trickled into high bay number 2 for installation: the despun bus. The lower Centaur adapter. The relay radio antenna. The booms. The scan platforms.

Come late July, they performed a second drop test of the probe. Using a revised parachute configuration, everything behaved as expected. Good to go on that front.

On August 14 the development test model saw friendly faces—ones that weren't there to hurt or injure. They belonged to the 1,500-plus visitors at JPL that day: family members and other guests of a diverse yet unified Galileo team building the world's first spaceship to orbit Jupiter. Everyone in attendance had an opportunity to sign large posterboard sheets. The stack would be photographically reduced, transferred onto small metal plates, then tucked aboard *Galileo* itself. A little facet of each signer was going to Jupiter.

Such events were the least that JPL could do for the employees' family members and loved ones. Casani sought to put on something every three or four months—a gathering or picnic with group events and usually games but always food, food, food. According to him, the idea started with coworker Bud Schurmeier. Back in the early 1960s, Bud had taken over the fiasco of the Ranger project and completely turned it around. His superior management tactics incorporated a supportive philosophy that Casani framed this way: "It really is a family effort. I mean, there's no getting away from it. You cannot perform as an engineer, or in that environment, without your family solidly behind you." People worked endless hours on these projects. Were always away. "And it took a toll. And they couldn't do it, you know, unless their families were fully supportive of it. I mean, you can easily see that a guy's coming home late every day, he's tired." Casani spoke of coworkers missing Little League games or family events or holidays and sometimes all of the above. "It's happened to everybody." It sure happened to him, plenty of times—including a European getaway he'd been forced to cancel at the last moment.

Bud Schurmeier had recognized this. And felt that an effective method of combating the situation was group recreation. "Probably every three or four months we would all go down to some hall in Pasadena," said Casani of the Ranger days. Spouses included. And after a series of such gatherings, the spouses—mostly wives at that time—began to feel as if they were part of the team. Instead of someone just waiting at home with cold food on the stove.

The effort continued even when in Florida during launch operations. Families would visit for a bit or even stay a while. Casani would task them with organizing prelaunch parties—finding a hall and decorating it and coordinating all the food. From where he stood, this made a palpable difference in attitudes. "They began to feel like they really were part of this," he justified. *Everyone* became part of the team. "They had some ownership. Not just having to put up with a missing husband or a missing wife for months at a time."

Casani sighed. "I don't think it's done as much now as it was," he commented. "The Lab's a lot bigger. People live farther away. It's harder to get people to come in."

Regardless, he offered an outlook of hope in his update on the saga of *Galileo*'s computer. A blistering, choking radiation environment had threatened the mission to orbit Jupiter by scrambling the very data bits that drove the ship. "And we had two courses of action. And I was funding both of them." The first utilized a Texas company the Lab had previously engaged. Its exotic silicon-on-sapphire process could take the pain, although it required a totally new integrated circuit design that hadn't been tested.

"And then we also had another approach goin' with Sandia Laboratories, 'cause they were doin' a lot of military rad hard stuff," elaborated Casani. "They'd actually packaged some [computer] memories for us in a rad hard situation on Voyager. And we knew they were good." Casani asked his contacts there whether Sandia could "simply" remanufacture Galileo's existing circuit designs—but in a rad-hard configuration that was otherwise a pin-for-pin match to the original.

Getting that answer took a while. But the Sandia guys eventually said, "Yeah, we think we can do that."

But Casani had his doubts. "We weren't too sure they could."

13

Mind the Gap

There is something noteworthy a rocket can do that the
Shuttle cannot. A rocket can be permitted to fail.

—Author Gregg Easterbrook comparing expendable
rockets to the space shuttle

The Summer Olympics came to Los Angeles in late July 1984, during which
John and Lynn Casani attempted a family vacation. All six piled into their Volkswagen bus and headed two hours northwest up the coast to Lake Cachuma
for a planned two weeks of swimming and fishing. One week in came a trip
into nearby Solvang for laundry. John and Jack dropped Lynn off at the laundromat, filled the bus with gas, handed the other boys money for ice cream,
then parked at a bar to have a few beers and watch the Olympics. Referred to
as "Charlie" until elementary school, Jack had never been a fan of the moniker. His given name was in honor of his grandfather, after all, and he much
preferred Jack.

They weren't even one beer in when the other boys ran over screaming,
"Dad! Dad! The bus is on fire!"

In understated tones, John picked up the story. "We ran out, and sure enough
the bus was a blazing inferno. I think I had overfilled the tank, and some
gas leaked out, hit the hot exhaust pipe, and boom. There it went." He stood
watching with Jack as it went full *Hindenburg* right down to the bare chassis.

Several bystanders grabbed fire extinguishers from a nearby hardware store
and, according to Jack, engaged in an "honorable but futile" attempt to save
the vehicle. "Pop paid the hardware store for the fire extinguishers," continued Jack, "but grumbled about the futility of the effort, as the bus was long
gone." Nobody got hurt, and the family had all their clothes, even. John rented
a station wagon and made arrangements to dispose of the bus, and once laundry finished, everyone backtracked to the campground to finish vacationing.

Jack returned to Solvang with his dad the following weekend because one thing *had* actually survived the fire. It was a six-foot-long wooden box—as wide as the bus and perhaps a foot tall—handmade by John and originally installed on the bus's roof to hold all the camping gear. It survived in part due to the heavy, marine-grade varnish job he'd painstakingly applied. "Pop wanted that box back. I was enlisted to support his wooden box recovery mission," Jack indicated.

With considerable effort, the men positioned the box atop the family's remaining vehicle, the '63 Volkswagen Beetle. It partially collapsed the roof. Never mind; they secured everything with multiple ropes and began driving home. *Well below* the speed limit. Even so, Jack soon expressed concern as to whether their jerry-rigged box would survive in the wind.

John responded, "How much load do you think those ropes can take before they break?"

"Well," responded Jack, "maybe a few hundred pounds each?"

"Since we know our speed, we can figure this out," informed John, rattling off explanations of fluid dynamics and Bernoulli's principle. "By using the dynamic pressure equation of one-half *rho* times velocity squared, all we need is the fluid mass density of air, or *rho*. Do you know the atmospheric pressure at sea level?"

"Fourteen-point-seven p.s.i.?"

"Very good! Now we can calculate *rho*." And calculations ensued.

The resulting figures indicated that the load as determined represented only a small fraction of the rope's estimated strength. "So we have a significant margin of safety," John assured his son.

"The engineering tutorial alleviated my concerns," Jack later commented. "But what really made me feel better is we were almost home!"

Also that July, *Galileo*'s major orbiter elements were test-stacked for the first time. Underneath them hung the probe, offering the team an uplifting sneak peek of how the as-built craft would really look. Afterward, men in fifty-five-pound, full-pressure safety suits rehearsed the accessing of various parts of the ship. Explained one of them afterward, "This was a test to see how well we will be able to perform our propellant loading operations at the Cape. We found that the scaffolding needs to be modified to provide us full access." Come August, the real *Galileo* moved to a separate facility for vibration testing in its launch configuration.

The development model bravely endured its test of Super*Zip. This stuff came into the mix partly because of all the extra boost a shuttle couldn't provide. After leaving the payload bay in orbit, *Galileo*'s backside would still be rigidly joined to the front end of the Centaur. But the two would have to part ways after the stage expended. Well, doing that would be easy with Super*Zip!

Both *Galileo* and Centaur had been accessorized at their respective ends with protruding circular flanges. They ringed the entire perimeter of each end and *almost* touched. Bridging the gap was a series of arc-shaped rectangular plates, 2.4 inches tall and 12 inches wide, made from a thin aluminum alloy. A V-shaped notch ran lengthwise through the middle of each. Thirteen plates arranged end to end looped all the way around the inner side of those flanges. Fourteen additional (and identical) plates circumnavigated the flanges on their outboard side.

Now add the secret sauce. In between these two pieced-together rings, situate a plump tube of silicone rubber. Through the middle of the silicone run a continuous length of explosive detonation cord. Secure each end of the cord to an electronic initiator. When the concentric rings are bolted up tight together, the silicone tube squishes to fill any voids.

After Centaur's shutdown, having pushed *Galileo* on its way, the cord would detonate—blowing itself apart at the ridiculous speed of twenty-three thousand feet per second. Instantly this would over-fatten the silicone tube and snap those aluminum plates along their V notches like graham crackers. Behold instant separation, and away would drift the Centaur. (Worked great on *Voyager*, BTW.) Early Galileo testing used two embedded detonation cords, which proved excessive. (One report categorized Super*Zip as "a dominate shock generation device in the *Galileo* spacecraft.") JPL stuffed test Zips inside a giant chamber simulating actual use conditions—near-total vacuum at some 36 degrees below zero. Things un-zipped as expected. For the real deal, they upped the potency of the charge ever so slightly and ran a final test with the actual flight spacecraft. Perfect recipe.

JPL kept marching *Galileo* down the road to launch day. Its orbiter flight chassis received more accessories in high bay number 2. The flight version of the probe showed up from contractor Hughes Aircraft. Vibration and acoustic testing of the orbiter wrapped that September. During October, General Dynamics in San Diego completed major welding of the Centaur's tank and began installing its plumbing systems. Great progress!

At a December all-hands meeting, John Casani played a videotape recording of the vibration tests. (How interesting that a break from work involved watching coworkers work.) The rolling tape showed the scene inside a specialized facility on campus, with the orbiter bolted to a massive rig capable of shaking the spacecraft with up to thirty thousand pounds of force. Everyone stood back and let 'er rip, and when it finished, they cheered like Kansas basketball fans.

To nobody in particular Casani sang out, "They're cheering because nothing fell off!"

Flight-qualified *Galileo* hardware assemblies began emerging from a testing chamber after cycles of punishment. Most everything had been thrown at the parts: rapidly dropping the air pressure to simulate launch. Ramping from minimum power state and coldest temperatures to highest power state and hottest temperatures. Crews working two shifts then began the for-real, final-final assembly with plans to run the works totally powered up 24–7 to verify operation, because when electronics fail, they tend to do so early on. Many called it the burn-in period. Any remaining changes needed to happen within the next four months before it went to Florida.

If the schedule held, *Galileo* would launch the following May aboard shuttle mission 61-G. Two members of its assigned crew visited Pasadena on August 12, 1985. A Monday. Wearing sport coats and ties, and therefore conspicuously overdressed for JPL, the astronauts met with Lab staff and toured facilities. They shook hands with assemblers and other personnel.

At one point, both astronauts donned white protective garments and caps, altering their appearance from businessmen to that of food-service workers, and entered the high bay. Casani briefed them using scale models. Nearby and up three wooden steps sat the real *Galileo* on its riser. Practically finished. Covering it were jet-black insulation blankets, secured by white lacing threaded in X patterns through grommeted holes crimped around the edge of each blanket. The boom arms hung in silence. If *Galileo* wobbled in flight, the unwanted motion could be stopped by changing the angle at which these booms were canted to the spin axis.

Galileo's Centaur went to the Cape in September. The delicate stage and its purpose-built shuttle cradle were successfully test-mated, rotated, and separated. Last time before the real thing.

As November rolled around, project scientists convened to more fully

20. A mustachioed astronaut tours the JPL high bay in preparation for carrying *Galileo* on his shuttle mission. Cabling technician Mary Reaves (*far right*) wears a communications headset. "Each spacecraft had its own problems," she said of her tenure on the flagship outer-planet machines. "And a level of cleanliness and handling constraints." Courtesy Mary Reaves.

develop the schedule of in-flight experimentation. "We have a rather clear group idea, I think, about what our overall objectives at Jupiter are, and how they relate to one another," explained Chief Galileo Scientist Torrence Johnson. "This is not a bunch of individual investigators, each pursuing his own thing, with no thought about the context that he's doing it in." Johnson exemplified the typical 1980s attire with a combo pack of massive eyeglasses and cop mustache. His blue button-down with faint vertical white stripes complemented his stereotypical black pocket protector.

The science group occupied a nondescript conference center. Some opted to wear ties, but all the men had on dress shirts at least. Very few women attended. One of the larger rooms saw people sitting in a U-shape arrangement of metal-frame chairs with ugly red vinyl upholstery that looked to have been salvaged from a Pizza Hut. Sport coats hung over the chair backs. Up in front, a dreaded overhead projector blasted an undersized square of a white screen that perched before god-awful cream-colored drapes. In one corner waited a Stone Age–era blackboard on a floor stand. Overall, the room itself was difficult to underestimate.

Everyone tended to their own massive stacks of papers; white foam coffee cups peppered much of the remaining open space on any table. People took turns clutching pens while making wildly gesticulating arm movements to more convincingly drive home some obscure point. Galileo observations had to be planned down to the individual minute. It made things stressful, but humor sometimes crept in. At one point, an atmospheric scientist lightheartedly suggested that the proceedings might finish more quickly if everyone from his discipline simply went home—allowing others to load the schedule with their own priorities. Later on after all the best slots were taken, he could make do with whatever scraps of time remained.

"We like that!" came the response.

"Of course you like it!" roared the atmosphere guy. "And if we trusted you, it might be the way to go!" The room broke out in laughter.

People fragmented into smaller groups, commandeering side rooms to hash through the more granular details. A lively session in room c-305 had scientists taking to the blackboard to make the case for one point or another. Arrays of taped-up pages covered the walls—depicting phases of Jupiter alongside relative moon positions over time, with various possible approach paths. The incremental nature of the diagrams suggested they could've been assembled into one of those animated flip-books showing the planet rotating. On a central table sat plastic drink pitchers and stacks of orange Dixie cups.

Some had been contemplating their Jovian wishes for years already. Despite the mini competitions for approach angles, instrument pointing, or data transmission time, everyone collectively shared Torrence Johnson's perspective—namely, harmonious collaboration would produce a coordinated master timeline of Galileo observations. And equally important, allow everyone to flee this soul-sucking conference center.

At 3:00 a.m. on December 19, 1985, after all the years of scientific jockeying, proposal writing, congressional ping-pong, and general inventiveness, the *Galileo* orbiter began the first leg of its physical journey—traveling from California to Florida. It went on a low-boy semitrailer, with the goal of reaching Cape Canaveral Air Force Station by Christmas. The spacecraft itself, cleaved from its high-gain antenna, hid inside a notably featureless giant white box strapped to the middle of the flatbed. It rolled behind a cab resplendent in white with blue and dark gold accent stripes, and whose sideview mirrors hung at the ends of comically overextended support arms. Ahead of the semi, an obscenely huge American station wagon led the way, bearing large rotating yellow caution lights atop. Centered between the lights hunkered a sign reading OVERSIZE in stenciled capitals—flanked by orange warning flags that flapped in the overnight air.

The truck's drivers went straight on through, pushing all night and day, stopping only for gas and food. None had been informed of the exact route in advance, lest the convoy fall prey to terrorism or protestors on account of the nuclear fuel which wasn't even in the ship. Or anywhere on the truck. Occasionally, a marked sheriff's patrol car, overhead lights cartoonishly ablaze, would swing hard into an intersection and block it so the lumbering semi-trailer could make an unobstructed turn.

For everyone involved, the trip was structured and finite, but arduous all the same. "To me, it's somewhat scary. It's a big responsibility. You tend to worry about it," professed JPL logistics man Tom Shain. He talked while commanding a beat-up camper van with hideous grandma drapes on the back windows. "You think of the repercussions if we get side-swiped drivin' along here, something like that. Well, there's an awful lot of people who worked an awful lot of long hours on this thing to, you know, for something like that to happen. So you tend to—that's in the back of your mind. It *is* mine, as I'm driving. Maybe that's one of the reasons I'm having a hard time sleeping."

Galileo's caravan arrived (safely) on December 23. That same day, news broke that an upcoming shuttle flight, carrying America's first "Teacher in Space" aboard *Challenger*, had been postponed by a day, until January 23. Several miles to the south, teams unloaded the semi into a prep hangar. Everyone remained unaware of how the recent days of overland road rumbling would be a major factor in crippling the spacecraft. They had no clue.

Challenger had been postponed again but was now supposed to launch on Sunday the twenty-sixth. Forecasted bad weather pushed it to Monday. The weather on Sunday turned out great after all; that night, the Chicago Bears trounced the New England Patriots in the Super Bowl. Monday arrived, January 27, with a new try for *Challenger*. This time, a handle on the outside of the shuttle's crew door took so long to remove that the launch again had to be postponed because high winds kicked up. Another try the next day. Centaur manager Larry Ross had a Monday conference call to attend but would then catch a mid-afternoon flight to Orlando for his big meeting. And if he was lucky, they'd all get to see the *Challenger* blast off too!

Monday night prior to launch, people from NASA's Marshall Center joined a rep from the shuttle booster manufacturer in a conference room near Cape Kennedy's iconic Vehicle Assembly Building (VAB). The big VAB is where shuttles were mated with their tanks and boosters.

The men in that Florida room began a three-way conference call with other Marshall people in Huntsville, Alabama, along with a tableful of booster managers and engineers over and away in Utah. For over two hours they debated whether to proceed the next morning. Booster engineers argued that the forecasted low temperatures—potentially 22 degrees Fahrenheit—would constrain the ability of rubber O-rings to hold in superhot gases. Some, but not all, of the Marshall people argued about a supposed lack of proof that cold temperatures had anything to do with how well O-rings worked. And in the end, the booster managers caved and approved the launch.

Next morning, the shuttle crew buckled in, and the count trickled down to just a few minutes remaining. Everybody in Larry Ross's meeting flooded the corner office to watch. Just south of them, a JPL team broke from *Galileo* labors in the hangar to head outside and try to catch it also.

At minus six seconds, *Challenger*'s three main engines came on. When this happened the entire conglomeration of boosters, external fuel tank, and shuttle actually bent forward more than two feet from vertical due to engine thrust buildup. "The Twang." This was normal. The entire stack flexed back to vertical at exactly zero seconds.

Instantaneously the twin solids lit up and frangible hold-down nuts the size of cantaloupes split apart and the whole thing started to rise. At that moment, several things happened that Larry Ross, John Casani, and everyone else watching could never see: Within less than a second of booster ignition, steeply rising pressure in each of the booster casing segments involuntarily swelled their middles by approximately four one-thousandths of an inch. Again, normal. This swelling caused the tongue-and-groove joints between those segments to rock open ever so slightly. In one joint, it happened just as scorching-hot exhaust gases raced horizontally between segments and through blowholes in packing putty, and turned downward into a gap that was tiny but yawning open all the same.

Sadly, all this was normal too.

The gases should've been halted right there by either one of two rubber O-rings, positioned one above the other, within separate notches along the joint's circumference. If things worked as intended, pressure from ignition would squish the top ring into the opened joint and block further escape of the gases.

But that morning's low temperature had indeed affected how pliable the O-rings were. They'd stiffened; the shmeers of frigid grease they sat in were viscous and sludgy. And the uppermost O-rings—too cold to perform an already difficult job—bore the additional burden of starting their downward journey into the gap from the uppermost side of their retaining notches. We're talking only fractions of an inch of travel and just over half a second of elapsed time. But hot gases don't care. In that near second, they screamed on past and overran the chilled lower O-ring and pushed through only a few short inches of travel until freedom. Blink and it happened. What burped from this joint then, as *Challenger* rose from its pad, were angry toots of brown-black smoke from burning rings and grease. The beast unleashed. A few seconds later it stopped, though, as combusted aluminum fuel reacted with the cold steel booster casing to form an impossibly impromptu weld.

That automatic roadside repair miraculously hung together until fifty-nine seconds after launch, when vicious wind shear hit the rising rig and reopened the wound. Leaking started anew and evolved into a blowtorch—mercilessly pointed right at the external fuel tank and booster attach strut. Solid-fuel rockets cannot be turned off once underway, and both still had another minute of run time.

At just over seventy-three total seconds into flight, the tank and strut gave way. The leaking booster pivoted on its top mount and impacted the forward end of the external tank, and the skies above the Cape began to rain propellant. Which ignited.

We'll never know exactly what *Challenger*'s pilot saw, but we do have his last recorded words. He said, "Uh oh." If he said anything after that it was lost to the wind.

In a flash, all the propellant combusted. A burn. A cloud.

And then *Challenger* was gone.

14

I Have a Goat

I loved airplanes from the get-go, and I had a passion—
well, it was very clear by the time I was in the 8th grade
that I was a born engineer.

—JPL's Bill O'Neil

O'Neil had been watching the *Challenger* launch live at JPL. "I was in the operations area that we had configured, within the last six months or so, for our flight operations of Galileo." Any operations area housed lots of big TV monitors for telemetry or press conferences or whatever might be needed that day. And on January 28, 1986, some of them were carrying a live feed from the Cape. "So everybody, virtually at the same time, saw that explosion. So that's how we all found out. We found out with our own naked eyes."

Not thirty minutes later he got paged over the campus-wide public-address system to phone Casani at the Cape. Galileo's project manager needed to talk about what they were going to do. The most immediate need actually seemed to be reassuring the troops. Wasting no time, Casani red-eyed back for an all-hands meeting and didn't bother reserving a conference room. With hundreds of project workers assembled in the main cafeteria, he stood on a chair looking haggard and disheveled while politicking: Keep to your work as if *Galileo* will launch on time. Stopping now gives our critics ammo to cut funding and leave us grounded forever.

"*Galileo* WILL fly," he promised the troops.

Therefore, in *Challenger*'s immediate wake, the general thrust of the Galileo project didn't change. As reported by mission designer Bill O'Neil, "The spacecraft, Orbiter and Probe, and everything was moving. Continuing to move toward an '86 launch. And that probably continued for two or three months. With ever-increasing doubt that it would happen." Looked as if they'd have to break things down and ship it all back home in August. So why not check what they could while still occupying Florida? Inside the Cape's Vertical Processing Facility, both a model and the actual flight *Galileo* were test-mounted onto its Centaur. Data connections were wrung out. Teams also test-fitted the ship's plutonium power supplies.

But the disaster had created a sort of emotional goulash with most everybody. During *Challenger*'s launch, cabling technician Mary Reaves had been watching from the roof of *Galileo*'s assembly hangar and couldn't wash the images of destruction from her head. "Your mind kept trying to put it back together," she glumly offered.

Five weeks after *Challenger*, astronomers the world over celebrated two uniquely improbable spacecraft from the Soviet Union. *Vega 1* and *2* had been launched in December 1984 toward Venus. Dropped landers onto its surface. Deployed balloons in the high clouds above. And *then* flew to rendezvous with, of all things, Halley's friggin' comet as it streaked through our inner solar system for the first time in seventy-six years. On their final approaches, both ships adroitly punctured the comet's billowing plume of dust and ice to confront Halley's nucleus face-to-face. The photos they took gave humanity its very first peekaboo at the inner workings of any comet whatsoever.

Superficially, the combination Venus lander/balloon flyer/comet kisser might seem bizarre indeed. What an unsellable hemorrhoid of a mission profile. But Halley's path in 1986 took it closer to Venus than to Earth. Multiple nations contributed everything from cameras to gimbal mounts to light sensors to dust detectors to little propellers that hung underneath each balloon and spun joyfully in the curious Venusian winds.

France played a huge role. For decades the Soviets had enjoyed a perhaps-unlikely partnership with that country's own embryonic space program, and the Venus–Halley concept essentially germinated with them. "The first proposal to do something related to Vega came from French, saying, 'Why don't we have a special telescope on Venus orbiter?'" So came the report from Roald Sagdeev. At that time, he headed up the Space Research Institute—the establishment closest to resembling a Soviet JPL—and bore the responsibility for managing his country's planetary exploration efforts. As the Vega concept evolved, mission planners began leaning more toward the idea of actually *entering* the comet's halo of dust and ice particles—to image the nucleus directly. Or at least trying to.

It was audacious—the kind of technological spectacular characterizing

21. Just prior to *Challenger*, Voyager's plasma wave team convened in Pasadena for the Uranus encounter. *From left*: Fred Scarf, Don Gurnett, and Bill Kurth. Kurth's finger is splinted from a sledding accident. Courtesy NASA/JPL-Caltech.

the rapid ascent of Soviet space activities. First satellite. First dog in space. First to fly past the Moon; then first to impact it. First to snag pictures of the Moon's back side. First man in space. First woman! First three-person crew! First spacewalk! First pictures from the Moon's surface! First pictures from the surface of Venus! Some claimed the Soviets also nabbed our first pictures from the surface of Mars, too (though that's *still* being debated).

But reaching the nucleus of a comet? Sci-fi stuff. Never done before. Too many variables: too much debris, too dangerous. As Sagdeev continued, "We started all the designs, you know, with all these problems. How to get closer to Venus? How to find the nucleus?" Their idea hopped the fence and took off through the meadow and found backing by the Soviet government. Other nations piled on in a scrum of participation. Ultimately the ambitious mission garnered widespread international attention, and the oddball ships even worked as intended in a real one-two-three. Vega's hat trick marked the final highlight of Soviet planetary exploration.

Gee, what about the United States? Ah, well, beyond one guy from the University of Chicago who supplied a Vega instrument, the United States didn't send a damn thing to Halley's comet. That divisive decision had been made years earlier in JPL's Bruce Murray days. While still at the helm during the late 1970s, Murray had shopped around a Halley visit, in part, by upselling a weirdly feasible propulsion concept known as solar sailing. It was pure theory. Oh sure, physicists well knew how sunlight mildly pushes against whatever it contacts. But nobody'd ever built a spaceship utilizing that concept. Murray's pitch went this way: unfold a ridiculously huge (and thin) sheet of material in space. Lashed to it is a bare-bones probe of comet experiments. Allow the sun's illuminating rays to propel the whole shootin' match out toward Halley's comet. Then be patient . . . it's slow-moving.

"The spacecraft could make a relatively leisurely approach to its target. With ample time to record the unusual physical processes taking place there." So argued Murray of solar sailing. To him it seemed an easy sell. Free propulsion? No fish would pass up such a huge worm on the hook!

Solar sailing drew its cluster of supporters. But Murray couldn't get the NASA brass to bite. It was too abstract. And politically, solar sailing competed with another experimental propulsion concept—massive solar cell arrays generating streams of charged particles to nudge a ship along. Way less far-fetched, with beefy dimples of support from Lewis Research Center and even JPL. Tons of money had already gone into it.

The solar cell thing got assigned to Marshall Center, where shuttle engines and boosters were top priority. The Marshall folks said they'd pick away on the idea the best they could, but development costs were projected to be as high as $300 million—not even including an actual spacecraft and instrument package. Analysts wondered if the scheme could even reach Halley's comet at all. To them, it seemed weapons-grade stupid. Marshall shelved it. As such, generally speaking, things were most certainly *not* going in Bruce Murray's direction.

"So *then* what he wanted to do, he wanted to trade Galileo! *Cancel Galileo*, turn that into a Halley comet mission!" An outraged John Casani picked up the saga, vividly recounting a low-low point in early 1982 with his friend and boss. Completely repurpose the mammoth Jupiter orbiter? No way in hell. The nerve of Bruce Murray to pull that thread! "I fought him on that. Head and toe."

Ol' Bruce wasn't letting go of his idea. Over a period of weeks, the two clashed as if rival gang factions. In life a common snippet of advice is to choose our battles, and the frat guy from Pennsylvania had definitely chosen to win this one at all costs. The divisiveness went supernova. Casani told his boss flat out, "Bruce. I'm not goin' along with this, you know? I think you're makin' a big mistake." Five long years had been invested in Galileo. Wipe out all that planning and effort just to see a comet?

"Let me say something about Bruce Murray." From his chair at the restaurant, Casani leaned forward, gesturing. "He and I were really good friends. I was very close to him." The men possessed the kind of relationship forged in fire via the stressful days, late nights, long weekends, and overall unrelenting hard work of creating brand-new things for the purpose of making brand-new discoveries. It went back to the 1960s, when the two had found themselves together on an intense timeline for the *Mariner 3* and *4* flybys of Mars—with Casani a systems engineer and Bruce Murray on the imaging team. From there a friendship had evolved.

Such a longtime bond offered John Casani the luxury of trying alternative approaches for persuading his friend to drop the Halley idea. Parables never worked with his kids, but so what. Casani entered dad mode and lectured Murray: "You're like that little dog that's taken a bone home and is crossin' a bridge. And he sees his reflection down there. He thinks what he sees is another dog, about the same size as him, with a bone. And he thinks, *If I'm smart I can get* both *bones and go home*. And of course, he dropped his bone in the river. And went home with no bones."

Bruce Murray stomped off. With no bones. One week became a few. He then summoned Casani to his office and said, "John! I have a new job for you!"

"What's that?"

"I want you to take over managing the Deep Space Network!"

Casani later regarded the offer as "very nice"—one that would've put him on an upward trajectory to even greater things at the Lab. A bigger paycheck. Assistant Lab director. Maybe even a reserved parking space. But integrity prevailed.

"Bruce, I can't do that," he scolded. "I know what you're up to. You just want to get me out of the way so you can cancel Galileo and have a mission to Halley's comet."

"Oh, no no no!"

"I'm not going for it!" Casani spread his arms wide before Murray in a religiously vibing sort of plea. "You're gonna wind up killin' Galileo and not gettin' Halley. And besides, what the hell you want a Halley mission for? You know there's already three countries goin'!"

Despite the ever-increasing impracticality of solar sailing to a comet, Murray offered no acknowledgment of his colleague's words.

"You're just gonna be the fourth Halley mission." Casani stabbed his finger at a nearby Galileo poster. "*This* is somethin' different."

Story delivered, a much older John Casani settled back in his chair and absentmindedly studied the wine in his glass. "Finally, he caved in on that." And according to Casani, the lack of support for an American comet mission was one key reason Bruce Murray quit the Lab only months later.

What was the point of that job offer?

"He was just trying to remove an impediment to him trading Galileo for a Halley comet mission. So I didn't take the job. Which was probably not the best thing for my career, I don't know, but I didn't anyhow."

Throughout the middle months of 1986, the lives and work of every Galilean seemed on indefinite hold. "We weren't sure *how* we were going to launch, *when* we were going to launch," remarked Bill O'Neil. With a glacier's patience, they waited. And to help manage emotions, a decision came forth to continue having picnics. One had already happened in Florida as kind of a last-minute thing while packing to leave after *Challenger*. A fourth would ultimately occur once *Galileo* finally left Earth. But what keeps everyone talking is the first of two that happened in between.

With it, Casani wanted to go big—and for solid reasons. "I thought it was a good way to get the people who were involved in the project to know one another, and to socialize," he justified of JPL parties in general. But this one would also attempt to erect a scaffold over the emptiness. His team could use a major release in the wake of the tragedy. And after the endless string of long weeks. For certain, any festivities needed to include their loved ones. Casani's vision encompassed a bountiful day of premier food and fun for every Galileo staffer and family member.

However, "it was gonna take *money*. And I couldn't get any money out of JPL at all." Pretty much the only way to fund his dream picnic would be by selling tickets in advance. But how in the world could he drive people to make sales?

Every week on the same day and time, John Casani helmed a Galileo status meeting. Its thirty-odd participants included ten or so key people representing JPL's major technical divisions. These reps bridged the pool of talent in their respective divisions with the project itself, serving as conduits for budgeting and scheduling and similar aspects. Bill O'Neil further categorized them as "technical experts" on division matters and super-smart people in general. Casani figured they could also handle ticket sales.

He announced this new responsibility during one of the very next project meetings. Before the assembled group, Casani disclosed an intent to hold the picnic—including his scheme for raising money. Every Galilean would be expected to hawk tickets. Aggregate sales would be tallied by the reps, then compared at each subsequent project meeting. It'd be a weekly contest, see? Among the divisions! With overall sales performance weighted against the percentage of Galileo workers in each. Just to keep things fair.

The rep with the highest sales that week would take home alcohol. And the loser? Well, the loser would take home the booby prize. Understanding that visuals often made more of an impact, Casani now gestured for his assistant at the back of the room to please walk the booby prize forward. Everyone turned to look.

What they saw plodding up to the front was off-white, a few weeks old, and bleating, and had just pooped a bit right there on the floor of JPL's conference room. It bore hooves and outstretched ears. John and Lynn had purchased it the previous weekend for fifteen dollars, cash, at a livestock auction outside of town. "We didn't know what the hell we were doing," John would later say of that outing with Lynn.

Squaring himself to the group, Casani declared, "I have a goat."

It had accompanied him to work that morning and been passing time inside a large box near his office. Staffers were bringing it vegetables from the Lab cafeteria. Don't panic; the goat would leave with John for the coming week. (Lynn had been keeping a diaper on the thing so it could roam freely about the house.) But during their next status meeting, every rep would have to report ticket sales. "And, uh, one of you guys will take home a bottle of wine. And the other guy has to take home the goat for a week." His terms were not negotiable.

Selling tickets was boring. "Why not make this interesting?" Casani's meeting was just about over; an attendee grabbed paper towels to swab the floor. One particular rep was outraged by the whole general concept. Later he'd also come to terms with how the Casanis performed a backyard neutering on the animal—guided by a book from the library because John didn't want to pay $125 to a local vet.

The ticket campaign lasted for weeks. Sales exploded. "All talk was about the goat," reported one Galileo engineer. Bottles of wine were awarded. The goat changed hands and moved around. Many were incredulous that such conditions were in play at a world-renowned institution. One week, that outraged rep lost and flatly refused to comply. But the picnic could now germinate with fiscal solvency. Teams of volunteers got in gear. Somebody negotiated the use of a nice park by the Rose Bowl, while others lined up games and prizes and food and special "prelaunch party" shirts. The event was huge. People spent all day in the sun and ate and drank themselves half to illness. Casani took his turn in the "Jupiter Plunge" dunking booth, where picnicgoers could whip waterlogged sponges at a target—aiming to drop everyone's favorite goat-hustling project manager into a huge tank of water. It marked only one path to that release everyone needed.

Casani's later retellings of this story always included a key detail: His goat cameoed at the picnic, to everyone's delight. It brings to mind visions of perhaps the saddest one-animal petting zoo that ever was. But multiple JPLers offered strong recollections of no goat in attendance that day—with John himself apologizing during opening remarks that it'd run off.

The goat hadn't run off. It was dead already. Eaten by some neighbor's Doberman, which smelled goaty smells and hopped a six-foot fence and dealt a prematurely blood-soaked ending to the ticket contest. Strictly speaking, Casani could *never* have admitted such a debacle on the day of the party. As one JPLer put it, "He'd have been booed off the picnic's lectern had he told the whole truth to the assembled crowd—exactly opposite from the intended morale boost!" And this party had indeed boosted morale.

Thirty years after it all happened, John Casani beamed at his table-side guest. "It was really, as you can imagine, a novel thing." He looked a bit little-kid embarrassed while acknowledging how times have sure changed. If the same thing went down today, expect a formal investigation, an addendum to HR policies, and perhaps one firing at least. Gingerly Casani mopped crumbs from his appetizer plate with a fragrant hunk of bread drowned in olive oil.

"So. Now you know the goat story."

American launch pads would remain silently still as no new lunar or planetary mission went up over the remainder of 1986—and through all of '87, for that matter. What a real bottoming out for space exploration . . . and for humanity in general when considering the April '86 disaster at the Chernobyl nuclear plant in Ukraine. Way to pile on!

Their whole project in limbo, Casani and O'Neil explored whatever fanciful notion they could think of to reach Jupiter. They asked the Johnson Space Center if additional kick stages could be duct-taped onto the existing air force booster. JSC said no way. Well, what about expanding *Galileo*'s propellant tanks? Good thinking, but that still didn't solve the problem. There were even discreet inquiries as to the feasibility of the Soviets' launching *Galileo*. Never panned out.

Trying to maximize his down time, Don Gurnett hit up O'Neil to enhance *Galileo*'s plasma wave antennas. O'Neil was visiting Iowa City and riding in Gurnett's car at the time. The pitch involved doubling the planned antenna length; Gurnett suggested that hinging them midway would enable the longer structures to be folded for launch. As they talked, O'Neil happened to notice that the antenna for Gurnett's car radio had apparently broken off at some point—replaced with a junky metal coat hanger stuffed into the body. An avowed car buff, O'Neil shuddered. "The big joke was, Don wants us to go to considerable trouble retrofitting the spacecraft, and he won't even replace the radio antenna on his car!" The whole discussion seemed extra moot after NASA definitively canceled Shuttle/Centaur in mid-June, leaving no means of reaching Jupiter.

In mid-July 1986 O'Neil and Casani were in a car together, driving back from some meeting in East Pasadena. O'Neil spoke up. "The shuttle can still launch *Galileo* with a vanilla two-stage IUS. Perhaps we should research everything we can do," he argued. "See where it can get us."

O'Neil was referring to the solid-fuel air force booster. Most everyone in the loop called it the IUS, for "inertial upper stage." That first word describes the type of guidance system on board—a rather under-the-hood kind of term not typically used as it was here. But NASA's original name, *interim* upper stage, referred to the idea of using this thing only as a stopgap until some form of advanced space tug (or other replacement) came along. Which never materialized. But the abbreviation stuck, and since everyone knew IUS already, a different first word was subbed in—whether it made sense or not.

Regardless, Casani thought O'Neil's idea to be worth checking. Once back on Lab, O'Neil forwarded the idea to Bob Mitchell, a manager overseeing trajectory design. Blue-sky brainstorming resumed, assessing every IUS option, albeit with encroaching futility. Nothing seemed workable. The idea of splitting the spacecraft bubbled back up. But this jigsaw puzzle had missing corner pieces.

On the Friday morning of August 1, 1986, Bill O'Neil picked up his ringing desk phone. "Hello?"

"Hi, Bill, it's Bob. I'm here in my office with Roger Diehl, and we think we have a path to Jupiter with a C_3 of fourteen." Bob Mitchell was on speakerphone so Diehl could listen in.

O'Neil thought maybe they were pranking him. "Are you serious?"

"We're very serious."

"You've got to be kidding." O'Neil completely forgot about the viewgraphs he'd been working on. "Well, you know, what's different?"

"We go to Venus first."

O'Neil sighed. "Oh. I thought you were serious."

15

S-TOUR

I was slow at finishing arithmetic lessons because I was too
meticulous about aligning my columns, and would erase and
rewrite them more neatly if they were not just right.

—Phil Roberts describing his childhood math homework

When asked, the miracle man who saved Galileo went so far as to say that
Mars had never before ended up being so useless. For his purposes, anyway.

Cloistered away in JPL's back-room discipline of orbital mechanics existed
Roger E. Diehl. He had papers from the University of Texas in Austin affirm-
ing his PhD in that very discipline—specifically, on the long-term motion of
Neptune and Pluto. Since 1975 Diehl had been roosting at JPL, plugging away
at what can safely be regarded as a niche. "I design trajectories to get spacecraft from the earth to almost any of the planets in the solar system," he began
of the career choice. And exactly what motivated him out of bed each morning? "Well, I guess it's always been my love and excitement of space and space
travel," came the response. "This was the closest that I could participate in
that." Even after decades in this field, the guy remained mesmerized by the
intricacies of how planets interact: "The way they're controlled by gravity. And
how you do things with the spacecraft, and you can go from one to the other.
I liked the fact that, you know, there were stimulating problems to work at.
And I enjoy mathematics."

He signed on to the Galileo effort way back in spring of 1977—months
before that initial round of congressional funding even came through—with
Diehl about to step away from Viking because its prime mission was ending.
The project still had him on for a bit longer, though, planning side observations of the Martian moon Phobos. Diehl explained that JPL missions always
ran to a predefined end. Sure, an extension could materialize if the spacecraft
was still operational, but usually on reduced funding. This translates into a
reduction of the workforce. People move to other projects.

"And in the case of myself, sort of being a new hire and not knowing anybody, I was relying on my group supervisor to have a position for me." That position ended up being on Galileo. "My group supervisor wanted me to do that, and I was interested in doing that, because Galileo looked like a fascinating mission to work on," he remembered. Diehl told his supervisor, "Yeah, great! Sounds good!"

Initially, the work was pretty straightforward. Diehl busied himself mathematically planning the orbiter's exact combination of loops through the extensive Jovian system of moons—all in accordance with fulfilling scientific objectives. He also met an amazing woman at church named Denise. They married and bought a new car together, a beige 1979 Subaru with a manual transmission, and Denise taught Roger how to drive a stick. Workdays at JPL stayed constant for a while. He shared an office with one or two other trajectory designers. All had battered old desks, complementing metal shelves along one wall that groaned under the weight of printed computer runs.

"You never knew when you had to go back and refer to something," he justified of the towering stacks.

At the outset, Roger Diehl's work of fashioning outer-space trajectories involved personal travel on decidedly shorter trajectories. He had to gather notes, exit his shared office in Building 156, walk a couple minutes down the hall past other offices, and finally duck into a small room. He categorized its furniture as "very plain, almost like surplus." Beyond its drab, Soviet-level trappings, the area housed a deafening line-feed printer along with keyboards and boxy computer monitors linked to JPL's mainframe computing network. This grand setting formed Diehl's arterial gateway to solving those stimulating problems. Merely showing up could be like visiting a popular restaurant: "There would be times where, if you had more people than you had terminals, you had to wait your turn." Anybody have today's paper?

Once seated, forget using a mouse to navigate friendly on-screen interfaces. The privilege of access was getting to peck away on a heavy mechanical keyboard to manipulate lines of green text on a black background. Doing taxes could be more visually engaging.

Diehl's order of business at the terminal would be to sign in with his employee credentials, enter a billing code for the current project, then select what he wanted to do. "They had libraries of programs," he commented of JPL's setup.

Many pieces of software represented a type available nowhere else in the whole entire world because they'd been created from scratch by a staffer.

For roughing out *Galileo*'s trajectories, Roger Diehl used a program called S-TOUR, which had been written by a colleague named Phil Roberts. "He was a member of the Advanced Projects Group," as Diehl put it—a small team investigating future mission possibilities that might be years down the road. Missions where the technology to realize them didn't necessarily exist yet. Their double-entendre motto "Any Body, Any Time" might today force a meeting with HR.

As Diehl continued his story about Phil Roberts, "He put together this program to be able to quickly generate multiple-orbit *satellite tours* if you were in orbit around a planet." Roberts had concocted it during the earliest days of J-O-P and specifically for that mission—knowing how the core flight profile involved repeatedly looping through the dynamic system of Jovian moons. In the eyes of trajectory people, they'd be flying a tour of sorts.

Diehl had nothing but praise for Phil Roberts. "He was very, very sharp. He was very fast too. I mean, he could sit down and have an idea in his mind and whip up a program to be able to analyze it." What Roberts had released into the wild conformed to Diehl's needs, as well as to the larger needs of his group, project, and employer. "A program that was designed just to see what one could do. And have that potentially be part of future mission proposals." The right tool for the job.

"Okay, let's say I'm at the Jupiter moon Ganymede. I want to get to Callisto." Diehl walked through an example. "I can tell the program, 'I want to go to Callisto next. Can I fly by Ganymede in such a way that the orbit will change, and I can encounter Callisto?' And so what you do, is you sort of sequentially build up a tour."

A minor sense of urgency always clouded the proceedings because JPL's computing center billed specific projects for how much computer time had been used. "Sort of a nagging nuisance," Diehl reported of the now-medieval practice. "You had to worry sometimes if you were spending too much money." Cheaper rates could often be obtained by hopping on during nonstandard times. "Early in my career there'd be many a time I'd come in for three or four hours, usually on a Saturday," he explained. "I could do so much more on the weekend for a given amount of dollars." On occasion, some heavy math procedure he needed to run could be set to auto-start after-hours at a preset time.

22. Roger and Denise Diehl in 1985. "My wife sometimes said she heard me
talk about trajectories in my sleep." Courtesy Roger Diehl.

Life for Roger changed again in 1983, when he took over supervising a trajectory group underneath Bob Mitchell (and had to report on how much overall project money went toward computer time). Roger and Denise, by then,
had a newborn and a one-year-old. Diehl moved to a private office in Building 301 to colocate with his group. The Galileo mission shifted into prelaunch
phase. The spacecraft went to Florida. *Challenger* happened. *Galileo* went into
storage, the oiled joints of its folded antenna dish slowly drying out. But such
things wouldn't matter if the whole ship ended up at a yard sale because of
no way to launch it. On any normal project, the mission and flight path are
defined first. How much science it'll carry. What it needs to *do*. Factors like
these drive how big the spacecraft ends up being and what combination of
booster stages will propel it to the final destination.

Diehl pointed out the unprecedented conundrum of these elements being reversed. "With *Galileo*, you had a *built spacecraft*. And all of a sudden you *did not* have a trajectory to get you to Jupiter." Finding some new path through space meant wading through an imposing number of constraints. The ship weighed a given amount; its propulsion equipment had been sized perfectly for a flight profile that could no longer be utilized.

Out of play, forever and for all time, was the Centaur-in-shuttle scheme. Any workable solution would have to make use of that underpowered air force "inertial upper stage" booster with its weak-ass solid fuel. Compared to Centaur, Diehl condescendingly referred to the IUS as "basically more like a firecracker" and mourned what he no longer had access to. Science risked losing an otherwise ready-to-fly Jupiter spaceship.

Technology naturally evolves, and JPL computer terminals gradually migrated from dedicated rooms into people's offices. Diehl himself welcomed a new terminal into his bigger digs in 301. It lived on a small table facing the door; while typing he could look up to greet anybody coming in. When not at the terminal, Diehl could swivel around 180 degrees to a larger desk up against the back wall. Plenty of room. Another group supervisor set up office in an adjacent space. And beyond both their doors sat two secretaries to run interference on the daily barrages of mail and memos.

Explore IUS-based options for reaching Jupiter. This new marching order came to Diehl via his superior, Bob Mitchell. "Since he was funding me as the group supervisor for my management time," Diehl explained, "I'm sure he thought, *Well, I'll put Diehl to work, too, since I'm paying for him.*"

Let the games begin. Since it lay beyond Earth and in the general direction of Jupiter, every search for new *Galileo* trajectories began with flying past Mars for a gravity assist. Now, understand that hopscotching through the solar system via planetary gravity is a common technique. Done properly, the intermediate planet transfers energy to the spacecraft and boosts it along. A sort of "gravity slingshot." Similar to having free rocket stages waiting out there to hasten the trip.

Faced with energy deficits for multiple aspects of the mission, Roger Diehl subdivided his problem. "In a way, I didn't have to think about what my deficit *was*. I knew I had two overriding constraints that I had to satisfy." First, he had to address that all-important unit of initial energy—the C_3. Original J-O-P documentation listed a C_3 of eighty with the shuttle and three-stage IUS sending the goods on a direct flight.

However, between the planned 1982 launch and the summer of '86, Jupiter had of course moved—rudely progressing through its solar orbit—and no longer offered such an easy path. In *Galileo's* new normal, there would still be a ride up to orbit on a shuttle. But its two-stage IUS, plus Jupiter's non-optimal position, equaled inferiority. Roger Diehl grumbled. "A trajectory with a C_3 of fourteen," he acknowledged, was the most he'd have to work with. Such a teeny amount! "So by definition I'm starting out with that constraint." Every S-TOUR session, he keyed that value as a top-end limit on the amount of initial energy.

And then Diehl's second constraint involved a need for the smallest-possible amount of delta-V. Think of it as in-flight potential velocity. Using any spacecraft propellant at all to reach Jupiter meant having that much less on hand for navigating the final satellite tour. Translation: possible reductions in achieving their science objectives.

These constraints are mighty tight. Crappy values to work with. And every option Diehl tried with Mars ended in disappointment. "I didn't have the performance to get to Jupiter," he said, dejected. The math never worked out. Slender Mars seemed unable to provide enough of a goosing boost to shove the massive machinery onward to Jupiter. "So I would have had to use additional propellant, which *Galileo* did not have, in order to complete the desired satellite tour at Jupiter." The man sounded worn out. Defeated by circumstances and bad decisions at the highest levels. And defeated by that shitty little booger-planet Mars. The arduous task bled into his downtime. "I'm totally convinced my brain worked on this problem when I was sleeping." Amusing yet frustrating all at once. "I think my subconscious or something was still not letting go of this problem."

Occasionally, when people have entered a self-imposed overdrive mode to work through some major passion project, routines and everyday basics can get left in the dust. But Roger Diehl strove to maintain a healthy balance between work and life. "I never got to the point of not eating dinner, or having a good glass of wine with dinner, and spending time with my family." Such things were important enough for him to remain aware of, and to prioritize, regardless of job strife. "And so when I worked, I worked hard. But I wasn't going to let the, you know, the enjoyment of things in life to suffer because of it."

On the Thursday afternoon of July 31, 1986, an exasperated Roger Diehl categorically washed his hands of the Red Planet. "I just came to the conclusion:

If I had to go out of my way at all to fly by Mars, I didn't gain performance." Notes of defeat peppered his comments. The attempts had included such meandering combinations as Earth–Venus–Mars–Earth–Jupiter, but even that one required a propellant-heavy maneuver after the second Earth flyby.

"Mars was not the solution at all."

After work that day he glumly shuffled out to JPL's expansive parking lot and found his Toyota Cressida, which promised everyday reliability yet few thrills. In it, Diehl wended home in frustration to Denise and the kiddos. He put on a good face for them but ached inside. The Diehl family's evening routine played out; the kids went to bed and those last sips of wine were savored and next up came Roger's longtime habit of deciding what the morning's initial computer runs should be. "Hit the road running," he advocated.

That night Roger Diehl hit something else instead—a paradigm shift. A new way to consider the problem. "I'm going to change my frame of reference. And I'm going to think about it as a tour of the inner solar system to get to Jupiter," he internalized. "And see what happens." The fresh idea helped liberate some of that daytime stress. "And I didn't care how long it took or how many encounters. If it took twenty years or whatever, I wanted to find something that, performance-wise, would work."

The next morning saw him land in his office a bit earlier than usual, around 8:15. Years later he would describe that Friday, August 1, this way: "Some days you just didn't know what was going to happen. And this day was sort of the ultimate surprise show." The areas surrounding his office remained empty and quiet in a meant-to-be kind of way that enabled focus. "Find a route for *Galileo*, from Earth to Jupiter, which can accommodate the spacecraft's weight. Along with the underpowered nature of its final propulsion stage."

He wouldn't be starting from *absolute* zero. Diehl had binders full of pork chop plots to reference. Within them, upcoming multiple-week periods were graphed against however much energy would be required that day—be it for a trip to Mars or Venus or Jupiter or wherever the destination. Every outcome was a function of celestial mechanics. "Injection energy," it's called. And the chops mathematically foretold that *Galileo* could go the distance if it accumulated sufficient injection energy before a final swing past Earth during some future December. "I knew it was *around* December, but I didn't know if it was late November, or—or, you know, I just didn't know how it would play out."

Having ruled out Mars, Venus remained to assist. And Roger Diehl harbored low-low expectations. "I launch to Venus, and then I'm going to use as many flybys as needed in the inner solar system to link up with an *Earth* flyby that would be at a time where, *if* you had sufficient performance, you would continue on a direct trajectory to Jupiter." Diehl smiled inside at the fortuitous decision he'd made prior to *Challenger*. A software engineer working for him had asked whether they should preserve the interplanetary option in S-TOUR because nobody needed it anymore. Diehl had told him to keep it just in case. And now here he was, sitting down to use something that almost got thrown away.

Enter credentials and log into mainframe. Access program library. Arrow down to S-TOUR. Open. Recalling this moment, Diehl got sentimental. "I almost felt like the program was part of me. I had run it so many times." The interface immediately asked its user basic questions about the journey. "I had to give it an Earth launch date, and I would have given it some sort of departure velocity." By the latter he meant that initial C_3 of fourteen, with the actual input keyed into S-TOUR being the square root of C_3.

Phil Roberts's software broke a flight down into legs. And as with every previous time Diehl searched, the first leg was always the same:

EVENT 1 LAUNCH FROM EARTH

Suitable launch energies to Venus would next be available a few years down the road, in July 1989. Diehl knew that from studying pork chops. And yesterday's final attempts had used July 1 as launch day, so he started there. Diehl typed in the date and hit enter.

EVENT 1 LAUNCH FROM EARTH ON 1989/07/01 00:00:00 TO GO TO VENUS

A pregnant shuttle in low Earth orbit had just birthed *Galileo* from its womblike payload bay. Air force IUS alight. Headed out. Now, reaching Venus from Earth actually necessitates a *reduction* in energy. "Because what you need to do is *fall down* to the orbit of Venus, for the Venus gravity assist," came the explanation. A bit of a contradiction, maybe, but hang in there. On the screen of Diehl's terminal, new lines of punchy green text detailed possibilities for how and when to go by Venus. "This is the initial step of pumping energy into the orbit," he continued. "There could be several trajectory options to Venus, all with the same C_3." Scanning down the columns of numbers filling his screen, Diehl saw how the July 1 option listed a C_3 of 13.957 along with his desired approach velocity and flight time. What an all-inclusive package. He tapped arrow keys to choose it, sending *Galileo* past the trailing side of Venus. S-TOUR updated with an assumption of the next destination, but it could be changed.

EVENT 2 FLYBY OF VENUS ON 1990/10/01 02:48:08 TO GO TO EARTH

"When I answer a question, it immediately prompts me with the next one. And when it has to do the calculation, it can do it in seconds." Diehl indeed wanted an Earth return—to check whether *Galileo* would finish arcing back around in some future December. S-TOUR gurgled and then upchucked columns of numbers indicating flight duration, celestial coordinates, angle relative to the ecliptic, and other factors that, on first sight, might've looked more like a bank statement than space-mission planning.

Over a year to Venus? Normally a ship can arrive there within four months or so. This option went the scenic route by tracing an empty orbit around the sun. But it conformed to the energy requirements and sent *Galileo* to the Goddess of Love for a helpful gravity assist.

Now the ship was comin' 'round the mountain. "I said, 'Okay, I've gone from Earth to Venus. Take me back to Earth.'" More keys clacked as Diehl answered questions posed by the glowing interface. Before it could present new options, S-TOUR had to know how many revolutions Earth should make around the sun. "There would be more than one trajectory that would take me back to the Earth, but it would take me back to the Earth at a *different time*." Diehl keyed in his December preference. From the bowels of S-TOUR bubbled up more columns of numbers and options. He scanned the various arrival dates. "I saw that one of the options took me back to the Earth in December," he smiled. "And that's when I jumped out of my chair."

EVENT 3 FLYBY OF EARTH ON 1991/12/05 04:47:48

Quickly Diehl's attention dropped to event 3's supporting data columns. Insufficient to reach Jupiter, but he knew that already. The loop back would see *Galileo* racing past Earth at just under nine thousand miles of altitude and in need of more oomph. Perhaps more could come from this—a nice, plump, two-year orbit to bring it back around *again* a couple Decembers down the road. Maybe that'd be enough for Jupiter, but he didn't yet know.

```
EARTH=VENUS=EARTH=EARTH
    RANGE         LAT         LONG       H ANGLE      PHASE
 257434370.       3.8034    118.3649      .0000       .0000


 EVENT  3  FLYBY OF      EARTH        ON 1991/12/05  04:47:48 TO GO TO MANEUVER
 BODY EQUATOR OF DATE

    PERIOD         RP          INC        NODE         ARG          TRUE
 63116865.    121122876.      4.2395      2.0958    =122.2844     =62.7773
   730.5193        .8097                            REVS =          .67438
    HP          THETA        VINF         DEC        RT ASC        PHASE
   14302.      172.8622     12.0528     =13.6597    251.5786      82.7391

 DES CROSNG   SUN   EQTR   ON 1991/12/09 AT 14:30:24   TF=   4.40458
   RANGE        LAT         LONG       H ANGLE      PHASE
 143035646.     =.0000      351.1311      .0000       .0000

 APPROACH TO  PERIAPSIS   ON 1992/01/17 AT 20:33:33   TF=  43.65677
   RANGE        LAT         LONG       H ANGLE      PHASE
 121122871.    =3.5832      212.0072      .0000       .0000

 ASC CROSNG   SUN   EQTR   ON 1992/05/29 AT 14:00:53   TF= 176.38409
   RANGE        LAT         LONG       H ANGLE      PHASE
 244433872.      .0000      251.7056      .0000       .0000
```

23. Roger Diehl's original 1986 S-TOUR printout displays every critical value involved in each leg of the trip. That "period" column specifies the number of days of travel, while "HP" denotes how high above a planet's surface the ship will fly in kilometers. "HP was always the first parameter that I would look at," Diehl indicated. Courtesy Roger Diehl.

From his terminal, Diehl now requested a custom parameter that the program wouldn't have automatically known to add. A tweak. "This was a case where I had to help S-TOUR," he explained. Properly aligning *Galileo* meant fine-tuning the aim point at the second Earth flyby such that the ship properly bent around for its one-way haul to the outer solar system. On the printout, it looked like this:

EVENT 4 FREE SPACE MANEUVER ON 1993/01/17 02:52:05 TO GO TO EARTH

Diehl didn't know the parameters of this maneuver. S-TOUR could easily compute it, but he couldn't give the program unbridled free rein to line up for Jupiter by any means necessary. He needed to constrain the software so it used minimal propellant to change direction yet retained the flexibility to search for Jupiter aim points. "The only thing I cared about was that it was so infinitesimally small," he said of event 4. And the result seemed about what he wanted—an option to head out for good in December. Good-bye, *Galileo*. "And when it intersects Jupiter's orbit, Jupiter will be there!"

EVENT 5 FLYBY OF EARTH ON 1993/12/04 16:51:33 TO GO TO JUPITER

Okay, this looked pretty darn nice. But what about the flyby altitude? Diehl's eyes zigzagged down the screen for this key nugget of info. Understand that s-TOUR treated a planet as if it were nothing more than a tiny pinpoint of gravity—located where the center of the planet really is—and ignored the planet's actual size. While sounding maybe a bit precarious, aiming for the center of a planet wasn't risky at all, as s-TOUR compensated. "It calculates the *bending* required," clarified Diehl of the program's computations, "and how far away from the center that you need to be" at the planet in question. Users therefore needed to double-check that s-TOUR's indicated flyby altitude would in fact be sufficiently *above* the surface of whichever planet the ship was arcing past.

In this case, the second row of data for event 5 displayed a positive number. "If that number had been a *negative* number, that would have told me I'm having to fly below the surface of the planet. Which obviously you can't do!"

Would this indeed send their beloved *Galileo* to Jupiter? Dutifully s-TOUR generated a new line of text:

EVENT 6 FLYBY OF JUPITER ON 1996/10/01 03:18:40

Why yes, yes, it would. Planet and ship could finally meet. In October 1996, to be precise. Just over ten years away.

"I knew I had a trajectory," sang Roger Diehl. "I was almost bouncing off the walls." Any solution for Jupiter used planetary flybys and trajectory corrections. But he had needed s-TOUR to collaborate on assembling the secret combination. "What's happening with each of these flybys: the farthest distance from the sun is being pushed out," Diehl clarified of the process. *Galileo's* flight path is essentially a giant oval. As the journey begins, that oval contains the orbits of Earth plus Venus. A gravity assist from Venus is used to elongate or "push" the oval into a broader shape. "The first Earth flyby pushes it out even more, and then the second Earth flyby pushes you all the way out to the orbit of Jupiter."

Roger Diehl had built an organic, mathematically proven road to Jupiter. And at that point was done needing Friday-morning peace and quiet. He snatched a curl of paper from the tractor-feed printer and rang up Bob Mitchell, saying, "I think I found a trajectory that'll work for us."

"Come up right away," Mitchell told him.

Diehl hustled over to Building 264 and dashed up to Mitchell's office and triumphantly extended the results into the hands of his superior. "I can remember him sort of pensively looking at the S-TOUR printout." They traded comments and questions for about ten minutes. Diehl knew he could trust S-TOUR. "Mitchell trusted *me* because he *also* trusted S-TOUR." Then Mitchell called Bill O'Neil on speakerphone with the news while Diehl stood there in the office. It was ten o'clock in the morning.

"This was momentous," Bill O'Neil would later say about the discovery. "JSC would not be able to deny us using only the IUS as upper stage, as they had already flown it several times." O'Neil scribbled notes as fast as he could, then tapped the on-hook button and dialed John Casani at extension 6578 without missing a beat, and gave a quick rundown.

"Send Roger over here," beckoned Casani's voice through the handset. "I want to kiss him on both cheeks."

But not right away. Casani was booked solid until the afternoon. The group met up later that day, and Diehl's full scheme was unrolled bare for objective analysis. "He loved the trajectory right away," celebrated Diehl of Casani's reaction. "He accepted it immediately." JPL practically auto-generates acronyms, and a new one emerged—VEEGA, denoting the eclectic Venus–Earth–Earth gravity assist. Casani rushed into a call to one of the hardware engineers and started listing numerous considerations: review the antenna design. Review the thermal design. Flying so much closer to the sun meant they were going to have to add blankets of insulation and maybe other stuff too. There was also the matter of an already-scheduled Galileo meeting for the following Monday, the substance of which might now drastically change.

Roger Diehl later categorized that Friday as "one of my most exciting days." Although one thing really bothered him—VEEGA's extra orbit around the sun. It served a purpose, yeah, but really stretched out the trip. NASA HQ and Congress both hate waiting; politics matter even in spaceflight. Diehl had been around long enough to know how the entire mission could still go in the dumpster because nobody would want to keep shoveling money into such a drawn-out mission.

Saturday morning, Diehl pledged to take the day off. But the idea of eliminating that extra solar loop kept dancing around in the unlit backroom recesses of his mind. He brought it up with Denise because he wanted to hit the office on Sunday and try snipping out the loop—even though they'd already committed to a church picnic. Diehl wasn't sure he could hold off until Monday morning to look for a more streamlined VEEGA. "You know, I just couldn't wait to get back to work," he admitted.

That Sunday afternoon began as anticipated. "We went in separate cars, because the plan was for me to go at the end of the picnic, and my wife would take the kids back home. But I just—it got to the point I . . . I just couldn't wait. I *had* to go in to be able to see if I could do this."

Denise recalled the event in a lighthearted kind of way. "He ate a quick lunch and went," she laughed of his early departure. "I remember staying for a while, because the kids loved the playground."

Reaching JPL's campus took half an hour. Diehl booked inside to his ground-floor office and plopped down at the terminal. "I understood the trajectory completely after finding that first one. And even though it wasn't an optimal one, I knew exactly what to do to take out that one empty rev before the Venus flyby." The process took about ninety minutes. And he did indeed find a workable option—one coming back to Earth a full year earlier. "All the flyby altitudes were nice and high. And I said, 'This is, you know, *this is a beautiful class of trajectories* that should work for *Galileo*.'" He said it with passion, the way someone might describe a Vermeer painting.

Roger Diehl couldn't help feeling that he had achieved a personal milestone. "In a sense, it was like my JPL dissertation. I had taken a problem, that we didn't think that there was a solution. And I found it."

Things now began happening even faster. At the very moment Diehl had first revealed his discovery to Bob Mitchell the previous Friday morning, Bill O'Neil had been frantically prepping for a Monday presentation at NASA Headquarters. "We never *didn't* have a plan," he alleged. JPL was going to recommend splitting *Galileo*. Again. Using a revamped probe carrier. Early Monday, O'Neil flew to Washington and did so, offering for his encore, "with appropriate fanfare," the miraculous discovery of VEEGA—despite its unattractive seven-year flight time. This announcement occurred just as Roger Diehl reported his improved "Sunday" trajectory to Bob Mitchell. Mitchell rushed the findings to Diehl's colleagues, who rapidly optimized VEEGA using different software and reported back as fast as they could.

Solution in hand, Mitchell impatiently eyed his phone.

Three thousand miles away in DC, Bill O'Neil finished explaining VEEGA and got icy stares. Geoff Briggs, NASA's head of solar system exploration, told O'Neil straight up that looping around in space for years didn't seem like a great idea. O'Neil struggled to respond, but the meeting time was up. People started gathering their things.

By prearrangement, O'Neil sprang from the room and called Mitchell to check for any news. Rapidly Mitchell said they'd gotten the flight time down from seven years to six. O'Neil flipped over a nearby scrap of paper and scratched notes on the back and dropped the phone on its cradle. "I quickly ran back into the room and reported this as the meeting was breaking up." He asked everyone to please sit for great news. They did so reluctantly. Briggs still objected. "The news did not impress our customers," as O'Neil put it.

But within an hour of the meeting's end, O'Neil received marching orders from John Casani to proceed with VEEGA anyway, no matter what Briggs or anyone else at HQ thought. Already Casani had in motion directives to price out what the changes would cost.

"We never looked back," said Bill O'Neil. They were going after all.

"Take me to Jupiter," is what Roger Diehl had told S-TOUR. "And I go, 'Whoa, it did.'"

16

Ten Years, Three Months, and Two Days

Ah, well, I would say yes. I mean, I wasn't chomping at the bit,
waiting. But I had finished the development of Galileo.

—John Casani on whether he wanted the job of assistant lab director

Having unearthed it from storage, *Galileo*'s processing team faced a unique situation—middle age. Their years-old yet unlaunched and unused ship, with probe in tow, had returned from Florida in late February 1987 after a grueling, five-and-a-half day slog on yet another flatbed semi. And ever since, it'd been sitting in the corner of a JPL assembly bay underneath a giant black drape. Leading a decidedly sedentary lifestyle. With the drape now off, everyone could bend forward at the waist and gently see to what extent things seemed to be holding up after all this time sitting around on their butt. Not visually apparent: damage wrought by trips on the semitrailers.

Panels came off and parts were examined. "A lot of these things were built prior to 1980," remarked one team member of the numerous custom assemblies and fittings and pipework. Serial numbers and delivery dates harkened back to the original 1982 launch plans. This thing wasn't even supposed to be here anymore. It was supposed to be nearing the end of its prime mission. Lengths of electrical cabling were now brittle and didn't flex as they needed to. Some metal parts showed corrosion. O-rings had fractured. Paint had chemically changed. Protective coatings might've shrunk. Assemblers had used conductive adhesive tape all over the place that didn't seem sticky enough anymore. Should this jalopy go to a museum after all?

The descent probe came off and went back to Hughes for refurbishing and new banks of customized lithium batteries. Teams of reviewers dissected the orbiter into its major systems, tabulating areas of concern for deeper digging. Every single moment of physical contact with the ship had the potential to dirty it just a little or overstress some critical part.

One thing made testing a real bear of a problem—only one *Galileo*. Many

24. In 1987 artist Henk Pander visited JPL and created this painting of Galileo
proceedings in the high bay. It is a work of oil on linen and measures 81 by 105 inches.
"The people at JPL were very good to me and allowed me very close to the
Galileo spacecraft." While the identities of the two male technicians are unknown,
the woman between them is Mary Reaves. Courtesy Henk Pander Estate.

existing spares had been committed to other projects, including the budget-
level *Magellan* soon heading for Venus. It hadn't always been like this. Back
in the Voyager days, two ships flew, but JPL had essentially built three. That
extra machine, their proof test model, functioned largely as a bed for testing spares but could've actually flown. So every time something unexpected
went down during Voyager testing, or even during the active missions, all the
techs could look at each other and ask: Well, how did the other ones behave?
Not the case here.

Galileo's frame-off restoration continued. Bolts and screws never meant to
unbolt or unscrew experienced just that. Long-united couplings and brackets and backplanes and motors and multipin connectors—all were separated
from one another and dispersed to various scrutinizing specialists. For certain, some good resulted. The West Germans improved their particle detector

by adding a sensor. The ultraviolet imager received its own new sensor. Also, engineers and technicians and scientists alike gained a better overall understanding of how spacecraft age.

But other modifications tampered with previously complete and working systems. During its outbound trip, *Galileo* would now face in a different direction than originally planned—nose-first toward Venus, with both high- and low-gain antennas pointing completely away from Earth. All-important communication required a new low-gain antenna, jutting from the ship's backside and spliced into the existing radio system. It would have to be designed, fabricated, connected to the onboard power system, and tested.

And speaking of, what about *Galileo*'s core power supplies—those plutonium-based electrical generators? Out of the box, the contraptions were only about 10 percent efficient to begin with. Ninety percent waste! Once assembled they were instantly on, couldn't be switched off, and the power output unavoidably ramped down over time. Mission designers would have to implement a number of power-saving measures to go the (much longer) distance.

Well, why not just get new power supplies? Couldn't JPL have simply requisitioned another set? Unfortunately, building new ones takes *years* and involves major negotiations with the U.S. Atomic Energy Commission. It's a nuclear waltz. And of course, replacement would also involve politics. "To have pursued new RTGs would have implied that we needed them!" squealed Bill O'Neil, who was more sensitive to the situation than most. "Which would have greatly fueled the fires of those who wanted the project canceled! We saved *Galileo* by minimizing cost and difficulty of getting it launched." He'd used an abbreviation for *radioisotope thermoelectric generator*—the official term for the power supply.

Reassembly of the probe continued. Reprocessing the orbiter itself was now squarely on track. Still, the antenna pins—nobody thought about checking the damn pins.

DEPUTY ASSISTANT LABORATORY DIRECTOR OF FLIGHT PROJECTS APPOINTED FEBRUARY 1, 1988

John R. Casani, manager of the Galileo project since 1978, has been appointed Deputy Assistant Laboratory Director, Office of Flight Projects, Dr. Lew Allen, Director of NASA's Jet Propulsion Laboratory, has announced.

"I had been on the project for ten years, three months, and two days when my boss was promoted. And they offered me his job overseeing all the flight projects." This is how Casani summarized the circumstances of his departure as Galileo's project manager. He'd be moving on into the upper tiers of management. So Casani would also be moving offices; time to pack up his current one in Building 264. "Which is the building just to the east of the cafeteria," as he located it on JPL's campus.

His outgoing office, not much bigger than anybody else's, contained a smallish meeting table off to one side, plus a whiteboard in case somebody needed to jump up and draw diagrams. The space had always remained one of sparse personalization without much to take down off the walls, save a few NASA recognition certificates. And never any mementoes from home such as family photos.

Hopefully his successor would embrace the culture that Casani had strived to bring mainstream—a culture of easy access. "You know, the only problem you ever have building a spacecraft, or running a company, or bein' mayor of a town, or president of the United States is communication." And poor communication, as he put it, represented "the root cause of all problems."

During his decade of Galileo tenure, Casani labored to be continuously available for whomever and whatever might be in need. A major catalyst was visibility. "The secretary sat outside the office. But I mean, I had an *open door.* Anybody who wanted to come and see me always got in." How remarkable: communication was facilitated by the basic positioning of a vertical panel. "And closing your door does not add to that, in any beneficial way. You know? It's, 'Well, he doesn't wanna talk.'" Casani straightened up. "No, I *wanna* talk. 'What's up?'" The person would come into his office and, depending on the scope of the discussion, maybe grab a chair at that meeting table. Casani would listen, focused, ignoring his phone messages, choosing affirmative responses. "First of all, I always wanna hear what they have to say, you know. And I'll ask questions. And what I'm trying to do is probe their level of understanding of the situation," he explained. "See how worried they are."

To help sustain communication, Casani habitually assembled key project people in his office at least once every month (not unlike those meetings with the goat contest). To the first guy he'd ask, "How you doin'?" And sometimes the conversation would unfold this way:

"Everything's fine."

Hmm, that's vague. Casani would ask about project targets. "You makin' it all?"

"Yeah."

"You behind schedule?"

"No."

"What are you worried about?"

"I'm not worried about anything."

"You're not worried about anything? That makes me worry." According to Casani, a lack of worry is a major red flag when dealing with large technological projects. He now had a new problem of his own: make this uncalibrated person sitting in his office lock into sync with reality. And confirm whether things really were okay or finally concede: Well, you know, yeah, I do have a problem.

An admission such as this—critically necessary for downstream success—would branch into Casani deposing the individual, working that person's knowledge of the situation to its periphery. And sometimes, if the person did indeed have a handle on things, resolution could be straightforward enough. Casani might ask, "Well, do you have a way to fix it?"

"Yeah."

"What's it mean?"

"I'm gonna need another $15,000."

"Okay, well, do what you think you need to do, and keep me posted." And as soon as the meeting ended and everybody left, Casani would update a notebook he kept for such things, adding the situation as a lien against his project. Now the budget had a hole in it shaped like $15,000 or maybe more, depending on whether the person seemed to truly understand the predicament. Sometimes it was a lien against the schedule—if another week might be needed to fix some electrical gremlin or thermal imbalance or whatever it was. Would they have a way of accommodating that downstream? Would it affect launch day?

Money wasn't always the answer—at least not until people had a better grasp on the situation. "I would ask enough questions to sort of satisfy myself that, *Okay, they've got their mind around this. They don't know the answer, but they know the problem. And you know, it seems to me that they're goin' about it the right way.* I wouldn't try to solve the problem for them." Rather, Casani saw his role as ensuring the individuals understood what resources were available for proceeding through to resolution.

"But I just wanted to know that they understood enough about the problem that I could feel confident that they're gonna go work it. And if something doesn't happen, they'll come back and tell me about it some more."

Leaving his old office for the last time, with his box of notebooks and note cards and NASA certificates in tow and balanced on one knee to catch the light switch, John Casani took a final look around and sincerely hoped its always-open door would remain open for those in need.

But Galileo's next project manager would ultimately not operate that way. "The guy that followed me, you know . . . he kept his door shut all the time."

July arrived. Early that month, Fred Scarf traveled inside Soviet territory for the first of two *Phobos* launches headed to study Mars and its moons. Following that, he made his way to the international Future Studies of Mars symposium at the Space Research Institute in Moscow.

Despite Cold War politics complicating joint East-West space exploration, the working scientists themselves always managed to quietly negotiate such roadblocks. In Scarf's case, an existing position with the diversified military contractor TRW made collaborating with the Soviets tricky—in the beginning, anyway. But Scarf had simply gone to work part-time at the University of California, Los Angeles. He used this academic footing to build a bridge across the Iron Curtain and get aboard Phobos with a plasma wave detector.

The Future Studies of Mars conference was scheduled to run until July 14. With two days remaining, Scarf heard welcome news about that day's launch of the second *Phobos* and the successful deployment of its plasma wave antenna. His enthusiasm filled the meeting room. But later on, while in one of the institute's lobbies, he suddenly collapsed on the floor. Scarf was rushed to Burdenko Hospital—a forty-minute drive even in favorable Moscow traffic. Somebody got in touch with his wife, Mimi. Fred died five days later. He was fifty-seven. Only a year downstream waited his appointment in Pasadena with Don Gurnett and the rest of the plasma wave team for *Voyager 2*'s Neptune encounter.

"I was out of country at the time, and learned only upon return," indicated Roald Sagdeev, whose own background was in plasma physics. He mourned

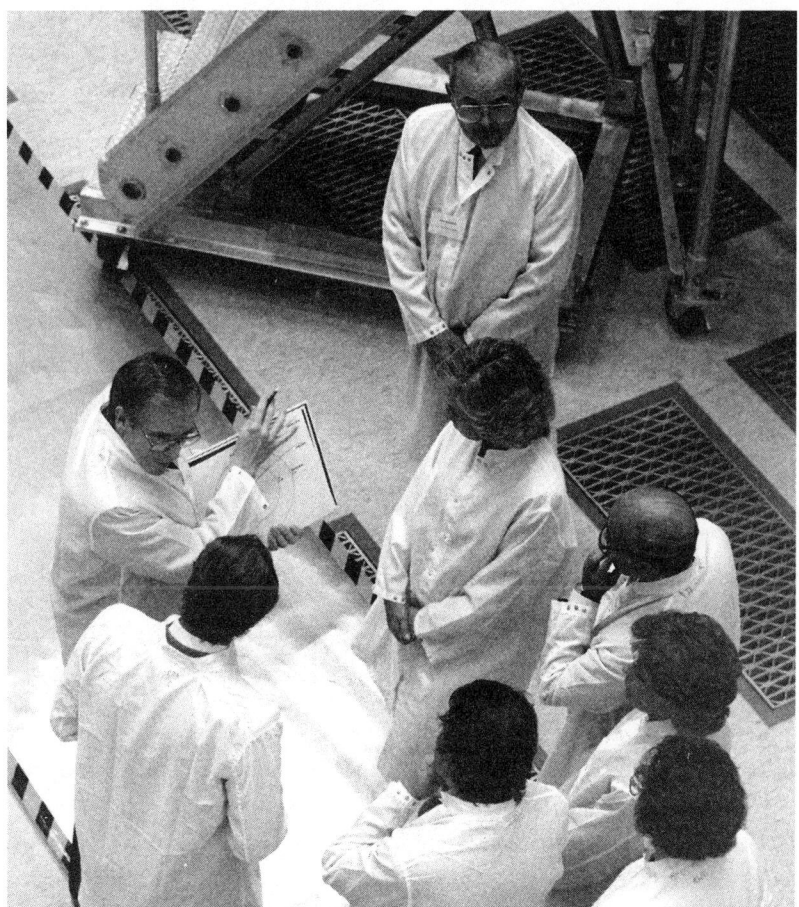

25. "My forte was explaining the magic of the project." Gesturing with a chart in one hand, Bill O'Neil speaks to King Juan Carlos of Spain during a tour of JPL's high bay in April 1988. Across from him is Spain's Queen Sophia. Explained O'Neil, "Royalty would visit because our two overseas tracking sites were on their land." Courtesy NASA/JPL-Caltech.

the loss of a kindred spirit, a collaborator, a friend. "He really was a leader in studying waves in space plasmas."

Gurnett had been in Cambridge, England, for a conference on auroral physics. The news reached him indirectly. "I think it got to this meeting somehow, and then I found out." Scarf would've been first author on their next round of plasma wave findings. "But as it turned out, he died before we got to Neptune." Gurnett now occupied the driver's seat for Voyager's final planetary encounter. Principal investigator. It was a position he wanted but not like this, and the Iowa physicist had to reconcile his emotions.

The final month of 1988 began. As had become entrenched custom, the Casani family prepared for their yearly New Year's party by updating last year's guest list, hiring security, and scheduling the chain-link fencing install. Lynn planned the food: hundreds of hamburgers and hot dogs. Eight gallons of chili. Coolers full of adult beverages plus some youth ones and even more adult beverages. John executed the PR with an in-person visit to the chief of the Pasadena Police Department. For years now he'd been paying his advance respects to grease the wheels just in case. But the 1988 visit played out a little differently.

"John," began the otherwise genial chief, "it used to be that you could serve alcohol on your premises to people underage, *as long as they stay on your premises*, which is what you've been doing. But they've changed the law this year. And now it's *not legal to serve minors alcohol* even on your own premises. So you probably shouldn't have this party."

Spontaneously, John Casani deflated like a punctured bouncy house. The party was in two days.

"I, uh, I can't. I can't change this." He swallowed hard and tried to summon more advanced negotiation techniques. Such vapor lock from a man who had repeatedly testified before Congress!

Warned the chief, "Just think what would happen if somebody got hurt or drunk or killed."

"Well, I—I—we'll have to see what happens."

Much of West Pasadena was, by now, quite familiar with Lynn and John's deserved reputation for celebrating. Although their festivities had begun smallishly, the couple's overabundant social adroitness would never permit that to remain the case. John explained, "This party at the Casani house got to be more and more notorious, and almost every kid in Pasadena wanted to come to it." Owing partly to the Casani boys' massive circle of friends, area teens knew they could eat, sleep, and drink, drink, drink in an environment that even their parents approved of. "I have a security guard," John would say to the adults. "And they have a guest list. And the names of all the children who were invited are on that list. And if their name's not on the list, they don't get to come in. And once they come in, they can't go out." What a sweet deal for

26. Lynn and John together in the 2000s time frame, with the ubiquitous glass of wine.
Said John, "I never went anywhere without Lynn, if I could avoid it."
Once, while in Europe for JPL work, John spontaneously called Lynn to drop
everything and come join him. Courtesy the Casani Family.

all 425(!) names that year, adults included. "So the parents were happy about
that. They knew the kids weren't running around the street."

Two days after meeting with the chief, the party began on schedule at noon.
Burly security guards had been emplaced. Guests arrived and were checked
in through the gate, and those staying overnight found their little corners.
Some crashed in bedrooms; some planned to camp on the front lawn. Others,
almost exclusively teens, actually slept on shallow-pitched areas of the sprawling roof. Typically the hot dogs and hamburgers gave way to roast pork and
chicken as the evening progressed, with a continental breakfast on deck for
daybreak. Once the Tournament of Roses Parade trailed off, around 10:30 in
the morning, "we organized a cleanup detail, always completed before game
time. The hardcore stayed for the game," John continued.

The family billiards table experienced as much use in that twenty-four hours
as it did the whole rest of the year. Bill O'Neil attended multiple Casani parties and summarized them as "food, booze, and wall-to-wall people."

"I say I'm going to quit doing this every year," vowed Lynn, during one party. "Then I get over it. We love it."

Out on the sidewalk, those teenagers pressed firmly against the wrong side of the chain-link fence also dearly wanted to love the party. But none were on the list. Their late-night pining and whining ultimately got the attention of police, and soon John—who by 1988 had given up any idea of sleep during the twenty-four-hour party—answered a knock at the front door. It was one of his security guards. "Mr. Casani, there's six or eight policemen out there that want to get in."

Nobody would've had an exact count of the underage drinkers present inside that house. But it was a lot. And John knew better than to resist the incoming tide of the law. To the guard he instructed, "Well, you better let 'em in."

What happened next is burned into his long-term memory. "They came in with full riot gear, plastic shields in front of 'em, and they had head gear and everything else. To find out what was going on." Rightly the cops assumed underage drinking. Not wanting to lie per se, John told them he'd address the situation immediately—but did so not by cutting off the kids or sending them away. No, he called the police chief. In the middle of the night. At home.

"Hey!" John cried out. "I got six cops, or eight cops, in my front yard, in full riot gear! You gotta do something!"

Briefly, silence.

"I don't know what I can do. But I'll try," came the chief's response. Fifteen minutes later the cops were gone, and the party never missed a beat.

That almost marked the end of it. "The next week," John concluded, "I got a letter from the Pasadena district attorney saying that Mr. and Mrs. Casani, if you have another party like this, you're going to jail."

17

Throwing Shade

Many of my contemporary colleagues who work primarily with data
have only a hazy perception of the immense amount of both scientific
and engineering work that is required to develop reliable instruments
and spacecraft for spaceflight missions.

—James Van Allen

In mechanicals of all types, temperature control is huge. Without it, car engines
seize up and computers shut down. Mechanicals *in space* also risk failure unless
temperatures are regulated with specialized materials. Engineers used to get
by with paint. Alternating stripes of white and black coated the outer shells
of early Explorer satellites, helping to even out temperature imbalances on
their insides. It worked fine on missions expected to last only a matter of days.

But spacecraft quickly evolved to the point of needing detailed thermal
analyses while still on the drawing board. How will systems be protected when
rotated into or away from the sun? How is waste heat going to be handled?
Every onboard component and science instrument factor in. Results lay out
expected flight scenarios in terms of heat loads, cold soaks, and overall temperature
fluctuations. It's an entire career track these days.

The particulars can be staggering. Look at *Galileo*'s retro-propulsion
module—that big knot of tankage and pipework and nozzles in the ship's
midsection. Mostly it saw use for positional tweaks but was also relied upon
for the big slowdown into Jovian orbit. Its propellant tank pressures and temperatures
had to be monitored obsessively. A few degrees or PSI of drift were
cause for action. Corrective nudging could be accomplished by shunting heat
and power around on the spacecraft. And if a shuttle had to abort home with
Galileo still in its payload bay, the temperature of this same propellant couldn't
be allowed to exceed a maximum 138 degrees. Ground crews needed specialized
cooling hardware so these tanks wouldn't over-pressurize and explode.
"I recall the worst-case scenario being an abort on ascent, requiring an emer-

gency landing in Ghana, *and* being unable to open the shuttle bay doors for cooling, *and* the runway being at the highest recorded temperature." So came the word from JPL's Bob Gounley, who spent years focused on this singular element of the mission.

Feel like being in charge of all this?

Materials applied to a spacecraft for thermal control are considered, at least by JPL management, to be their own subsystem—same as power or navigation. Separate job code, separate budget. And responsibility for properly swaddling *Galileo* lay in the hands of an individual who, as a youth, had been painted by Norman Rockwell, became an Eagle Scout, later got kicked out of the army for flat feet, never graduated college, and survived a car crash that killed his entire family. But who started at JPL in 1955 building missile containers for two dollars an hour. This was the guy. The guy translating *Galileo's* thermal analysis into an application of physical materials. While commanding a budget of over $2 million to do so.

Why oh why, with all the opportunities in Pasadena, had he chosen the Jet Propulsion Laboratory?

"I needed money. So I went up to JPL and applied for a job, and they hired me."

Triple-retired Hugh von Delden described his principal career this way: "Design the blankets. Pattern the blankets. Figure out how many—what the blankets were gonna be made of, and what configurations. And the final assembly, is all my job."

Von Delden further explained how the Lab recognizes a hierarchy based on the principle of the cognizant engineer—a capstone individual, topping the chain for each technical aspect or subsystem of a project. "And that person is the *only* person that has all the answers from cradle to grave, as we called it. Never passes off in the middle of something," he informed. The job never gets shoved over to somebody else. "It is the same guy that has to answer all the questions." As his ultimate calling played out, von Delden would find himself cognizant engineer of the "temperature control hardware subsystem," as it's known, for every JPL spacecraft going back to *Mariner 4* in 1964. If a ship ever caught a fever, it wouldn't be on his watch.

Hugh von Delden researched almost as much as he fabricated, often meeting with such vendors as DuPont, the supplier of Kapton film. He'd review JPL needs and occasionally request custom finishes. Von Delden also headed up all materials ordering. Every six months he faced audits to check how his budgets and materials estimates were holding up against what the department actually consumed.

To create *Galileo*'s original round of protective blanketing, Hugh von Delden's team had begun soon after the basic hardware design settled in the late 1970s. First, they broke down the craft's multifaceted shape into some three hundred manageable sections. Next, each blanket pattern was roughed out by wrapping sheets of everyday brown kraft paper over a spacecraft model. Wrap, cut, rewrap, nip-'n'-tuck, then wrap again until things fit. Each real blanket derived from this process and ultimately received a final fitting to the real ship. Mostly they worked out of Building 18, the "Shield Shop," although von Delden's office, by the time of Galileo, had moved to Building 158 on Surveyor Road near the middle of campus. The Shield Shop itself boasted some of the best HVAC systems on Lab because high humidity could affect the more exotic finishes.

Any complete blanket encompassed a perhaps surprising number of layers. Ten to twenty thin Mylar sheets, aluminized on both sides, alternated with synthetic Dacron netting. Exactly how many layers of Mylar covered which area depended on predictions supplied to von Delden's team. "It's the thermal analysts who had to tell me how good the blankets had to be, or what performance they needed." Over top of the Mylar and Dacron sat a layer of aluminized Kapton, with straight Kapton atop that and a final outer layer of Kapton-coated indium tin oxide.

Kapton tore easily enough, but von Delden praised it. The material not only insulated but also eliminated static buildup on any one part of the ship. And it didn't reflect sunlight, which could potentially disable spacecraft instruments. "I don't really know how it's made," laughed von Delden of the amazing film, going on to mention how static discharge problems had led him to request carbon-filled Kapton from DuPont in the mid-1970s. He used it on the *Voyagers*. Since then, however, his preference had swung to indium tin oxide. This exotic coating helped resist Jupiter's punishing radiation—as well as safeguard the underlying blankets against ravenous atomic oxygen lingering in low Earth orbit.

All these materials went together at the hands of trained fabricators using commercial-grade PFAFF-brand sewing machines meant for heavy fabrics (think denim). When placed on the ship, most blankets simply attached to other blankets or sometimes to themselves. Von Delden likened it to how a shirt envelops the body and fastens to itself. Most everything tied together with crisscrossing number 12 Dacron cord in white or black, depending on the location and other requirements. Dacron offered stability at higher temperatures— versus something like nylon—and wasn't as stretchy.

The final product had to armor as well as insulate; destructive cosmic particles move lightning fast. After hitting the blankets they'd vaporize and become harmless, but they needed room to do so. That meant installing little standoffs between the blanketing and the ship's body panels. Standoffs that had to be designed, fabricated, test-fitted, redesigned where necessary, approved, then reattached and measured, and forever hidden. Fully assembled, *Galileo*'s crinkly black tinfoil appearance gave no indication of such complexities hiding beneath.

"So what you see is not exactly what you get," affirmed Hugh von Delden.

Galileo's instruments themselves—the whole point of going—also needed protection. As von Delden continued explaining, "One of the jobs I had to do is go around to the different science people and tell them what their blanket was gonna look like." For some instruments, the blanket wouldn't be able to tie back onto itself. "Okay, you've got to provide these attach points," he'd instruct. And the hardware an experimenter thought had been finished was suddenly not quite finished.

Once fully diapered, the orbiter traveled up the hill to the north end of campus to Building 150 and JPL's giant, eighty-five-by-twenty-seven-foot space-simulation chamber, featuring interior dimensions of twenty by twenty-five feet. Goodbye, atmosphere. Hello, xenon lamps arrayed by the dozen and broiling at the intensity of up to two suns. Only after weathering such punishment would von Delden learn if the variously applied blankets and other treatments were protecting as intended. One of his proudest accomplishments involved blanketing a *Voyager* instrument that had to gimbal around in space. "A thermal rotating joint is hard to make," he insisted. "And the last I talked to Ed Stone, he had had over I don't know how many million cycles on his instrument, and it was still working. Still going back and forth. And—and the blanket was still on it! Performing as designed!"

By early 1986 Hugh von Delden considered his *Galileo* thermal work to be almost complete. It had involved at least one European trip to collaborate with West German engineers on their retro-propulsion module. Even beyond those strict temperature-control requirements, the module also had to be protected from micrometeoroid impacts that could pierce a tank and blow the works. "And, uh, they wanted to know how I was gonna do that." Hmm, great question—standoffs can't be bolted to a propellant tank! Whatever he dreamed up had to be okayed by prickly West German engineers.

The full picture still had some blurry spots. But von Delden concentrated his efforts on a skeleton of fiberglass struts, to be attached at the extreme top and bottom of the central tank enclosure. Like staves on a barrel. The concept traveled to Goleta, California, and a small engineering company with roots going back to the mid-1970s in, of all things, tennis racquet design. This was going to be handled by a tennis racquet company? Well, they'd moved on to deployable spacecraft booms and knew fiberglass, and von Delden regarded the company as worthy of providing *Galileo's* struts.

When anchored in place, each strut bowed out to leave, between the tanks and itself, a gap of four inches. "Which was the distance that our guys at the Lab came up with for meteoroid control. So the blanket was spaced four inches minimum from any tank surface." Attached-yet-uncovered struts wobbled a little from side to side—but held fast when the blankets were tightened down across them.

The arrangement needed confirmation, so it met the firing squad. Into a testing chamber it went, rigged with a specialized gun firing rounds at the blankets. "They could get a projectile up to like 15,000 feet a second, which was about half of what a meteoroid really travelled," von Delden went on. Supplemental layers augmented the blanketing for additional protection. And these tests conclusively validated the choice of fiberglass. Impacts to a metallic strut could eject little pellets of metal and breach the skin of a propellant tank. Fiberglass, however, only splintered and vaporized, posing no threat.

"It was a tricky deal, but it seemed to all work!"

The remainder of *Galileo's* blanketing tested out also, and von Delden relocated to the Cape in early 1986 for final checkout. His shared Cocoa Beach apartment happened to face the launch facilities; they could see Pad 39B pretty clearly. On January 28, he had the morning off and wandered out onto the balcony with a roommate to watch the *Challenger* go.

"So we saw the whole thing."

Despite being the Lazarus kiss that it was, VEEGA necessitated a fresh assessment of what temperatures *Galileo* would now need to withstand. Compared to 1986's flight profile, thermal analysts were staring down a barn burner—a nearly threefold increase in peak solar irradiance buffeting the craft during its scorchy loop through the Venusian neighborhood. Such exposure could easily overheat things, warp them, damage parts, and burn out electronics.

Mitigating this is what ultimately dictated a list of changes to the insulation blankets. But also needed were two completely new shields for protecting *Galileo* during the sunniest parts of its journey. One to safeguard the folded high-gain antenna's delicate ribs, mesh, and other hardware. And then another for the ship's whole main body to hide underneath while nearest the sun. The toughest umbrellas that ever were. *Galileo* had never been designed for aftermarket accessories. It had no spare brackets or hitching posts or luggage rails or anything of the sort.

"That was the biggest change for me," acknowledged von Delden of these additions. "I mean, the blanket modifications weren't bad at all. But coming up with this brand-new structure, and a way to attach it, and what to build it out of, and blah blah, on and on!" He'd joined the conga line of trajectory designers, mission planners, and so many others who had to redo already-finished work. Thankfully, *Galileo*'s revised, 1989 launch meant this didn't have to be a rush job.

The antenna shade ended up being straightforward enough to anchor right atop the unit's central mast. Think sombrero. Just broad enough for the collapsed ribs and mesh to cower behind, like a scared kid protected by mama's skirt. But the main shade would have to mount between the base of the antenna and the primary bus structure. Looking over his thermal requirements, von Delden knew the shade itself must contain two specific materials layered atop one another. Each would have a base of Kapton but with different finishes. Flat black had to coat the upward-facing side; otherwise, any direct sun hitting it "was gonna reflect back up into the antenna and probably melt things." But the side facing down, toward the ship's body, required a coating to migrate heat outward to the perimeter of the shade and away into space.

When finished, this shade closely resembled the size and shape of a backyard trampoline. Measured twelve feet in diameter. Well, what could support it? Von Delden figured on ringing the ship's body with tubes, similar to spokes on a bike wheel. But he wasn't sure about mounting them, and nobody was going to let him poke around for options on the flight spacecraft.

The project never had enough funds to build a proof test model. And Galileo's development test model, traumatized survivor that it was, didn't resemble flight hardware closely enough. But they did still have a thermal test model. "The bus on that was an accurate bus, uh, except for cabling and that," indicated von Delden. Studying it, his team identified a short length of midsection in which holes could be drilled and a ring of eight fixed nuts installed. They'd have to punch through existing blankets in that area, but it'd work.

"And then we went right to the flight structure and drilled the holes, and it was a little tricky." To guard against metal shavings and debris, one person carefully held a vacuum nozzle close to the bit while drilling was underway. Thin-wall aluminum tubes threaded into the nuts; the distal end of each tube intersected a support brace about two-thirds of the way from center, which dropped down to supplemental mounts on the bus proper. The shade itself came together from two half circles of material, each with stitched-in sleeves enabling the fabric to slide over the tubes like curtains on rods. Crisscrossing lengths of Dacron cord lashed the halves to one another, and baddda bing, there it was.

An oversize generic shipping case enclosed *Galileo*. Large, vertical white case doors swung closed on silvery reinforced hinges. From there, the case transitioned onto a pallet of wooden two-by-fours and disappeared underneath a massive brown tarp. The entire package then went onto an orange-colored, lowboy tractor-trailer whose contrasting, white-colored cab proclaimed 3-WAY CORP. on the side in large script. A uniformed guard stood idly by as the assemblage rolled out heading east.

The security and discretion and vanilla appearance had much to do with seventeen tiny heating capsules affixed at various places inside the ship. Each about the size and shape of a C-cell battery. Containing plutonium, the capsules were embedded deeply enough in the working parts to require installation while still at JPL. "Even if the blankets were really good, they were still gonna get too cold, so they needed more heat," explained von Delden of the capsules' presence. The units output one watt apiece and made for an extremely simple method of warming targeted areas. But their radioactive nature frustrated environmental activists, who feared catastrophe if bad things happened during launch. Hence both the nondescript nature of transport and the security.

By this point in his career, Hugh von Delden understood prelaunch activities at the Cape Canaveral Air Force Station. He'd been going down there since

27. Hugh von Delden poses with the overhauled *Galileo* spacecraft in its
updated configuration for the VEEGA trajectory. Topping everything is the new shade
for the (when folded) high-gain antenna. Cupping the antenna from below
is the second new shade for *Galileo*'s body. Courtesy Hugh and Pam von Delden.

1964, and things ran a little, as they say in the space business, "off-nominal."
When performing clean-room tasks on or around the spacecraft, everyone
had to wear electrically grounded, antistatic "bunny suits" with accompanying food service–style hats. Such a getup reasonably protected against dirt or
debris or even stray hair fouling the delicate ship. But every second *not* work-

ing in the clean room meant going Saturday barbecue casual. "Down there, everybody just dressed in beach outfits and shorts and whatever." No need to bring other clothes; change in the airport if you could.

For a quick bite, the Cape's lackluster vending machines offered unappetizing and nutritionally questionable sandwiches dating almost to prehistory. People could bring food, but why? "Very seldom did I make a lunch, mostly because you never knew when you were gonna eat." Prolonged integration tests and last-minute whatevers could upend anybody's schedule at any time. You could always hit one of two on-site cafeterias. But if the day's events permitted, "Cocoa Beach wasn't far away at all," von Delden pointed out. "We could knock off for lunch at 11:30 and be back at 1, you know, which was fine, and go into town to have lunch."

On many a Florida Cape workday, Bill O'Neil's schedule called for visiting *Galileo* inside its special hangar containing the clean room and various amenities. There, he could look up and glimpse other visitors as well—tourists, who filed through a small observation room to witness moments of prelaunch prep. Having never been up there himself, O'Neil checked out the room one day and happened to notice something with queasy implications. "On the wall of this little observation room, there's a box with a big button on it," he started. "If there's a fire, you can press that button." Doing so would activate an immense hazard-protection system and deluge cascading sheets of water throughout the building's interior. It'd cause a flood worthy of the Old Testament, exposing *Galileo* to the most destructive of rinse cycles. O'Neil's eyes bulged at what amounted to an accidental discovery of something so alarmingly bad. The button didn't even have a cover on it. Off-the-street randos came through here all the time.

"*Anybody* standing in this room could set off the deluge system!"

He ran off and found John Casani, who screeched, "My god! That spacecraft's worth a billion dollars now! It'd take a hell of a lot less than that to rebuild the building! So we shouldn't have that system enabled."

Quickly O'Neil arranged to have the fire-protection system turned off. And that was fine and dandy and one less thing to worry about until 7:30 a.m. on the day to fuel the spacecraft. The task procedurally required higher than normal levels of caution. On that morning, O'Neil entered the hangar to find a number of technicians "running around bitching and swearing." He asked what happened.

28. Joined to a solid-fuel air force stage, *Galileo* finally sits inside the cargo bay of space shuttle *Atlantis* during closeout operations. More than seven years had passed since it was first supposed to be launched, and more than three since the *Challenger* accident. Courtesy NASA/JPL-Caltech.

They told him, "Some character turned off the deluge system, and we can't start."

"Well, that character's me." They all looked at him. "And then I got in trouble." The system had to be reactivated, and everyone got back to work.

Insulation blankets waited to go on until most every other prep task was finished. "It was shipped down there naked," remarked Hugh von Delden of the aging machine. "They have to go through a system test down there, and direct-access cabling wouldn't allow blankets to be on." Once receiving the go-ahead, his team of four to six people worked around the clock for three days straight, laboriously lacing the blankets onto *Galileo*'s hardware in a precisely defined order. "When you tied 'em together, you wouldn't pull 'em real tight, and you'd just pull 'em until it was nice and sturdy," he described. And that about wrapped things up. Their creation mounted onto its supposedly safer-than-Centaur solid-fuel stage; then the whole shebang went into the belly of a shuttle.

"Huh. We were ready to go again," said Hugh von Delden.

In Florida to see *Galileo* launch were the Diehls and their kids. They had made the event into a family trip, including special shirts. But two shuttle-related delays, over a six-day span, pushed the "big send" far enough down the calendar that the Diehls had to leave while *Galileo* remained stuck in its shuttle on the pad. All that way only to miss it. On the cloudy Florida morning of October 18, 1989, they were at the Orlando airport, aboard a return flight and waiting to taxi, when the shuttle's twin boosters finally ignited, didn't leak, and climbed up into the sky, bearing the work of thousands.

Mused Roger Diehl, "I saw the launch from inside the airplane, with my nose pressed up against the window."

That afternoon, astronauts ejected their passenger from the shuttle's payload bay. An hour later the IUS underneath *Galileo* lit up and sent the works toward Venus. Two stages of solid fuel burned and Super*Zipped away, and the Jupiter machine finally had some time to itself. Atop JPL's Building 264, a huge sign had been unfurled.

GALILEO—WE'RE ON OUR WAY

18

Not-So-Tasty Rib Tips

You've been the person, or the group of people, that have put
this thing together and tested it. And turned it on every day,
and turned it off every night, went home, came back. And then
all of a sudden you launch it and you have no role.
It's another bunch of other people that are running it.

—John Casani on the transition to flight operations

Days after launch, *Galileo* began receiving ground commands. A complete
health check occurred, starting with the probe. It seemed to be in great shape.
Such interrogations would happen every year of cruise as a matter of housekeeping. One orbiter instrument was powered on and soon detected high-energy
ions from distant solar flares. The protective cover for the particle detector
opened on cue. The retro-propulsion module's small maneuvering thrusters
were test-fired. Late that month, two other instruments came to life for inflight evaluation. All were fine opportunities to evaluate *Galileo*'s onboard reel-
to-reel tape recorder—used not for audio but for data storage and playback.

In early November, a two-day series of maneuvers almost perfectly aligned
the ship for Venus. A second and final tweak happened late that December.
"Galileo's solar shades are all performing well," praised a status bulletin for
the project. "Keeping the spacecraft within the acceptable temperature range
as it approaches the intense heat near Venus." Any desired functionality for
Venus encounter—including the operation of ten science instruments—would
be shaken down by the end of February. Those plans didn't include *Galileo*'s
high-gain antenna, which was to be kept folded under von Delden's custom
shade until long after Venus.

The year 1989 became 1990, and that February, *Galileo* picked up its gravity
assist right on schedule, ratcheting up the speedometer by some five thousand
miles an hour. All instruments worked well; just after closest Venus approach,
the camera misbehaved a little but no biggie. Data spooled onto *Galileo*'s tape

recorder exactly as intended. All captured science data would be transmitted to Earth that coming November. Sending home one single image of Venus, at the current low data rate, took over three and a half hours—and eighty-one images waited in line. Opening the big antenna, with its big-big data rates, couldn't happen soon enough.

The ship made its closest approach to the sun in late February. No temperature issues arose there either. The end of that month also marked the end of tenure for Galileo's current project manager, who moved to a completely different Lab effort. Filling his role? None other than Bill O'Neil, stepping up from mission design. This former Wisconsinite (from the minuscule town of Hartford, population four thousand-ish) readily admitted to a rebellious childhood. He was a fifties greaser punk. "Oh, I guess I don't know where the rebellion came from. But I—I just didn't get along well with the teachers. And I got in a fair bit of trouble in high school." Now he headed the world's most expensive planetary mission.

During March, *Galileo* demonstrated a remarkable ability. As brainstormed, proposed, lobbied for, funded, promised, designed, redesigned, fabricated, tested, shelved, recovered, overhauled, then finally delivered into space, half the craft could rotate while the other half did not. Yes, indeed: that much-ballyhooed, dual-spin mode had proven itself. It actually worked. It even changed modes relatively quickly; within fifteen minutes, the mechanicals could relock into "all spin"—and would do so in accordance with planned mission events.

Dual spin represented some of the blackest, black magic sorcery ever attempted on a spacecraft—a diabolical single point of failure in a machine specifically engineered away from single points of failure. To understand it, behold a cumbersome construct known as the spin-bearing assembly, within which two halves of a spacecraft could freely rotate yet still exchange data signals and electrical power. But how? Cables running between the two would quickly wrap around themselves and lock up the works. Wouldn't they?

Engineers in Arizona, at subcontractor firm Sperry, interlinked the ship halves with a joint akin to the hub of a bicycle wheel—a spindle rotating inside of a sleeve. Atop *Galileo*'s lower, despun section perched a tall vertical spindle with a column of metal rings stacked on it from bottom to top. This fitted up and into a cylindrical hub nestled at the center of the spun half—a hub lined with an equal number of larger-diameter rings. Every power and data line between the spun and despun sections ran to contacts embedded within these rings. When assembled, the rings aligned in concentric fashion and almost touched. All the Sperry company needed was some way of bridging that gap.

In this weird world of ringing rings, the initial approach became more rings. Sperry added a hundred of them, each about the size of a wedding band, rolling in shallow channels between the main spindle and the hub rings. At first it seemed great because the rings were super lightweight and rolled with little friction. They were also super easy to make. Rings!

But major issues quickly gunked up this approach. The rings intermittently broke contact, bleeding noise into data signals and corrupting their quality. No way could a spaceship go all the way out to Jupiter only for every morsel of homebound data to contain more artificial noise than a Dead Kennedys song. Don Gurnett would freak.

The rings also intermittently broke—physically fractured—because one hundred rolling rings all squished together into channels represented a landmark case study in mechanical loading and material fatigue. How easy it must have been for the authors of JPL's original 1976 "Orbiter Description Document" to wave dual spin through the gates with "the despun section of the Orbiter is coupled to the spinning section through the despun bearing assembly." Okay, got it. No problem. Not even mentioned was the way multiple propellant lines would have to feed up through the middle of the spindle and complicate things even more.

Sperry tried bunches of possible fixes, including greasing the hell out of the rings. And grease did help reduce noise levels but wasn't a genuine *fix*. The concept failed. Every supporting ring had to come out. Flurries of shop talk ensued. To fob up a better approach, a one-off "tiger team" convened.

Ultimately, Sperry overcame the problem by cleanly separating power from data signals and creating a specific type of contact for each. Transferring power between halves would now be accomplished with a more traditional brush-and-slip-ring arrangement. It was tried and true, used in electrical generators and the spinning drum of your grandma's VCR. The stacked rings on *Galileo*'s despun shaft received electrically conductive blocks made from graphite and carbon. These "brushes" physically rubbed against a stack of rotating metal rings with embedded electrical contacts located inside the spun half. It was less than ideal because the brushes generated dust and wore down over time.

Not nearly the same lifespan as the original mega-ring deal but the best anybody could come up with.

Then the data side of the equation would be handled by rotary transformers, which were also commonly used on VCRs. Here, a small New Jersey industrial company fashioned two concentric metal doughnuts from long lengths of thin wire and separated by the most minuscule of air gaps. With one doughnut on *Galileo*'s hub and the other on the spindle, signals passed between because of magnetic fields generated by electrical current in their windings. Some forty-eight of these transformers were stacked into a tall cylinder for shouldering the task of moving data between the craft's two halves. Rotary transformers had unlimited life. No wear-induced noise. Or debris. And no greasing required!

Major flight-path tweaks occurred over four days in April 1990, plus two in May. *Galileo* added a paltry seventy-eight miles an hour in speed as it transitioned course from post-Venus to Earth return.

In late May *Galileo*'s distance from the earth stopped increasing. Now on the rebound, it screamed back toward a first Earth encounter to happen months and months down the road. Come mid-August, the core Project Science Group met for two busy days. They hashed through plans for studying Earth itself during the impending flyby and traded early results from the science instruments. This mission was finally happening.

"We have just had our first Galileo science press briefing," announced Bill O'Neil eagerly in a follow-up newsletter. "Scientists presented and described their first look at the data *Galileo* obtained at Venus." Owing to the still-folded main antenna, that data had been interned on the ship's tape recorder until close enough to our planet for reasonable transmission times over a slow link. Everything came down just fine. Atmospheric scientists were no doubt going to wet themselves over the photos showing equatorial cloud patterns.

Continued O'Neil, "The broad spectrum of information about Venus obtained by *Galileo* in this very limited, 'first' encounter is awesome. It is a wonderful confirmation of the ability of the *Galileo* orbiter's instruments to thoroughly investigate the Jovian system."

Earth swing-by approached. The particulars of this maneuver demanded the absolute latest navigation data, so by the time it came through, the thirty-four-person team in charge of building command sequences had only one week

29. In this artwork by Michael Carroll, *Galileo* darts past Earth. The probe waits inside its blunt, cone-shaped heat shield (*left*). The high-gain antenna points rightward and remains stowed. © Michael Carroll, http://stock-space-images.com.

to prep everything for transmission. Any final toots of the thrusters had to be dead-on accurate. "Earth's gravitational assist will amplify all deviations from a perfect trajectory," commented one member of the project—adding that small adjustments, made earlier, could save five hundred times the amount of fuel as compared to implementing corrections later on down the line.

Between mid-October and early December, one instrument retracted its protective sunshade. Ready for action. More science data flowed in from recent parts of the cruise. "We're really learning how to operate these instruments," praised a team member. "It's been invaluable for us to collect this data." Some housekeeping steps were run on the retro-propulsion module to flush its thruster feed lines. More instruments powered on, including Iowa's plasma wave hardware. A different instrument struggled with overheating but was under control. A third instrument needed a software patch after exhibiting flaky behavior. Overall though, pretty smooth sailing relative to the mission's scope and complexity, and the ship darted past Earth on December 8 while picking up a speed boost of 11,600 miles an hour.

Months passed. Finally, the sun-to-spacecraft distance grew large enough and safe enough for their communications' pièce de résistance to open wide. Blooming season had arrived. On April 11, 1991, *Galileo* began executing a sequence of commands to unfurl its umbrella-like high-gain antenna and—*uh oh, what's goin' on? That's not right.* With every single aspect of the spacecraft constantly under microscopic monitoring, ground controllers rapidly noted how the deployment was sapping way more power than expected. As if things were trying to move but not making enough progress. It hadn't opened all the way, they knew in part, because of what physics told them: Full deployment would decrease the spin rate of the spacecraft by a calculated amount, and that hadn't happened.

Well, shoot. The antenna had never been a problem, right up until it was.

JPL formed a big-brain crisis response team that same day. Spent weeks plowing through the situation. And by late June, Galileo's project manager had enough conclusions in hand to address the troops. He called an all-hands meeting.

Dressed in a dark suit, Bill O'Neil stood on stage in a darkened von Kármán Auditorium and leaned into the lectern's microphone. He talked at a rapid clip. He seemed excited. "We have overwhelming evidence that the flight antenna is in *this* configuration."

Next to him and filling most of the stage's back wall, side-by-side projector screens hosted rectangular images of the Galileo high-gain antenna. In the left-hand shot, it proudly posed in full-peacock mode, 100 percent unfurled—just as Lew Allen and others had seen over the course of deployment tests. Those tests had always gone great.

But the right-hand image portrayed a lopsided, half-open antenna, slumped in asymmetrical blight. Almost embarrassed by its own catawampus state. None of the ribs had fully opened, with those traveling the most having made it only 35 degrees out from vertical. Opposite them, a trio of pouty, uncooperative ribs sat upright in complete defiance. Fully stowed. Like the North- and South-going Zax of Dr. Seuss fame, they refused to budge. These stuck ribs were preventing the whole antenna from opening further, and four possibilities had made the initial cut:

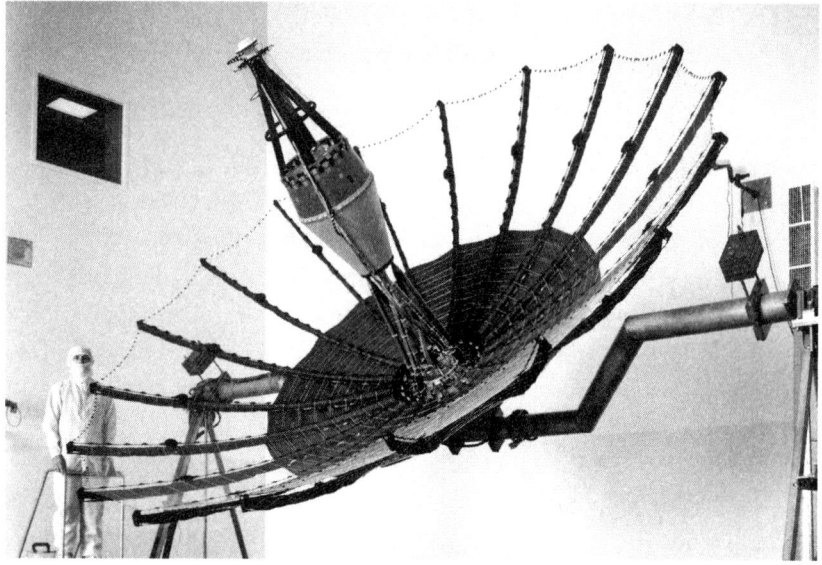

30. Open wide: the intended, fully deployed configuration of *Galileo*'s folding antenna.
Courtesy NASA/JPL-Caltech.

1. The Shield Shop's tip shade had snagged in the antenna mesh.
2. The antenna's mechanical nut-and-thread deployment system hung up or broke.
3. Tips of the three ribs were lodged in their sockets.
4. A secondary rib-retention system had failed to let go.

"Not a chance!" defended Hugh von Delden of the first option. Indeed, more poking about laid it to bed. As a summary report phrased it, "No configuration of tangling the tip shade in the wire mesh could be found that would restrain the ribs at the stowed position." Also eliminated: the mechanical drive system responsible for unfurling the antenna. Four possibilities had been cut in half.

Maybe some kind of trouble with the rib tips? A spring mechanism had secured them until soon after *Galileo* reached space. Although it unlatched, the ribs stayed closed for years, and something could have happened. But they had data; telemetry indicated the spring mechanism properly released. And the measured torque on the drive motors didn't match the values necessary for the spring mechanism to be at fault.

That left the secondary rib restraints. Below their tips, each rib was also held in position at its midpoint by two protruding and identical pins with rounded ends. Pin ends fit into receptacles mounted around the perimeter of the antenna's central tower. And each of the receptacles featured one of two different shapes—either a concave one to match the pin end or a V-shaped groove. This double approach intended to guard against both pins on the same rib exerting identical separation loads. It was a hedge against trouble.

Luckily the project had a spare antenna handy. Both the spare and flight models had had the crap beat out of them. Lots of vibration testing. Which naturally rattled every last bit and bolt and part on the units, including those pins and sockets. Under close inspection, the V-groove sockets, and only those, were discovered to have progressively worn little flat spots on their pin partners. Could this be the culprit? Worn pins?

The investigation expanded to Ohio and NASA's Lewis Research Center, where the anatomy of the mid-rib restraint went literally under a microscope. It was a simple enough system. Until deconstructed! The sockets originated from an exotic nickel-chromium alloy known as Iconel 718—an aerospace standard. Dependable stuff. But the pins had been made from an entirely different alloy—a titanium one, finish coated with dry lubrication. All fine and well from an engineering standpoint . . . until a few specific things happened. The very first time the flight-version antenna ribs were stowed under tension, at manufacturer Harris Corporation, high levels of contact stress began to deform the ends of those pins that'd been mated with the V grooves. This business of contact stress contributed to abrading the pin coatings—a process severely exacerbated by Harris's cross-country, jostling, over-the-road transport of the complete antenna assembly from Florida to California. Inside a shipping container, the antenna lay on its side but was anchored only by the interface bracket meant for its eventual mounting to the spacecraft chassis. That meant the whole damn antenna rode the whole damn way cantilevered over to one side, groaning in silence. Any pins facing straight up or down suffered the most. At JPL the antenna emerged from its shipping container with multiple pins already at risk, and hadn't even been installed on the spacecraft yet.

Next came vibration testing, which further accelerated pin wear. This test regimen, as executed by JPL, coincidentally oriented the antenna exactly as it had been during transport by Harris. More damage occurred to the at-risk pins and coatings. Guess which ones ended up stuck? The likelihood of successful antenna deployment was falling faster than the reputation of the *Pink Panther* movie franchise.

In JPL assembly bays, the antenna was test-opened. These tests failed to unmask the situation because the level of pin-socket friction, *in an atmosphere*, remained low enough for things to still work. Meanwhile, extremely thin coatings continued to wear off the ends of the little tiny pins and roughen their surfaces.

In a vacuum chamber, the antenna was also test-opened. No problems were found, because the pins and sockets happened to be stationary relative to one another. But the now-rough surfaces of the nearly bare parts greatly increased the amount of friction between them—to a level seven times higher than normal. The time bomb really started ticking loudly. Nobody heard. No testing regimen on the books would've found this problem.

So *Galileo* went to the Cape in 1985, with more jouncing and bouncing for days on end. During this, the antenna's second cross-country trip, those pins rattled in their V-grooves like chattering teeth. After *Challenger*, the ship returned to JPL; cue shake-a-thon number 3. In 1989 came, sadly, a *fourth* cross-country voyage, again from Cali to Florida. By the time October 18 materialized, vital coatings were mostly gone from pins mated with V-groove sockets.

Launch forces naturally shook the hell out of *Galileo*, and once off Earth, the vacuum of space degraded these critical parts even further. Translating into more force than ever before envisioned to yank them apart. Antenna ribs had begun deploying, but three stuck fast due to the pins lodging in their V grooves, and this situation proved to be more than the motors could handle.

That's how they'd gotten into this position. People weren't sure how to get out of it. Some tried to focus on the bright spots—such as how *Galileo* had scored our first flyby of an asteroid, back in October 1991. The two were only a thousand miles apart. Returned images showed Gaspra, as the asteroid was called, to be a hostile body of irregular and cratered surfaces awash in dust. Such discoveries helped mask some of the pain.

Efforts to unfold that antenna consumed the next year plus, as *Galileo* predictably drew closer and closer to home and a final gravity assist for Jupiter. The year 1992 began and progressed and entered its final months. Singer Sinead O'Connor used her appearance on *Saturday Night Live* to publicly shred a photo of Pope John Paul II. Windsor Castle caught fire. Then, only 188 miles above Earth—below even the typical shuttle orbit—JPL's hobbled ship arced past for the second and final time on December 8, ticking up its speedometer by some 8,250 miles an hour. Not far below, the American television network NBC announced that its popular show *Cheers* would end the following year. Many called it one of the greatest TV shows of all time. But *Galileo* did not care. Wounded yet alive, and certainly beloved, at long last was it heading directly for Jupiter.

19

Ground Truth

Here was a satellite with an atmosphere more massive than Earth's.
A primitive world frozen in time, where chemical reactions
taking place today may resemble some of those that led to
the origin of life on Earth.

—Tobias Owen, astronomer and coauthor of
the original Cassini proposal, on the lure of Titan

Fast forward just a bit to Christmas Day 2004. Different spacecraft. Different planet. Different satellite tour.

During *Cassini*'s third orbit 'round Saturn, points of attachment on its side blew their fittings with literal Swiss precision. Pyrotechnics fired and fasteners released and compression springs unsprung and little guide tracks went *spizzz* through rollers and instantly a plump, trampoline-size, gold-colored disc peeled away in a gentle spin. Handed off to the laws of motion. Deep within the disc a trio of timers began redundantly counting down as the whole thing just lazily coasted along through open space, rotating seven and a half times a minute with absolutely no way to adjust course. Destination: perhaps *the* most fascinating world in the solar system beyond ours. Titan.

Separation data fluttered back on radio waves and to a parental welcome-home embrace by the giant antenna dishes of the Deep Space Network. From there, signals routed to the European Space Operations Centre in Darmstadt, Germany, and then via a live telephone link south to Zürich and the modest offices of engineering firm Contraves. It was closed for the holiday, not to mention being four o'clock in the morning. Inside, displays of data next encountered the eyes and brain of fifty-three-year-old mechanical engineer Platon Tatalias. A man who'd focused the recent era of his career on just exactly how *Cassini* and its passenger would faultlessly split apart.

In the office completely by himself, a few colleagues followed along remotely as Tatalias monitored the news. "I had instructions from our CEO to call him and the public relations office any time." Over the open phone line, a NASA voice heralded major steps as they occurred. "After some seconds the voice declared that *Huygens* released," indicated Platon Tatalias, naming the thing he and his teammates had worked so hard to perfect. And what did this man feel, knowing they'd executed the divorce so precisely on cue?

"First, relief. But what about the trajectory?"

Mother ship *Cassini* imaged the receding probe. The giant craft then performed cleanup maneuvers prior to being retargeted for a flyby of icy moon Iapetus. The tail end of this lengthy maneuver would swing the orbiter back past Titan to serve as a radio relay for *Huygens* as it entered the thick atmosphere and began descending toward who knew what.

More information came through on what the separation velocities were estimated to have been—based on *Cassini*'s rebounding motion at the instant it happened. Results were then double-checked. Inside Contraves, each second of time dragged. Finally came the answer to years of effort. "We had to wait about half an hour to hear first news that the *Huygens* should be on the correct path." Great to hear! Especially considering this split-second, you've-got-one-shot kind of moment. The straightforward delivery of Tatalias's comment is stereotypically indicative of someone hardened by the discipline of delayed gratification.

Tatalias dialed a nearby phone. "I informed the CEO and the public relations officer. Some hours later we got a visual image from NASA, taken from *Cassini*, which showed *Huygens* like a small ball far away." With the trajectory already vetted, this otherwise unremarkable picture functioned as mere punctuation—but as a lifter-upper all the same. It confirmed things. With ripsnortin' mathematical certainty was the fascinating moon of Titan going to be visited by a machine from Earth. Only three weeks away.

"That made everybody happy and confident for the rest of the mission."

Calling on Titan might have *eventually* originated from American efforts. JPL, for starters, had a Saturn orbiter + Saturn probe + Titan probe study underway going at least as far back as 1979. But the eventual reality of *Huygens* traces directly from the motivated machinations of exploration fanatic Wing-Huen Ip, and if anything's for certain, it's that Ip was a guy who got around. Born in Nanjing, China, he grew up hundreds of miles away on the edge of the South China Sea, in Macau. Ip's undergraduate days occurred at

31. In this artwork, the *Cassini* orbiter has just released *Huygens* for entry, descent, and (hopefully) landing on the moon Titan. This image is illustrative only; the positions of the various subjects are not necessarily correct. © European Space Agency.

the Chinese University of Hong Kong, where he earned a bachelor's in physics. Beginning in 1969 came graduate studies all the way across the Pacific Ocean in San Diego, California. There, he crossed paths with some resident experimenters on *Pioneer 10* and *11*—both destined for the outer solar system. Ip loved these exciting times. "That was my very first brush with space experimental data, not to say a first with Saturn!"

At nearly the same time, one of Ip's professors accepted an open position in Lindau, West Germany, at the Max Planck Institute for Aeronomy. A few years later Ip rejoined him, switching continents yet again. Both *Voyagers* launched and struck out toward Jupiter; Ip couldn't wait to learn what they found. And the comet! During this same period, in the early 1980s, Europe's unified space science community had begun gearing up to study the legendary Halley's comet during its impending solar system fly-through, some eight years down the line. Ip lived in an unglamorous basement apartment underneath the institute's cafeteria and first heard of rings around Jupiter from a BBC report covering Voyager discoveries.

Two years later he moved back to America for Johns Hopkins University and an opportunity to process Voyager data from Saturn. Ip focused mostly on the ringed planet's icy moons and not so much on Saturn itself. Regardless, "that was a great learning experience for me." He started wondering about what might come next after Voyager shut down. The Americans already had their giant Jupiter orbiter in the works, although its launch day always seemed to get pushed back for weird reasons. "I came up with the idea that after the Voyager mission, a Saturn orbiter should be the next logical step."

As 1982 marched along, busy Ip happened to see a brochure at a scientific conference. The booklet came from Ames Research Center and read *Saturn/ Titan Probe*, featuring cover art of a *Galileo*-ish mother ship near Saturn, having just released a *Galileo*-ish probe covered in fancy gold foil. Ip recalled the Ames pitch: arc past Saturn, drop a probe into Titan's heady atmosphere, then map the moon's surface with radar on the way out of town. Okay, so America had also been thinking along similar lines? Was this good news or bad?

Beyond Titan, though, the whole scheme didn't seem much of an advancement. How come Ames only suggested flying by Saturn? Both *Voyagers* were already doing exactly that, based on super-old plans from 1965. Where's the progress? The new milestones? Why wouldn't Ames want to orbit out there and conduct a really deep study of the whole neighborhood?

No matter what he thought his own next steps might be, Wing Ip felt spurred to action when the European Space Agency (ESA) issued a request for new mission concepts that very July. "Being young and naïve, I thought it interesting to propose a Saturn *orbiter*, with a probe to Saturn or Titan." Still a student, he'd never attempted something so grandiose. Ip started asking around for advice because how in the world could he ever assemble such a thing as a brand-new space mission proposal?

That September he partnered with a French atmospheric scientist working for a hundred-year-old Paris observatory. Our sparse knowledge of Titan, based largely on *Voyager 1* data, made people like him crave returning. "The atmosphere of the satellite appears to be a kind of laboratory," suggested the Frenchman, "in which a complex organic chemistry has occurred for several billion years. And on a planetary scale!" His fantasies for a Titan probe meshed almost perfectly with what Ip had in mind. Their biggest problem seemed not at all what to do—rather, how to manage the language barriers.

Collectively the pair slapped together a proposal, beating the bushes for two dozen European scientists who joined the bullpen as advisers and potential experimenters. These people specialized in everything, from planetology to space physics to exobiology, and understood that the whole concept might well go south. But not trying was an automatic no. Success would heavily depend on tight cooperation between ESA and NASA because there was no way Europe could do it all alone. They needed America's money, worldwide tracking network, and expertise—specifically with probe tech because Europe had never done that before. But the two space agencies hadn't exactly been getting along: NASA's recent walkout on an international mission to the sun, without saying so in advance, had resulted in some unhealed wounds.

So when the bound formal pitch went to Paris and ESA Headquarters in November 1982, Wing Ip understandably held out little hope. "Remember that the original Cassini proposal was written by a small group of scientists, with very little experience in atmospheric probes!" He stressed that their end product resulted from the group's collective effort. "Not just me!" Holding a copy in his hands, Ip reflected on its unconventional Nerf-orange cover. A circular cutout on its top half cleverly framed a title page underneath, which read *PROJECT CASSINI: A Proposal to the European Space Agency for a Saturn Orbiter/Titan Probe Mission in Response to the Call for Mission Proposals Issued on 6th July 1982.*

The thirty-three-page, English-language document impresses with its scope and detail—especially in light of the rushed assembly by a multinational group for whom English was surely not the primary language. "The combination of a Saturn Orbiter and a Titan Probe as outlined here is believed to be the best choice for a post-Voyager mission to Saturn," read part of its introduction. "The basic element in our proposal, in any event, is to establish necessary and sufficient conditions for an in-depth investigation of the Saturnian system in the next decade."

If you're going to peddle a big-money space mission, you have to justify it. The proposal's inner sections highlighted the most intriguing unknowns of Saturn's atmosphere, interior, and ring system. Plasma waves got a paragraph. The lesser (and known) moons all got their mentions. Sure, the *Voyagers* had discovered a great deal. But they essentially raised more questions than were answered!

Titan claimed a whole section of its own, with the prose driving home the relevance of this moon in particular. "Studies of the synthesis of complex organic molecules in Titan's atmosphere, which may have a composition similar to that of the Earth primitive atmosphere, is another issue of fundamental importance to the understanding of processes which have led, on our planet, to the formation of life." A quartet of major Titan goals were specified, none of which involved outright landing. Basic data on "the physical nature of its surface (liquid or solid)" is about as committed as they dared.

Europe could furnish the bare-bones orbiter—the text calling for *Galileo*'s dual-spinning configuration. Maybe. And then America could hopefully provide most everything else: the orbiter's plutonium-based power source, the entirety of the descent probe, and comprehensive ground-based tracking via the Deep Space Network of antenna dishes. "Since the Titan probe could be derived from the Galileo probe in Jupiter, its realization by US is probably financially advantageous," argued the proposers. Their included line drawing of the Titan probe looked very much to be a cut-and-paste of *Galileo*'s with all the names scrubbed off.

Of course, merely submitting a great-sounding proposal with a fancy cover doesn't guarantee approval. Europe's space agency relies on the same sorts of academia-based advisory boards that America's does. And after serious review, ESA's Solar System Working Group saw great potential for this Cassini endeavor. It asked the ESA science director to approach NASA HQ about collaborating in the manner suggested. A dual venture looked full of promise, despite that nasty ghosting back on the solar mission, and Ip's brainchild slowly worked its way up the chain. A beautifully tall flower in brisk wind struggling to bloom.

In America the proposal also reached the battered desktops of planetary scientists and commingled with that earlier mission concept from Ames. Europe confidently had their *Giotto* ship coming together for an eventual launch to Halley's comet. What about making a second *Giotto* that could orbit Saturn and check out its moons? Suggestion endorsed. And then what did everyone think of having *two* probes: one for Titan, surely, but another into Saturn's atmosphere? The Galileo probe design could be reflown. And while they were there, why not keep the radar mapping of Titan? That's what ESA heard at an international planetary science research meeting chaired by Fred Scarf, who led the proceedings "with great charm and efficiency," according to one glowing report. Another participant categorized these get-togethers, which intermittently occurred over the next few years, as "sometimes argumentative," though the planetary scientists all wanted the same basic ride to Saturn's neighborhood.

This patchwork mission approach is what walked into the door of a workshop in Snowmass, Colorado, during the summer of 1983. Held by the Outer Planets Group—a fourteen-member subset of NASA's Solar System Exploration Committee. Full-time planetary science professionals from such institutions as Ames and Caltech and Stanford and the University of Arizona spent days presenting and debating and refining future mission concepts. They were like kids who wanted a say in the next family vacation. Guest presenters— some European—also came with their own suggestions, and worth noting is how these sorts of meetings always happened in scenic locales.

What seemingly made the most sense would be to package the orbiter and dual-probe concept into one single mission. But NASA costing guidelines led to their whole mom-plus-two-and-radar-too being initially proposed to HQ as fully separate endeavors.

Also discussed in Colorado was a hardware concept JPL had in the works called Mariner Mark II. Supposedly the spacecraft of tomorrow, for use deep in our solar system. Heir to the throne of the Voyager platform. JPL promoted it as a generic machine—*customizable for most any mission!*—and this ambitious "Mark II" concept ultimately reversed the roles. If it all went ahead, America would now supply the Saturn orbiter itself, with ESA on the hook for the probe.

This flip-flop made for a bit of a sea change. But overall, development did not resemble Galileo's up-and-down theatrics. As the years progressed, both sides of the Atlantic ran their respective follow-up studies. Formal collaboration began in the late spring of 1984 and continued for over twelve months. Wing Ip kept tabs on the developments and stayed busy with other endeavors—one of which took him to Los Angeles at Fred Scarf's request. "Fred was an American member of the joint ESA-NASA Cassini mission study team, and we met frequently during that period," explained Ip. Although this latest visit had nothing to do with the dreamy Saturn orbiter. Rather, "he invited me to give a seminar at UCLA, because he asked me to join his plasma wave experiment" on a proposed mission to a comet.

Afterward they met up at Fred's house before heading out to dinner. It marked a rare period for the well-traveled Scarf to actually be home from his continual efforts at international cooperation in space. As Ip recalled, "He used to joke that he had no more concept of jet lag, since he was in the air all the time between Los Angeles, Moscow, and elsewhere." Like everyone on the hopeful team, they retained their excitement as Cassini momentum built during the closing years of the 1980s. Wing Ip swore that people on both sides of the Atlantic promised him expensive champagne if a new Saturn mission happened in their lifetimes.

"These spacecraft are our cathedrals," observed one of them. If only this particular cathedral to Saturn could cross that sacred threshold of approval.

The penultimate trial called its witnesses during October 1988 in Bruges, France, at a rented-out movie theater nominally showing *Who Framed Roger Rabbit*. Over three hundred people showed. Cassini faced four other equally valuable mission proposals received by ESA. It went in as a strong contender, though plenty were lobbying hard for the "Vesta" mission to comets and asteroids. "This has been a long, arduous selection process," moaned one European space official. "But if the result is that ESA comes away with a good new-start mission, then it has been worthwhile."

Late the following November, ESA green-lit Cassini, and the Titan probe formally received its own name. They called it *Huygens*, after the Dutch astronomer who originally discovered Titan. The U.S. Congress issued its own approval a year later, including funds for Mariner Mark II—but not until Cassini-Huygens took on a stowaway in the form of a comet rendezvous submission added to sweeten the deal. They'd aim to launch in April 1996 on a traditional booster provided by America, despite the shuttle's late-September return to flight after a vacuous thirty-two-month gap.

20

The Least Unacceptable Solution

I always try to anticipate what could go wrong. Where could
something go wrong. I always used to say, "I want a horse in every
chute." So that if one horse gets out and it falters or something,
I've got another horse that I can release.

—John Casani on his project management philosophies

"You don't understand the way it was—the way it *still is* for space instruments.
There is a great deal of prestige tied up in being in charge of one of these big
investigations. It's not only money for your institution."

The tone of Don Gurnett's comment stung just a little: *The answer's so
obvious . . . why bother asking?* He'd taken issue with a question about the significance of being principal investigator on a mission such as Voyager or Galileo or Cassini. It's so much work. And classes still need teaching. Families
still need loving. Vacations still need taking. So why knock yourself out to
get aboard? Just leave it to some other colleagues at the next university over.
They're studying the exact same thing anyway, right? Why does it matter who
actually measures the plasma?

It matters because the person making measurements is the person *making
discoveries*. Opening a new window on the natural world. Being the person
who first experienced something nobody else had. "I find that rather historic," summarized Don Gurnett. Discoveries are going to happen on every
mission. Nobody knows exactly where or when, but they will. And the people riding on *Cassini* would be forging new chapters of history—with their
names at the top.

A call for proposals went public in a timed, simultaneous effort between
NASA and the European Space Agency. Riders wanted! The calendar said
October 10, 1989. *Voyager 2* had skimmed Neptune only a few months before-
hand, then headed for the end of the solar system. Shoehorned inside, the
University of Iowa's plasma wave instrument would be reporting back on

exactly where that end really lay. Making history at thirty-six thousand miles an hour.

For Cassini, Don Gurnett proposed an expanded version of what he had on *Galileo*, and he eventually received a welcome aboard as principal investigator alongside some eleven other orbiter instruments. New NASA contracts said so. Next would come the laborious building of hardware. "After you're selected, there's usually an accommodation phase where you kind of work with the engineering people," Gurnett explained of the steps to come. Preliminary designs existed; they had a decent handle on the anticipated size and weight of each hardware element. But nobody yet knew, strictly speaking, where all the bolt holes would go. Although announcements of opportunity contain supreme amounts of detail, "they usually don't specify exactly how instruments are mounted."

For all the high contractual language and heady multisyllabic terms, Iowa's deliverable came down to a box in the ship's belly, cabled to external antennas, and the latter are perhaps surprisingly more challenging to integrate. What, a simple structure such as an antenna? "If it sits in the field of the view of an instrument," Gurnett continued, "it's a problem. If the antenna vibrates out there, that can affect the attitude control system. And so we have a lot of engineering problems with things that we do, mostly having to do with the antennas. And to say nothing of the electrical noise from a spacecraft. We want to detect an *actual signal*. We don't want to detect tape recorder motors running and stuff like that."

Location issues affect many onboard devices. Planetary spacecraft often carry magnetometers for detecting whether celestial bodies possess magnetic fields. These instruments are so sensitive that spacecraft electronics themselves will contaminate measurements. Ever wondered why some space probes have those long-ass boom arms extending away at ridiculous distances? Blame the magnetometer people. Their stuff must operate at antisocial distances.

At least Gurnett wouldn't have to travel very far to check the progress of hardware fabrication. "Every single instrument I've ever flown has been built at the University of Iowa, here," he affirmed, while sitting in his Iowa City living room. Van Allen Hall, on the corner of Jefferson and Dubuque Streets, represented the tactile and emotional home of the university's Physics and Astronomy Department. It opened in 1964, shifting the department uphill from MacLean Hall and closer to downtown—only five walking minutes, by the way, from the inviting, greasy-spoon atmosphere of the Hamburg Inn No. 2. (Recommended: Presidential Breakfast with kielbasa and an extra egg.)

The hands-on guy that he was, James Van Allen recognized the obvious advantages of building instruments in-house and gradually cultivated a world-class machine shop to serve the fabrication needs of his expanding department. It's still the case today. Gushed Gurnett, "I can't say we don't subcontract some things. But by and large, we're proud of building the instruments here. And that's—that is kind of a heritage of Van Allen."

With that, Gurnett had a plane to catch.

In early December 1990, he traveled to Pasadena and a Cassini science team kickoff meeting at JPL. Within the hallowed chapel that is von Kármán Auditorium, an otherwise corporate and aesthetically stale main-floor nave had been reconfigured with a line of cheap folding tables up front. Rows of chairs filled the back. The stage-left wall displayed *Voyager 1*'s *Family Portrait*—a composite image of our solar system assembled from multiple shots taken the previous February. That spacecraft was now so far away that in the portrait, Earth took up less than a single pixel.

Gurnett parked at one of the tables. He had on a coat and tie with a striped shirt underneath. He also had a teammate sitting nearby. "And somewhere along the line I added Bill Kurth, who was my student in Iowa," came the identification of this teammate. "Yeah. And I don't remember exactly when that was." Coffee cups already littered most every horizontal surface.

Now a coinvestigator, Kurth had remained in Iowa physics after completing his PhD, and he clarified that he himself doesn't shoulder the added stress of managing students. "Um, I'm a research scientist. I am not faculty. I do not teach," he indicated. "I've been on soft money since the day I graduated!" While not specifically citing prestige, or history, Kurth's participatory motivations certainly bore those flavors and had as much to do with the simple wonderment of it all. "Being in the position to see things for the first time, whether they're an earthshaking discovery, or just more of the same? There's something gratifying about that, and, um, I really get a kick out of that."

The general welcome to the day's events came by way of Dennis Matson, Cassini's project scientist at JPL. He didn't have an experiment aboard. Rather, Matson would function more as a coordinator and referee for the various science investigations and the people behind them. An impending and rather drastic change to the orbiter, which nobody yet knew about, would later add a frustrating new dimension to Matson's world. Other speakers took turns at the lectern. JPL director Lew Allen offered some high-level remarks on his excitement for the mission. Gurnett listened in silence. Across from him sat Hasso Niemann, on board to study Titan's atmosphere . . . assuming the *Huygens* lander made it in.

Everyone had been working their tails off to reach this point. Not even a month had passed since Cassini's official funding had gone live—courtesy of President George H. W. Bush. Public Law 101-611, dated November 16, 1990, also went by the title "National Aeronautics and Space Administration Authorization Act, Fiscal Year 1991." It provided billions in line-item funding to the American space agency and contained language noticeably evolved from the all-shuttle rhetoric of documents past. The law's introduction, in a sort of preamble titled "Findings," highlighted Congress's determination that "the United States space transportation system will depend upon a robust fleet of space shuttle orbiters and expendable and reusable launch vehicles and services." Quite the concession.

Within this public law of a document, paragraph S focused on the "Comet Rendezvous Asteroid Flyby/Cassini mission" and set aside an amount "not to exceed" $1.6 billion "for development, launch, and 30 days of operations thereof, to remain available until expended." Money would be unlocked in stages—with some available immediately, and more to come after key design milestones had been reached. Paragraph S further specified that a cost-containment plan be submitted twice a year until the money ran dry.

This single-sounding mission actually was two. "Cassini" per se remained the Saturn orbiter plus sidekick *Huygens*. Nothing had changed in that regard. Whereas the comet mission, abbreviated CRAF, aspired to actually rendezvous with a comet, then hang close—within several miles—for detailed and long-term rubbernecking observations. Project leaders envisioned what they called "adaptive exploration," with early findings influencing the direction and scope of downstream investigations. Gurnett's tablemate Hasso Niemann had an experiment on CRAF too.

Cohabitating these two missions aimed to reduce costs and streamline operations. Take out the fluff. They'd share a common management structure and data-handling system. They'd also share an identical spacecraft design. Earmarked for both was the fancy-schmancy Mariner Mark II chassis, which was at last a bona fide reality. Able to carry any science instrument, be it for comets or Calypso, and absent any Galileo gotchas such as folding antennas. Or dual-spin halves, for that matter.

Waltzing next onto von Kármán's hallowed chancel was the esteemed Cassini project manager, dressed in an uninspiring charcoal sport coat with off-white shirt and pale maroon tie. He parked behind the over-varnished wooden lectern, which mutely offered comforting support for the procedurally dry task ahead. To the right of the man's Brylcreemed hair stood an American flag with gold fringe around the edge, festive on a pole stand. From the lectern's top corner a gooseneck microphone dutifully jutted upward. And into the mic he offered, "Seriously, I'd like to welcome everybody. Nice to be here. A lot of old friends, a lot of new faces, which I hope we'll be friends before we're through with this process." He made a joke about how those who didn't know him would find out "a lot" over the next couple years. It earned a couple polite laughs plus diplomatic silence from those who didn't know what to think.

To the man's left, one of those despised overhead projectors shined a viewgraph on the screen. In a practical, *oh-that'll-work* font, it proclaimed PROJECT OVERVIEW J.R. CASANI DECEMBER 11, 1990. That's right; Reverend Casani himself had descended from the echelons of assistant lab directorship and the Flight Projects Office (FPO), returning to the pulpit and plebian workaday life of space-mission project management. The overall workload was taking its toll. His deteriorating physique, resembling that of a hamster, shifted inside worn dress shoes.

Quickly Casani replaced the transparency with another showing a bar graph. This one read:

CRAF/CASSINI FUNDING PROFILE COMPARISON
1988 OMB SUBMISSION VS EXPECTED POP 91-1
(AMOUNTS IN $M)

The abbreviation "POP" stood for Program Operating Plan, JPL's internal process of mapping out intended mission expenditures. POP numbers didn't always, ah, *harmonize* with the amount of money already submitted to the Office of Management and Budget.

As Casani continued talking, "This will require doing business in a fairly drastically different way than we've done before." His ensuing presentation related to funding challenges. Substantial ones. And not to mention a bit of a shock, considering they were only as far as the kickoff meeting with budget shortfalls already staring them down. Casani charged the group with keeping expectations in check. Accepting limitations. Making the most of resources they *would* have. The global outcome of this process would be what Casani described as "the least unacceptable solution."

Discord immediately avalanched upon the group because *Cassini* signified the world's only Saturn orbiter for the conceivable future. Investigators wanted to secure every last second of observation time or bit of data they could beg, pinch, wheedle, or cajole. Alms for the poor travelers?

"There's going to have to be compromises," reiterated Casani, leaning forcefully into the mic. The entire mission had but one shallow pot to sip money from. People needed to work together, take turns, and sip lightly.

He relinquished the lectern to a man collaborating on *Cassini*'s radar system, Charles Elachi. Who tried a joke. "Ah, good morning. I understand from Dennis that the first speaker gets everything he asks for!" Elachi stared out over the crowd during his post-joke pause. Absolutely no one laughed. The radar system Elachi was pouring his soul into would be used by *Cassini* to map Titan's surface—a dreamed-of goal that'd been front and center since the mission's earliest cellular divisions nearly ten years prior.

Tobias Owen, collaborator with Wing Ip on the project's original proposal, couldn't stop himself from conjecturing on that surface . . . right on out to the end of conjecture. "What lay beneath that smog? Speculation ranged from a global ocean of ethane to a rugged landscape carved out by precipitating hydrocarbons to a nondescript, gooey surface resembling a melted and refrozen chocolate ice cream dessert."

Most people can appreciate the coolness of mapping another world for the very first time. Especially a world that might harbor life. The idea is almost out of a video game. Not like any terrain found on Earth but real terrain nonetheless. People can relate. But what about something as obtuse as plasma waves? Things that can't even be seen? Can the everyday Joe appreciate such a thing?

"Why is this important to mankind?" echoed Don Gurnett, aloud, of perhaps anybody's first question regarding this particular passion of his.

Gurnett cited energy—humankind's primal lust for energy. One natural phenomenon the world already exploits is the spontaneous decay of radioactive elements. Consider uranium. As its large atoms split, they release heat that can be applied to water and generate steam. At high-enough pressures, steam can drive turbines, which rotate and in turn drive electrical generators.

Build something similar (carefully), and voilà, you've got yourself a nuclear fission power plant. In the absence of phenomena such as tsunamis or Soviet incompetence, nuclear power plants are relatively safe.

An even more efficient method—nuclear fusion—is demonstrated by the sun. Super-hot plasmas combine and, in doing so, release massive amounts of energy. Controlled fusion is a potentially limitless source of energy . . . although we aren't yet able to reliably control it. As Don Gurnett continued, "To achieve controlled fusion, you have to get the plasma up to a temperature of about 30 million degrees. Really hot. And you can't put that in any kind of ordinary bottle." So the approach thus far has been trying to magnetically confine hot plasma. In a chamber fitted with *really* strong magnets. And what closely mimics this is the existing phenomenon of plasma waves.

Another big gotcha with controlled fusion is that waves spontaneously generate within the plasma. It can happen in a split second and destabilize everything. "This is just almost identical to the development of chorus in the radiation belt," mentioned Gurnett. His rationale had now fully unfolded: If you're struggling to control hot plasma in the lab, and working natural examples of this occur all over the solar system, then *examine them.*

"So, actually, I think many of the great advances in plasma physics have been made by studying *space* plasma."

The instruments created by Gurnett, Scarf, Kurth, and the whole plasma wave team represented a distinct evolution of the species. On Voyager, the challenge had simply been getting aboard, and initially they hadn't. So Gurnett had tendered the most alluring pitch he could: "I made a proposal of a real light instrument," based on a proven design, and promised "that we could build it in time. And *that* might be a winning combination." It turned out to be just the ticket. "But by—by doing so, it made our instrument out-of-date from a standpoint of what *could have* been done."

But going for the bleeding edge carries inherent risk. Gurnett maintained that when it came to technology, he leaned conservative. For their Cassini instrument, he specified an older and less robust computer chip—yet previously flown and reliable. "There is one thing that I have learned over all of these years: The instrument has to *work*." Imagine if he and Scarf had begged their way aboard Voyager only for the thing to conk out. "If you screwed up and failed to deliver an instrument on time, or it didn't work, your career was toast."

That got him talking about *Galileo*'s high-gain antenna—still embarrassingly stuck half open and therefore useless. Already the ship had tooted some 110 pounds of irreplaceable propellant while attempting to dislodge those jammed-up ribs. The relay radio antenna couldn't send any data at all to Earth because it *had no transmitter*. So more and more, the dyspeptic workaround appeared to be strict reliance on their wimpy low-gain antenna atop the mast. Although not subject to the detailed pointing requirements of the high-gain, its rate of data transmission measured some 3,350 times lower. "A light bulb alongside the broken searchlight that was the high-gain," mourned JPL's Bob Gounley. What a slap to the spleen. Despite flying by a second asteroid in mid-1993 (and discovering that asteroids could themselves have *moons!*), the mission's scientific objectives faced a rocky uphill path. Gurnett struggled for words. "That was terrible," he moaned of the antenna problem. "I . . . I just, it . . . it was depressing. I mean, I can remember thinking that, *God, you gotta be on a spacecraft mission that WORKS.*"

In no way had John Casani planned on stepping into Cassini project management. He'd liked his spot. Its title had a distinctive ring to it and looked great on a business card: "Assistant Laboratory Director for Flight Projects." Superb! Only a few of these assistant director positions existed at JPL, making for something of an exclusive club. "All the flight projects that were going on at the time were under my supervision," he noted. And from this perspective, one specific project had been experiencing numerous bumps in development—the comet rendezvous. Supposedly, the community of comet and asteroid researchers hadn't been able to agree on exactly which ones to rendezvous with and/or fly by. Among other sticklers.

"We had put forward several different proposals over the years, and they all sorta didn't work," intoned Casani, "because of this *instability* in the community." To eliminate bureaucratic duplicity and increase the likelihood of actually reaching a comet and/or asteroid, the effort had been forcibly mated with Cassini. But this did not make for any kind of miracle cure. The supposed meshing turned into more of a cross-threading of intentions. Casani had mostly good words for the JPL man originally running things. "He was really a smart guy, but the project wasn't going so well," came his opinion. "There were problems with it." And Lew Allen summarily requested that Casani visit the director's office, because sometimes a topic needs to be discussed in person.

"John," began Allen of the delicate issue, "you know, things haven't been going too well with this."

"Yes sir, I know that."

"Well," Allen continued, "I'm going to appoint you project manager."

Casani puckered. "Lew! Is this a demotion? Are you demoting me?"

"No. I'm promoting the project." A nonanswer, for sure, and quite the spin-doctored phrase. Worthy of the Oval Office. "And when you get it sorted out, then you can go back to what you're doing. But right now I want you to get this project on a straight and narrow path." A jilted John Casani fled his boss's office, newly burdened with what certainly represented a demotion. Allen didn't name Casani's successor outright, and never did, so a deputy took over as assistant lab director in the Flight Projects Office, and everyone got back to it.

In July 1990—months before the kickoff meeting in von Kármán—a request went out from NASA bigwig Lennard Fisk. At that time the associate administrator for the Office of Space Science, he'd sat down and written a letter to the Space Studies Board (SSB)—formerly the Space *Science* Board. Still a major wing of the National Academy of Sciences, the SSB was in a unique position of bridging government space programs with the scientific community undergirding them. (Or at least, they *should* be.)

Fisk had been weathering a few stormy months. That April, the much-ballyhooed Hubble Space Telescope had finally left the ground. "I remember I was flying home from Europe, and I'm thinking, *Fisk, you've got this job knocked. You got the Hubble Space Telescope up, you just sold three new starts, everything is wonderful, you are terrific!*" As if he'd rigged the lottery. The following Monday, Fisk's triumphant return to the office got blown to smithereens because Hubble's main mirror captured every image out of focus. Media outlets went into a piranha-like frenzy. As only one example, Hubble made the July 9 cover of *Newsweek* and not in a good way. "NASA's $1.5 Billion Blunder," it screamed.

Fisk kept one eye on the engineering aspects of a Hubble fix while simultaneously facing a similarly bad issue: He needed help balancing NASA's checkbook. In that letter to the Space Studies Board, Fisk referenced CRAF spending bloat and petitioned the board to please review possibilities for downsizing the mission. Every good board needs a chairperson, and here, Louis Lanzerotti—a Harvard-trained research geophysicist—fulfilled that role. "The

32. An early 1990s photo of the Casani family that was one of John's absolute favorites.
From left: Josh, Drew, Jason, Jack, Lynn, and John. "It was after a rugby game in
my freshman or sophomore year," indicated Jason. Believe it or not, Lynn and John
weren't done adding boys to the family. Courtesy the Casani Family.

established procedure at that time," he clarified, "was that for questions such
as Fisk posed, a request for a Letter Report would go from NASA to the SSB.
The SSB would direct the request to one of its relevant Committees as appropriate. The relevant Committee in this case was COMPLEX, chaired by the
very capable and dynamic Professor Esposito."

Underneath the umbrella of the Space Studies Board, its *Com*mittee on
*P*lanetary and *L*unar *Ex*ploration functioned as one of five standing committees. Each enveloped a separate discipline. "We were aware of NASA budgetary concerns, but I don't recall how far ahead we had discussed this issue,"
explained the aforementioned Professor Esposito. As quickly as possible they'd
address CRAF. "We prided ourselves on providing timely advice," he mentioned, although Fisk likely arranged for his letter to coincide with an already-
scheduled COMPLEX meeting that summer. Committee members gathered in
the sleepy yet scenic resort town of Breckenridge, Colorado, snuggled against
the Rocky Mountains.

Since 1977 the Massachusetts Institute of Technology–trained Larry Esposito
had been working professionally in planetary science at the University of Col-

orado in Boulder. Already he was part of the science teams on Voyager and Galileo, as well as Cassini. Just recently though, his *Voyager 2* instrument had sadly crossed the rainbow bridge. It had examined how sunlight scattered in the atmospheres of Jupiter and Saturn, and had been switched off that April due to *Voyager's* increased distance from Saturn. Regarding CRAF, Esposito surely would've wanted to see the mission fly, yet he remained necessarily pragmatic. "Every NASA mission is in danger of cancellation," he pointed out.

COMPLEX met in mid-July, as planned. Reps from NASA HQ—specifically the agency's own program manager for CRAF/Cassini—also joined. The comet mission's project science group attended and put forward two major options for saving money: One involved a laundry list of miscellaneous trims, shaving down lots of little things although not drastically enough to cover the shortfall. The second option outright deleted the mission's hero instrument—an oversize, five-foot-long, golf tee–shaped penetrator to be driven down into the comet and run experiments.

Regardless of how timely COMPLEX strove to be, the group took the long route of hashing through numerous permutations—weighing everything against an earlier report of its own that had specified fundamental strategies and goals for exploring comets. Important to follow their own advice. Not six weeks later, the promised letter report went out and traveled back through the chain of the Space Studies Board to Lennard Fisk's desk and via this path did the associate administrator learn of the committee's verdict.

If anyone didn't want to see CRAF affected, it was Lennard Fisk. His space roots began with Earth's first satellite. "In 1957 I was fourteen. *Sputnik* made me want to be a scientist. *That* was the event." From there he endured a slow climb through Cornell, the University of California, and then NASA's Goddard Space Flight Center. "I was an expert on cosmic ray propagation," he explained, using a phrase not many people say. "In the early Eighties I got on the slippery slope of administration, where you just go further and further into it." This continued until 1987, when Fisk accepted his associate administrator position at NASA HQ.

Structured as it was, NASA housed multiple associate administrators. Each had a specialty: one for piloted spaceflight, another for legislative affairs, yet another for communications, and so on. In the case of Lennard Fisk's office, "you were the CEO of science," as he defined it. A straightforward enough job. "You had to sell things." By that he meant programs—selling them to the NASA administrator, to Congress, and really to the world. "New starts," they were called. Lately the starts had been selling almost better than Teddy Ruxpin. Fisk's three new ones, the subject of so much initial excitement and self-satisfaction on that plane trip, included CRAF/Cassini. He'd also muscled through a giant x-ray space telescope and a comprehensive program of terrestrial studies known as Mission to Planet Earth.

In reviewing his letter report, Fisk learned that COMPLEX endorsed neither option initially presented to it. Rather, the committee had developed its own. Two million smackers could be saved right off the bat by deleting the penetrator's heat shield cover. Future penetrator expenses could be capped in accordance with recommendations from Principal Investigator William Boynton, who guesstimated $10 million in reductions over two years. He envisioned saving money by slashing the number of coinvestigators and by using phone calls to replace a more expensive in-person meeting every year. Supposedly none of this would affect the science return. "He has also agreed to descope as necessary to stay within budget," reported the report. And then beyond the penetrator, four other CRAF instruments could be axed completely.

Next Fisk read the following: "Selecting which experiments to delete was of course difficult, because all the experiments have the potential to yield unique and important information." The report further lamented "a painful loss of scientific return" by eliminating what'd been suggested. And its closing drew forth a mood of dejection—noting that too many trims can kill a mission. "COMPLEX expresses its dismay that such a major reduction in mission scope has had to be undertaken so early in an approved mission."

Project Manager John Casani had been keeping one eye on penetrator costs as they mounted higher and higher, more than doubling, to the point where he ended up in DC for another meeting at headquarters. The JPL contract with NASA itemized a specific dollar amount for the penetrator; any overages would have to be somehow absorbed by the Lab. "Whoa!" Casani had shrieked when presented with the numbers. "This is out of control!" He wanted to cut the cord on that thing and get out while they could. His recommendation persisted despite an eleventh-hour visit to Casani's Washington hotel room the night beforehand. There'd been a knock on his door. And when Casani opened it, in came Mr. Penetrator himself, William Boynton, looking decomposed yet clutching hope that Casani might change his mind.

Boynton had in tow his young new hire as maybe a sympathy ploy and even presented a six-pack of beer to help lubricate the deal. Casani appreciated such gestures, and welcomed a brief discussion with the two, but his mind was already made up. "We cannot continue with this. This is not working. I mean, this has to come off the spacecraft," he told HQ the next day.

Saddened yet seasoned enough to deal with it, Louis Lanzerotti chalked the situation up to how the system operates. "I have been around long enough to recognize very well that the budgeting life is fraught with puts and takes, and the political and science and engineering winds of a given time," he commented. "It is taxpayer dollars that are always being balanced."

The summer of 1990 gave way to autumn. Sans penetrator, not much of a CRAF mission remained. Lennard Fisk stared down a tough decision. "You can't afford what you even sold," he lamented, and selling programs remained a cornerstone of his job. He was out of money to distribute among his new starts. "That's where I lost CRAF. I got asked to choose one or the other. I chose Cassini."

That decision didn't mean safe harbor for the Saturn orbiter. Not even the deletion of a whole entire mission could protect every last one of *Cassini*'s gloriously robust features and accouterments. "Still too expensive," griped Fisk. To have a chance in hell of leaving Earth, parts of the mission, or ship, or some other aspect would have to go. But which ones? No matter the funding profile, Saturn is still three planets away, and anything going there would need the requisite chassis, propellant tanks, antenna dish, and so on.

People began looking at *Cassini*'s twin boom arms. As designed, one could articulate in three directions for instruments needing to point: cameras and spectrometers and so on. "You can point it *anywhere*," gushed an engineer of the Cassini design. The other boom featured a rotating turntable at its very end. Mounted instruments could thusly sweep the local environment, which the fields and particles crowd very much wanted to do. Don Gurnett and Bill Kurth loved this arrangement. With such appendages, Cassini engineers had wisely avoided the quicksand of creating another dual-spin spacecraft.

The *Voyagers* don't spin. Each employs what is essentially a stabilized platform in space. Despite that, Gurnett still got the data he wanted from them, and why he did came down to a coincidence of housekeeping. Every few months, a calibration procedure twisted the ships through a complete loop on the axis of their high-gain antennas. As Gurnett framed it, "They roll the spacecraft, and that turns out to be a *great* advantage to me," because the plasma wave antennas rotate as well.

Gurnett happily dug into the significance of this motion. "Well, plasma, you have to measure what's called the *distribution function*: how many particles are moving in any given direction. And frankly, these plasma instruments, you really need to measure *all* directions." If the ship is otherwise fixed in space, never rotating, then direction finding is nigh on impossible. "All you can say is, 'I hear it but I don't know where it's at.'" Addressing the deficiency means introducing more hardware. Gurnett theorized having, ideally, three hundred separate instruments looking in as many unique directions, and he wasn't kidding. But a spinning platform trumps even this. "Well, suppose you made a measurement every ten degrees of rotation around the spin axis. So then you can get 36 measurements during a rotation. So you can get 36 times ten, that's 360—just turns out to be 360 different directions."

No turntable on *Cassini* would mean no instruments sweeping through space—a massive handicap for any fields and particles experimentation. But everybody working the mission had their own problems. As John Casani pointed out, "We were under a lot of cost pressure at this point in time. And NASA was pressing us, uh, saying they were gonna cancel the program 'cause the costs were too high." Fisk was on his butt. So was snarly top dog Dan Goldin, NASA's administrator, who barked about killing the Saturn mission entirely.

All this work on Cassini and CRAF both, just to abandon everything?

During a critical meeting with Lew Allen, the formidable JPL director brought up Cassini's alarming expense forecast. John Casani later explained, "Our expenditures at the time were not out of line with the *current* funding. The problem was that our *projection* of funds needed for future work exceeded what NASA was anticipating." The discrepancy hadn't been tallied to the nearest penny but was on the order of several hundred million dollars. "Headquarters had declared they could not afford what we said was going to be needed." And wanted the situation addressed.

This prompted Allen's meeting about costs. To John Casani he charged, "What are you gonna do about that?" His question spoke to the core issue of under what conditions the project might be able to soldier on. They were on the verge of losing Saturn as the entire mission buckled under the weight of its own ambitions.

In the years since this meeting, John Casani's resultant actions have been discussed and debated, celebrated and vilified. Decades later he framed his rationale with additional perspective: "I did not want to involve the science community in the formation of our response. For a couple of reasons. First of all, because time was of the essence. And second, because when I recommended to HQ to drop the penetrator from the CRAF mission for budgetary concerns, and HQ accepted, the entire science community went on a full-court press to persuade HQ to reverse their decision. Instead, HQ decided to cancel the entire CRAF mission." No way did he want that to happen again.

"Well," responded Casani to his boss, in measured fashion, "very simple. We're gonna take those two booms off the spacecraft. Those are what's causing the problems."

Lew Allen cocked his head. "Well, what—what does that mean?"

Strictly speaking, it meant rigidly attaching every instrument to the body of the spacecraft itself. No articulating appendages. No turntables. "You'll see out in one direction. But we'll rotate the spacecraft, and that will allow them to sweep out the full sky around them." Such an arrangement would provide most everything the science teams wanted. Casani smiled at his boss in an attempt to sell things as well as Lennard Fisk could.

Lew Allen foresaw all kinds of problems. It sounded too drastic. "What about the camera? The remote-sensing people?"

"We'll put a very simple platform right on close to the spacecraft. That can only move in one direction."

"Well, how are you going to point?"

"We stop spinning the spacecraft. And then we use the roll position of the spacecraft as one degree of freedom. And the camera on the spacecraft only has to move in the other degree of freedom. It will be a very simple, simple camera."

Lew Allen stared straight ahead.

Preached Casani, "Yes! That will absolutely work! We can do that!" As he later justified, "I *knew* all this would work because we had already developed, and flew, the required guidance and control functions successfully on the *Magellan*." They'd have to rewrite plans for transmitting Cassini science data, and that would suck but was certainly doable.

The meeting ended, and Casani's to-do list overran its margins. "I immediately terminated all work having to do with the two science booms, as well as the rotating turntable." New work orders initiated a "simple platform" for the remote-sensing gear. It wouldn't be rigidly affixed per se but would offer a single degree of freedom only. Casani further directed mission designers to ensure support of the fields and particles requirements by modifying how the ship would move in space. Instead of a spinning turntable, the whole darn thing would spin and acquire essentially the same data.

In the wake of these changes, any guesses as to the general response? Avoiding the catch-all term "unhappy," Casani instead leaned on "disenfranchised." As in, what he'd just done to the entire corps of scientists already working this mission. A whole class of people assuming the ship would remain configured the way they'd been promised *all the way through* to the part where it reaches Saturn. Instead, here came Mr. Big-Pants John Casani, lopping off body parts as if running the Belgian Congo and forcing everyone to take turns with instrument use like kids waiting on a playground slide. And when the next one's turn came up, their wingless amputee of a ship would have to groan about as if wearing a permanent neck brace.

"You shouldn't anthropomorphize spacecraft," Casani joked to his restaurant guest. "Because they don't like it."

During communication sessions, the ship would naturally have to aim its antenna at Earth—no problem whatsoever in the full-figured olden days of yore. But while doing so, the science instruments wouldn't necessarily point toward anything of value. That's wasted mission time. While the camera was running, the direction everyone else faced would be at the mercy of the shot. Many pictures would have to be stored on the spacecraft's internal recorder. "When it was time to play the pictures back, we would have to point the high-gain antenna toward the earth. And keep it pointed that way. And so while we were doing that, we couldn't spin the spacecraft, and so the fields and particles people now lost a lot of their mission observing time." So detailed Casani of the new inefficiencies awaiting investigators. "And they were furious about that."

Indeed, they were. This was a disaster. The most polarizing decision of his career. Multiple members of the science teams took their project manager to task for wiping out functionality and horrendously overcomplicating the process of making observations. How could he bitch-slap everyone that way? Casani denied nothing. Took full responsibility. He acknowledged limitations and communicated understanding. But at the same time, apologized for naught. His job was to get *Cassini* off the ground. Yes, no question, the mission would now cost more to operate after launch. And absolutely, three diverse camps would now have to plan every last move way in advance, often in competition with one another. "But it's either that or you have nothing," Casani told them all, summoning his father's no-nonsense attitude. "You're gonna be working your ass off, or you won't be working at all. So take it or leave it."

Iowans Don Gurnett and Bill Kurth first learned of the mauling during a group project meeting at JPL. The now-gone turntable would've rotated eighteen degrees every second and spoon-fed them excellent coverage of the local environment. No way could the whole ship rotate that quickly; they'd now be lucky to get even one degree per second. Talk about a demotion! Even so, Kurth stated that their robust triple-antenna design remained "mostly detached" from any fallout related to the deletions. "We could 'see' our waves regardless of spacecraft attitude," he insisted, and their work would be spared the agonizing hassles of repeat coordination meetings. The remote-sensing folks with the cameras and stuff would undoubtedly have it tougher without their scan platform.

Most every scientist harrumphed their way back home, and despite all the calamity and resulting negative air, the mission remained alive. Blood still flowed through its veins. And hey, Casani surely had no interest in making life harder. But something had to give, and he'd gone with cutting the booms. "We saved a lot of money by doing that. And it demonstrated to headquarters, you know, that we were really serious about controlling the cost on this thing." Also, operations at Saturn wouldn't occur until some ten years in the future. As such, "NASA's ability to accommodate it was less challenging."

Sure, Don Gurnett would've preferred the turntable. But like his colleague, he wasn't too worried ultimately. "We can do great direction finding with three antennas!" Even so, for all the anticipated time and money that everybody was going to have to invest in micrometrically coordinating every last maneuver, Gurnett figured the U.S. government could just pony up enough dough to bolt the damn arms back on, and the whole mission would return better science to boot. "They took the scan platform off and saved some money. And I think it was actually more of a face-saving thing than anything else," he suggested. "Because they were able to say, 'They reduced the cost for Cassini,' and Congress approved it."

In the wake of these draconian downsizings, both John Casani and Lennard Fisk acknowledged how their whole paradigm had to change. What a different

33. In this cartoon, John Casani figuratively douses the lights on JPL's Flight Projects Office. The closure wouldn't last. Courtesy the Casani Family.

playing field these days. Any future flagship mission, even if surviving Congress, would never survive unmolested all the way through to launch. Cost projections never went *down*. So why try to sell another huge one-shot when it'll just run over budget and put everyone through the wringer, again, with demoralizing redesigns and compromise?

Fisk vowed to change his way of doing business, commenting, "So when we talked about selling new things, we only sold small things after that."

The trajectory of the Jet Propulsion Laboratory had, throughout its history, gone in one direction only—bigger. Bigger missions, bigger ships, bigger workforce. The last time John Casani had been able to hold an entire spacecraft in his own two hands, it'd been *Pioneer 4*. Now they were supposed to go back to thinking small? Ventured Casani, "People outside the Lab, and NASA Headquarters, everyone said, 'That's not a job that JPL can do. They're only good at building big spacecraft.'" He started hearing rumblings of the work going elsewhere. "And I was saying, 'No, wait a second, we can do small spacecraft.'"

Hundreds of miles north of JPL's campus, the Ames Research Center had hired in a guy specifically to recapture its past success at running cost-effective and science-driven missions to outer space. Most notable of the lot were *Pioneer 10* and *11*. Uh oh, competition.

Alarmed, Casani marched into the JPL director's office. Greeting him was a new face yet an old one all the same. Longtime Labbie Ed Stone had functioned as Voyager's project scientist since the mission's very beginning in the early 1970s. These two men knew each other well. And to Stone, Casani implored, "We've got to get rid of the Flight Projects Office." Every one of their flagship-class missions—Voyager, Galileo, Cassini—fell under its purview. And for many working in space, the FPO had come to represent an unnecessary layer of bloat and bureaucracy. Wasted dollars. Casani wanted to nuke and pave the whole doughy, stodgy FPO concept and replace it with something trim and healthy and quick on its feet. To Stone, he championed a need for the Lab to broadcast a strong message that JPL had the interest, the mindset, and the capability to accommodate small-scale missions. Stone told him to go for it.

John Casani smiled. "And so that's what we did."

21

Saving the Appearances

In the book entitled *Sidereus Nuncius* by Galileo Galilei there is
nothing contrary to the Holy Catholic Faith, Principles, or good
customs, and that it is worthy of being printed.

—Heads of the Council of Ten, *Sidereus Nuncius*, March 1, 1610

Arguably, the first space mission to orbit Jupiter should've been called Van Allen.

But to understand why it wasn't, as great as that name might've been, let's skip backward four hundred years or so to a narrow point in time where a major discovery about Jupiter first occurred. Despite the intervening years, we know exactly when this happened because the person behind it carefully dated his writings:

> On the seventh day of January of the present year 1610, at the first hour of the night, when I inspected the celestial constellations through a spyglass, Jupiter presented himself. And since I had prepared for myself a superlative instrument, I saw (which earlier had not happened because of the weakness of the other instruments) that three little stars were positioned near him—small but yet very bright.

The individual in question—a highly educated, forty-five-year-old Tuscan gentleman named Galileo Galilei—had a lot going on at the time. Already he was an established mathematics lecturer at the University of Padua. Tutored students one-on-one. And ran what amounted to a private boardinghouse for multiple students. Unmarried but with a fertile mistress, he had three children dependent on him for support.

During the summer of 1609, amid all this swirling activity, something came to Galileo's attention that began steadily consuming any scrappy remainders of the man's free time. It was a new type of apparatus that today is generally referred to as a telescope. Throughout his writings, Galileo called it a spyglass. What it did was nearly magical. People had, throughout history, attempted to magnify distant landscapes and objects by various means. Think of the advantage in wartime alone! Familiar with optical theory, Galileo previously had experimented with arrangements of large, shaped mirrors, which failed to magnify much of anything. Crazy to think it could be done for real with two glass lenses in a tube.

Galileo was floored by the device—but not so much by the overall quality of the lenses available commercially at that time in spectacle shops. At best they offered triple magnification. One of the more versatile people in the world, Mr. Galilei simply endeavored to make his own. Key to improvement was not larger diameters so much as the quality of the grind and polish. His first one, completed that summer, gave way to progressively better and better instruments. Galileo's "objective" lens—the one at front, farthest from the user—measured less than 1.5 inches across and thinner than 0.08 inch in the middle. Producing it required a grueling effort and untold hours of elbow grease. *Grind-grind-grind, polish polish.* Repeat. If anyone ever set the standard for persevering through trial and error, it was Galileo Galilei.

Come autumn of that year, his two best finished lenses (each shaped differently) were bound within strips of wood inside a leather overwrap. And Galileo now possessed an improved, absolutely one-of-a-kind science instrument offering a power twenty-one times over that of the naked eye. Instead of spying on neighboring armies—or potential future mistresses—he trained it on the moon, because why not? Others had already done so. But none with the insane level of magnification that Galileo now possessed. How odd: Under such magnification, the lunar surface failed to exhibit its historically assumptive characteristic of glassine smoothness. Rather, its appearance had the same sort of unevenness prevalent on Earth. Successive nightly observations proved it so.

That alone was profound. But surely the sky held multiple wonders, and Galileo Galilei, after carefully documenting his lunar observations via notes and drawings, began planning a new round elsewhere. The obvious subject would be a planet. During that particular time of year, every known planet happened to be in an unfavorable position for viewing. (Mars, as only one example, was too close to the sun for any useful assessment.) Except Jupiter. Eagerly he turned his attention to it and first glimpsed what he regarded as three stars in alignment near the planet's equator. Initially, that's what they seemed to be—stars, forming a coincidentally straight line with Jupiter. That and their surprising brightness constituted his initial attraction.

This had to be a fluke . . . right? Jupiter *just happened* to be passing through this line of three stars and *just happened* to line up with them as it did. Right? The next night, when Galileo again trained his superbly tuned spyglass at the heavens, he expected to observe progress in Jupiter's orbit as the giant planet gradually lumbered away from the starry line-up. But the planet had moved in the wrong direction. And those stars remained close. Hmm.

Understandably this prompted more and more observations, and a week of them accrued. These nights were taxing as Galileo squinted through his tiny lenses, pausing to wipe fog off the glass with a cloth or briefly rest his eyes. Imperfections in the lenses could be somewhat mitigated by limiting the amount of light entering the far end; Galileo did this with a sheet of cardboard the size and shape of a mini doughnut. It drastically limited his field of view but provided clearer images. And unexpected ones. The relative motion of Jupiter to these not-fixed-at-all stars told Galileo he must be on to something. Super weird was how the stars changed position—moving to the other side of the planet sometimes—and not all three appeared every night. But they always stayed loosely together in more or less a straight line.

And then on the night of January 13, he saw four stars. Four! More sketches in his notebook. Diagrams. Comments.

Two days later, Galileo concluded that his four stars weren't what he originally thought. No, they could only be moons. Moons of Jupiter. How electric of an idea, that other planets might have dedicated minor bodies twirling about themselves. Plenty of people at that time still held fast to a belief that our teensy Earth signified the center of movement in the universe. There could be no such thing as moons orbiting other bodies. Well, that idea had just been smothered.

A lone Italian math professor held no monopoly on planetary observations; anybody at all with enough resources to make their own telescope could see the exact same thing on the exact same nights. And be gearing up to take credit for it! Wasting no time, Galileo swiftly compiled his notes and commentary and drawings and diagrams into a hastily prepared manuscript, which was published only months later. (Indeed, he had the writing underway even before discovering Jupiter's moons, and pages were being printed during the course of his further observations of those moons. Most—if not all—of the 550 initial copies handled a particular word swap by gluing in paper strips with the change on them.)

He wrote the book in Latin. A quick introduction of how telescopes work shifted gears to a larger section on lunar observations. Followed by a summary of the visual differences between stars and planets. This transitioned into the largest section, which reported the Jupiter findings. It included sixty-five diagrams tracing moon movement, lest Galileo be accused of failing to pile-drive his conclusions right into the ground. Altogether his book was brief, slanted in tone for political pandering, and lacked any sort of tied-in-a-bow ending. Depending on which Galileo scholars are queried, the title of his book translates to either *The Starry Message* or *The Starry Messenger*. (Galileo himself allowed the disagreement to persist.)

This modest publication signified the dividing line between medieval and modern science. With it, Galileo faced a perhaps not obvious challenge of explaining the reality of the cosmos in a way that, to use the vernacular of the time, "saved the appearances." That is to say, whatever radical ideas his text proposed had to coincide with the observed motions of the various heavenly bodies. The two had to reconcile with one another. This formed a problem at very basic levels. Imagine the objections from your average Tuscan on the street: "Mister Galilei, if the moon really does have mountains, then how come its perimeter is smooth like a circle? Shouldn't it be all jaggy around the edge?" Addressing these questions, Galileo offered multiple explanations. First of all, he proposed, the moon may well possess a heavy atmosphere that blurs things. (On that he was wrong.) Secondly, a multitude of lunar mountains— some closer to Earth, some farther—no doubt stack up in such a way as to present a smoother edge to our eyes. (Ding ding!)

For his discoveries to be validated, Galileo knew the winning path involved others performing their own observations. But nobody else had a spyglass as good as his. And they were suspicious things. (After all, you use them to *spy*.) People had vague ideas of how they worked. Maybe the instrument itself was somehow creating those moons Galileo thought he saw? When an advance copy of his book went to the grand duke of Tuscany, Galileo took the trouble of sending one of his best spyglasses along. (Plus instructions.) Laboriously he created additional ones and supervised their delivery (also with instructions) to other key people. Galileo toured the lands, discussing his findings and inviting people to look through the groundbreaking instrument he'd conveniently brought along. Such efforts consumed the remainder of his life, and ultimately the Catholic church—patently unamused that someone would speak in opposition to holy scripture—confined the man to house arrest.

If only something even better had been available than a twenty-one-power spyglass. A more advanced instrument. Or perhaps multiple ones, all of higher precision. Able to figuratively position Mr. Galilei's senses more closely to Jupiter than what could be done with a tube of DIY glass lenses on a windowsill in Padua. All that would take centuries. But imagine the pure wonder that likely would've overcome this man in knowing that future times would witness a diverse collection of scientific measuring instruments arriving at the very system of planet and moons he had struggled for so long to understand. Instruments borne by a vessel "adjusted to the heavenly ether," as it were, quite unafraid of the empty wastes, and even named after The Man himself.

It began with the *Galileo* probe—a 750-pound product of scientific fantasy for hundreds of years and now nearing the end of its months-long coast. Six hours ahead of first contact, the probe's onboard timer began cueing a sequence of pre-entry observations that, sadly, did not include photography. An initial finding—tons of helium atoms, moving at surprisingly high speeds. They bombarded the probe as it slammed through a band of radiation ten times stronger than any surrounding Earth. The news raced through wiring bundles to an onboard radio system, which passed it up to a set of embedded antennas on the probe's back end. Data flung off the antennas and were gobbled up, Pac-Mac style, by the dedicated relay antenna slung underneath *Galileo* as it soared overhead. Information then trickled home. Slowly.

The probe barreled in, shallow trajectory, heading for the planet's middle. Next came the distal wisps of Jupiter's atmosphere and the most thermally demanding stage of the mission—"deceleration." A unique heat shield had to defend the probe's internals from temperatures twice as hot as the sun itself. The probe experienced some twenty-five thousand sizzling degrees while plunging at more than a hundred thousand miles an hour. Its exterior glowed white-hot. Layers began vaporizing. Roughly 44 percent of the probe's overall weight had been allotted to just this shield alone. And during deceleration, half of it flashed away.

Inside of two minutes, peak deceleration 228 g-forces, Jupiter's atmosphere slowed the probe to just over a thousand miles an hour. There came the drogue chute. Off went the heat shield's back cover. Main chute now—some fifty-three seconds late. What remained of the front heat shield tumbled away. Probe fully exposed. Fitted with simple aerodynamic vanes, but absent wings or other control surfaces, the bulbous machine dropped lower and lower into the planet's atmosphere underneath its single main chute. Jupiter possesses no solid surface, and the probe would run for as long as it could withstand the increasingly treacherous pressures and temperature. Or until its batteries died.

Floating downward through white-colored ammonia ice clouds, the machine hummed with life as a five-inch arm extended, four small mirrors adorning the end of it. A small laser beam illuminated clouds rushing past the mirrors. This measured the size of cloud droplets and gave clues to their shapes and would help determine whether the particles were solid or liquid. Until this occurred, no direct evidence had ever existed for the presence of clouds below Jupiter's uppermost layer. Wind speed measurements also began, clocking nearly 450 miles an hour. As the probe fell, the wind speeds never slowed. A separate instrument slurped in atmospheric samples for analysis, confirming the presence of methane and hydrogen sulfide and puzzling amounts of ammonia. Later analysis would reveal that, in general, the planet's interior is much drier than originally thought.

Jupiter squeezed. Pressure built. Temperatures climbed. Turbulence swirled. The probe had been designed for worst-case pressures in excess of ten times that on Earth. It accomplished this in part due to a counterintuitively vented enclosure that further reduced weight. Another onboard instrument, a sophisticated lightning detector, had come from Louis Lanzerotti—that guy from the Space Science Board who'd worked to scale back CRAF prior to its cancellation. Lanzerotti's instrument failed to catch any visual flashes, though the probe's cumulative results—synthesized across every onboard instrument—noted some fifty thousand high-energy lightning discharges. Maybe that sounds like a lot, but it is a fraction of what's recorded here on Earth.

Elsewhere on the probe's exterior a small, cone-shaped appendage twisted on its base. Quickly. Twice a second it looked up-down-up-down, and the light it caught during each "peek" entered the instrument through a small window made of solid diamond. From there the light bounced off two mirrors before striking a handcrafted plate of detectors that, fundamentally, measured the amount of radiation streaming from the planet against how much reached it in the first place. This kind of study was key to deciphering how Jupiter's clouds are layered, how thick or sparse they are, how they circulate, and how much water vapor they might contain. Like the others on board, this instrument had been custom produced to necessarily high standards. Besides

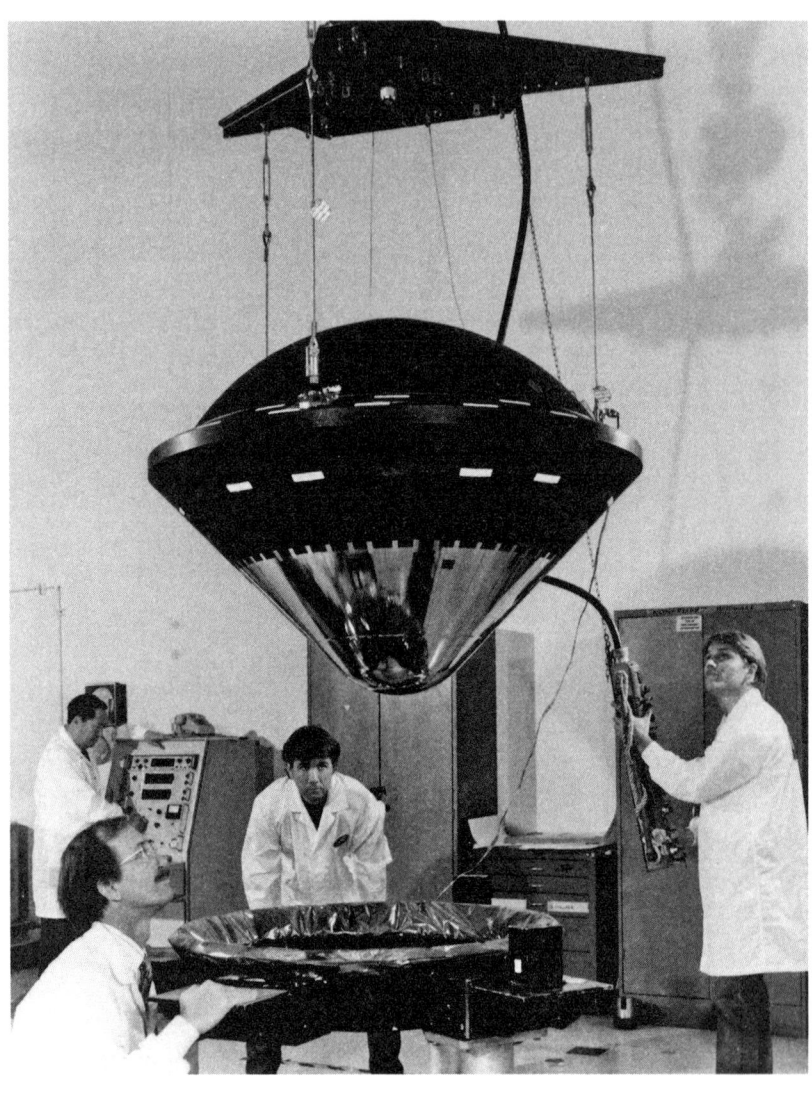

34. Enclosed within its heat shield, *Galileo*'s descent probe is handled by technicians. The shield's downward-facing business end bore the unfortunate responsibility of withstanding three minutes of up to 25,000 degrees Fahrenheit, which would burn away much of it while entering Jupiter's atmosphere. How come nobody ever named the probe? Courtesy NASA/JPL-Caltech.

the diamond window, it also contained platinum, beryllium, tantalum, and gold. Its constituent materials drove the delivery price to $6 million for this one device alone.

Despite such materials and exquisite workmanship, past forty minutes of activity, the various systems capitulated to Jovian pressure and heat and began winking off one by one. Minutes later the probe transmitter overheated and failed, but a backup instantly took over. Ten minutes afterward the backup also yielded, and signals cut out. A lone survivor, measuring atmospheric structure, marked the only instrument lasting the full period of data transmission. But every one of the devices had otherwise worked perfectly. And at well over 200 degrees Fahrenheit—far in excess of a design spec calling for only half that. The greatest recorded pressure was twenty-three times that of sea level on Earth.

We will never know exactly what happened to the probe after it shut down. But its parachute certainly melted, followed by its aluminum fittings. Bit by bit, as temperatures and pressures mercilessly swelled, the various components would've liquefied and evaporated: the titanium shell. The platinum and beryllium. The gold. "Ten hours after entering the atmosphere there would have been nothing left to see," read a subsequent comment in JPL's *Galileo Messenger* newsletter. "And the Probe would have become a part of the planet that its sister Orbiter will be watching so closely." Almost fifty-eight total minutes of probe data had been received and relayed by the *Galileo* orbiter. Each speck of it took nearly an hour to hit the receiving antennas back home on Earth.

"Data from an actual probe into a giant planet's atmosphere." So marveled one scientist on the project.

The *Galileo* spacecraft, having fired a long engine burn to miss Jupiter, did so by over 130,000 miles. And then, not suddenly at all, on December 7, 1995, the singularly unprecedented VEEGA trajectory fulfilled itself as *Galileo* blazed through the Jovian system, flying initially by Europa, then bending around Io—two of the "Galilean satellites" seen by The Man himself in 1610. In Io's case, the ship cleared it by just 550 miles in a carefully planned maneuver to crank the spacecraft's flight path and speed. Future orbits would encounter the other two Galilean moons of Ganymede and Callisto. A total of thirteen Jovian moons, at that point in the mission, were known to exist. Over the next twenty months that number would balloon.

Way the hell back on Earth, JPL's campus had been overrun with photographers and news crews and other journalists. Not to mention project workers and their loved ones. The King family went. Clyde's son James suggested that his father typically avoided these kinds of events, skirting all the ballyhoo and attaboys and round-robin back thumping of colleagues. Rather, the elder man celebrated in his own personal way. "His response was more of a quiet pride. Something that you kept for yourself," James said. "You didn't need to broadcast it. You set a task or goal, and even if it was a hard one that seemed impossible, accomplishing it gave him more satisfaction than getting a bunch of loud praise."

The Lab's video department had just concluded a ninety-minute TV show that bookended the approximate time of orbit insertion. *Galileo at Jupiter* featured an in-studio host with impaneled experts and occasional cutaways to the mission support or navigation areas. Considering the predictable technical fumbles and dead air with such a production, it unrolled rather smoothly.

For the events of the day, Bill O'Neil had donned a dark suit and tie, and outwardly resembled a senior FBI agent. It'd been a long day already. "People work ungodly hours," he later said of Lab life. "I used to quip that I didn't work full-time on Galileo. I worked *all the time* on it." He was standing in an operations room when the TV crew wanted him on. Fellow project engineers and managers closed in for the shot. With deliberateness, O'Neil muscled a phone handset while waving to a camera feeding the broadcast. "We are just ecstatic already because we have confirmed that, indeed, the Probe was successful. It radioed a signal to us. The most difficult entry ever attempted has been accomplished. So it's just a wonderful afternoon."

Two other people had explicit reason to celebrate. One of them, Roger Diehl, harbored obvious personal reasons. Without his persistence and brilliant use of S-TOUR, a workable flight path for *Galileo* might never have been found. The second person with explicit reason to celebrate was the guy Roger phoned after VEEGA had been initially accepted—the man who'd created S-TOUR in the first place. "I was very pleased," beamed Phil Roberts of the warmest news in the world flowing over the line that day. His little computer program had come through in the clinch! Okay, so S-TOUR didn't have the reach of Windows '95. But it saved a billion-dollar space mission. By that point Roberts had already moved on from JPL but greatly appreciated his old coworker taking the time for such a momentous update. "We had mutual congratulations on the phone."

As *Galileo* began tracing a long, initial orbit for its prime mission, both

Roberts and Diehl could reflect on how S-TOUR had been responsible for so much more than just reaching Jupiter. That wasn't even why Roberts had initially created it. As first envisioned, his program represented *the* tool for a mission designer who's *already in* the Jovian system specifically, has over a dozen moons to contend with, and needs to quickly reduce mountains of flight path permutations down to a scientifically bountiful favorite. Indeed, without the work of Roberts, *Galileo* may well not have been able to use Io the way it just had—altering course and beginning to decelerate, conserving propellant in the opening orchestrations of a years-long mission profile. What *couldn't* Phil's programs do?

When Phil Roberts began work on a Jupiter-orbiting spaceflight, the year was 1974, the mission hypothetical, and the expectations rather uncertain. The JPL group he worked in, advanced projects, studied all manner of potential missions. (The same group produced Voyager's Grand Tour.) Roberts had been at the Lab for about five years and recalled the environment of the time: "Jupiter was high on the list of planets to orbit. It was also desired to study the Galilean satellites. But it was deemed important not to crash into any of them and leave earthly microbes." So all the concept studies kept their distance. One colleague's proposed "tour" of the Jovian system repeatedly looped past the moons at an impressive variety of angles . . . but never venturing closer than about a quarter-million miles to any one of them. Scientifically speaking, it almost wasn't worth the trip. But did play to caution: Flown this way, a spacecraft at the end of its mission—with nearly empty tanks—ran a low risk of impacting any virginal moons.

Phil Roberts appreciated his colleague but nevertheless saw irony in the man's effort. "You could say he fought against the gravity of the moons while trying to get a good look at them."

Developing mission possibilities involved running a computer program. "There were *many* programs," noted Roberts of this period—some created in-house by fellow JPLers while others came from outside contractors including Lockheed. You kind of had to ask around. Generally speaking, the landscape of trajectory-planning software was fractured and specialized. Akin to writing a memo with one brand of word-processing software, then having to find another for writing a to-do list and still another for writing a letter. Oftentimes people would embark on some exciting new assignment only to get tripped up by needing vital yet nonexistent software.

Even with the right tools available, plus expensive IBM and UNIVAC computers the size of refrigerators, plenty of labor-intensive steps remained. "Mission planning involved many 3-dimensional calculations," explained Phil Roberts. "And these were worked out with vector algebra. And *then* software was written to carry out the necessary calculations." People sometimes became programmers out of necessity. It all seemed so darn cumbersome. But Roberts, who'd begun writing programs back in his undergrad days at the University of Kansas, sensed that things could surely be simpler. So he wrote a "library" embracing all this vector algebra that everyone continually depended on. It functioned as an add-on for any program, existing or new, and saved having to recode the math each time.

It streamlined things almost instantly. "This library allowed me to write code about 3 times faster than before. With fewer mistakes." Roberts was thrilled. The efficiency afforded him extra time to contemplate what else could be restructured or somehow improved. To investigate a possible mission to a comet, he modified someone else's existing trajectory program that didn't account for comets at all. This led to the writing of complete software programs. His reputation around the Lab improved.

"My mother taught me that nothing was more important than my reputation."

He built another library—way more expansive than the vector one—that unified bunches of scattered datasets on the position and velocity of planets and their moons. Everybody referenced that kind of information, but nobody had ever consolidated it into a single source that nearly everyone's programs could utilize. Roberts had solved the problem of access. And not long after, began exploring the possibilities of a Jupiter-orbiting mission. On this, he worked primarily with one specific colleague. "He was better with calculus," Roberts suggested of the man, "and I was better at geometry. We would discuss the limitations of what gravity assist could accomplish. I recall asking him why I can't accomplish a certain orbit change. He said, 'Of course not, that would not conserve the Jacobian.' I wasn't sure what *the Jacobian* was." Phil Roberts was no academic slouch; he'd followed his undergrad at Kansas with a Caltech physics PhD. Nevertheless, exchanges such as these crystalized in him an awareness of there always being more to learn.

Navigating from Earth to Jovian orbit would be easy enough—people had known how to do that for years—but the tricky bit came down to dealing with all its moons. Each one orbited differently, was sized differently, and imparted differing amounts of gravitational influence on each other. To accommodate each known moon in software, Roberts recognized that each would need its own "routine," or discrete component, of code. Thousands of lines in total. All grouped together into a "master routine" addressing every peculiarity of the Jovian satellites. To be clear, that wouldn't represent a complete program per se. It would be only an element of one, similar to how Microsoft Word has a dictionary that by itself can't be used for writing. So Roberts integrated his master routine with an existing program called KPLOT, used for depicting the arrangement of planets and moons at any desired point in time. The two components interacted with each other just fine, but Roberts needed to check for accuracy. "I got a photo of Jupiter and its four moons taken through a telescope. I put the time of the photo into KPLOT, had it plot Jupiter and the moons as seen from the earth, and overlaid the plot on the photo. It was a match!"

Early work on Galileo had Phil Roberts trying to make the most of how the spacecraft would approach its target. Following probe release, the orbiter needed to execute an avoidance maneuver to prevent hitting the planet. "The greater the miss, the less the radiation exposure. But the more fuel would be needed to *slow* the Orbiter and get into orbit." Roberts noted how the spacecraft would, at this point, be in the vicinity of Io, for which a close pass had been highly desired from the beginning. "It occurred to me that as long as we were passing Io, we could use its gravity to slow the Orbiter some. And then finish the slowing with the rocket burn." This led him to create a specialized program, JOPROBE, strictly for planning the Io maneuver. It put the ship where everyone wanted and saved fuel to boot.

"My first coup in trajectory design for this mission," he cheered.

Success with Io led Phil Roberts to think about a new, from-scratch program to suss out the excursion through Jupiter's complicated system of moons. A really user-friendly program. Guiding people as if a travel agent for Jupiter. Fast yet comprehensive. His core version of it, nearly a thousand lines of code bearing the utilitarian name Satellite Tour, considered Jupiter to be "the center of things." To fulfill his intended user-friendliness, "the S-TOUR program was designed to be interactive, allowing you to sit at a terminal and work your trajectory design from one moon to the next. At each flyby, all the possible trajectories to reach other moons—or to return to the same moon—were calculated." Roberts went on to explain how these options were accompanied

35. During a 1973 JPL open house, Phil Roberts shows off a bit of his handiwork. "I am demonstrating a plotting machine that would make ink plots of any planet which visitors could take with them," he explained. The inkjet printers of their time, plotters used multiple-colored pens that mechanically traced lines on paper. Courtesy Phil Roberts.

by helpful characteristics, such as what views of Jupiter would be available between a current encounter and the next. As if S-TOUR was a Choose Your Own Adventure in space.

"You basically explore your way along!"

S-TOUR proved immensely powerful. Roberts never bothered writing a manual. He personally oriented anyone who asked, handling follow-up questions as they arose. The usefulness of Io persisted. A hypothetical spacecraft orbiting the Jovian system would need to do so in the same plane that all the Jovian moons orbited. But coming in from Earth, the ship would arrive slightly out of plane. "The Io flyby could help fix that," noted Roberts. Not bad for a moon about the same size as ours. And S-TOUR further revealed how subsequent orbits could be progressively tightened into much more scientifically useful periods of one to two and a half months. What a perfect series of maneuvers! "With no fuel burn! And it could get the orbit into Jupiter's equatorial plane, where the other moons can be reached."

The resulting satellite tour could happen at decidedly intimate distances, offering many rich opportunities for science investigations, because S-TOUR enabled a projection of the entire phase of orbital operations. All the way through to exactly where the spacecraft would be as its maneuvering fuel began to run low. Bonus: An interplanetary component—advantageous for planning the outbound voyage from Earth—could easily be added to the software by having S-TOUR simply reconsider "the center" to be our sun and not Jupiter. This bolt-on had nearly been deleted before Roger Diehl used it to find VEEGA.

"S-TOUR was the highlight of my career at JPL," beamed Phil Roberts.

22

Wiggly-Line Science

With its high-gain antenna jammed and its data being squeezed back
through a tiny dish, with its components aging, whether *Galileo* in its
compromised state could complete its primary mission was in doubt.

—Spaceflight writer David Clow

Hobbled yet resolute, *Galileo* blazed along. Somebody got ahold of John Casani and asked him to summarize the lessons of the project so far. He responded, "Tenacity certainly has its rewards"—a statement worthy of promotion to the mission's motto. If the ship hadn't been named *Galileo* (or *Van Allen*), well, *Tenacity* would've been a solid contender also.

That name certainly would've applied to what went into addressing the broken tape recorder. During routine use in mid-October 1995, ahead of entering Jovian orbit, the unit had gone into rewind mode and seemingly never left. It rewound for fifteen hours—way longer than necessary for traversing the entire length of tape many times over. Was a capstan slipping? Had the tape itself snapped? The latter would be a deal-breaker. If this thing was going to be broken forever, *Galileo* didn't have many onboard options left for storing data. Maybe they should've put on a roof rack?

One thing they *had* given the spacecraft was a computer. Possessing dedicated onboard memory. Some of which could be given over to temporary data storage. And that would've worked okay for probe descent and for the lightweight fields and particles readings. But images or other high-rate remote-sensing data—basically anything from an "-ometer"—would rapidly saturate the limited amount of memory. Darn it anyway. A multitude of engineered precautions had been taken to guard against tape recorder malfunction—from the composition of the tape itself to *filtered grease* to the way the entire unit had been sealed in a fixed, stable atmosphere of nitrogen and helium. The specialty manufacturing firm that built it had offices just forty miles away from JPL, in Anaheim (and were definitely on the case, just

36. Reelin' in the years: a spare Galileo tape recorder on display at JPL. Containing approximately 1,850 feet of tape, the twin reels are stacked atop one another. Tape heads and capstans are to the right of the reels. Photo by and courtesy Joseph Chiu.

to be clear). This was not some cheap-ass tape recorder from Kmart. Was it really done for good?

After studying the situation for nine straight days, which undoubtedly involved serious lapses of personal hygiene, technicians breathed deeply in anticipation and commanded *Galileo*'s tape to move forward for ten seconds and learned that it worked after all. Nothing had snapped. Those ten quick seconds of playback affirmed that the tape itself had never budged a millimeter during its supposed hours of rewinding. It'd merely been stuck. But now ran. Why was it working again?

Investigators had two spares on the ground—one each from Galileo and Magellan. When the latter was plugged in and commanded to rewind, it stuck right away. Damn. Both spares were of the same model. They even used the same manufacturing lot of tape as that currently orbiting Jupiter. Yet with differences. The Magellan spare hadn't had as much tape run through it prior to final cleaning and sealing at the factory. That became a major clue because any audiovisual pro old enough to have handled tape understands what happens to it during use. As tape is physically pulled across a bank of read and write heads, under tension, little particles are scraped off and build up on the heads over time. *Galileo*'s own tape had ingloriously stuck to one of these debris blobs. During *rewind*. Running it *forward* had freed the works. Eventually the blob would shear from the head, and the sticking problem would basically fix itself. But just to be safe, the recorder would be strategically run in the forward direction for a short distance, every time, before any rewinding.

A series of carefully incremented actions proved that *Galileo*'s onboard unit still reliably recorded and played back. And could be utilized after all. Since one length near the head of the tape had been continuously burnished for some fifteen hours, and maybe weakened because of that, it was wound onto the supply-side spool and never used again. With twenty-five turns atop just to be safe.

For project manager Bill O'Neil, the most painful aspect of this entire episode was his abruptly dropping the hammer on one particularly anticipated event—super-close-up Europa and Io photos as the orbiter raced by. "I had to say there was no way we could do the approach imaging," he affirmed. Jarring starts and stops and high-speed operation would overstress the tape recorder. Good grief—no future orbits would bring the ship anywhere near as close. "It was the worst decision that I ever had to make." When his decree went forth, it was not subject to negotiation.

"There was no point in debate as far as I was concerned."

Naturally, that failed to go over well. "The scientists were up in arms," recalled one member of the science team. O'Neil's decision had unstapled their nerves. "The only place in the solar system with active volcanism. The only pass we have. And we're not going to take it?" Closely flying by Io, in particular, had been anticipated for years on end, with *Galileo*'s approach deliberately shaped for the proximal encounter. It was bad enough how those twenty-five extra tape winds contained additional images from Jupiter approach that would never be seen. The risk of breakage wasn't worth trying to retrieve.

"You realize how much of your life has been invested in this experience. And through some accident it's gone in a puff of smoke. Or because of a tape recorder."

Away from Europa and Io soared their wounded machine. Already it'd made history and not just with the two asteroid flybys. Remarkably had *Galileo* directly imaged fragments of a shattered comet whomp-whomping into

Jupiter. This occurred in July 1994, on the planet's "away" side, meaning Earth-bound telescopes could witness the aftermath but not the impacts themselves. By sheer chance, though, the orbiter happened to be tracking through a prime vantage point. Thousands of astronomers had followed the events and clamored for pictures and rejoiced in the unprecedented spectacle of massive fireballs and shock rings blasting across the planet's surface like artillery. And they got to see it only because of *Galileo*.

"In all of history, no one had ever confirmed seeing an impact on another planet," offered astronomer David Levy, who codiscovered the comet.

By the time of Jupiter arrival, the ship had ingested nearly half its permitted amount of radiation. *Galileo*'s initial, grandly looping, seven-month orbit would skirt the greatest intensity of the radiation belts and reach a max distance of twelve million miles from the host planet. Going even that far out wouldn't take the ship completely beyond the largest single structure in our solar system—Jupiter's magnetic field. It's one more set of statistics making this our planet of superlatives: largest diameter, fastest rotation, most energetic radiation belts, most moons, largest and strongest magnetic field. Regarding the latter, it occurs because a gigantic internal repository of metallic hydrogen is churned along by swift rotation—once completely around every ten hours. This generates a vast cocoon of magnetic energy that surrounds the planet and is called the magnetosphere. Earth has one, too, but it's twenty thousand times weaker.

Now, if that's not crazy enough, here's where things really cross the center line. Io is close enough to Jupiter to be caught in a never-ending gravity battle. Such force is bad enough as is. But just past Io is Europa, pulling on its sibling from the opposite direction. Perpetual tugs of war bulge Io's middle up to three hundred feet. All the friction makes things hot. Volcanoes continually well up like adolescent acne, with one to two tons of sulfur and oxygen being stripped away from Io every second. The debris swells Jupiter's already-huge magnetosphere to even larger dimensions. It completely encircles the planet's radiation belts, plus the orbits of all four Galilean moons, and trapped particles can actually form little atmospheres around those moons. If this feature was visible to us standing on Earth, it'd be as much as three times the size of the sun. "An unprecedented physical laboratory" is how some researchers have described Jupiter's magnetosphere, helping to explain their fascination.

Galileo's lazy, introductory orbit further afforded time to regurgitate every scrap of probe data off the tape recorder—data leading to new scientific puzzles. The biggest so far seemed to involve water. Why hadn't the probe encountered as much as the *Voyager* flybys indicated was present? Were our major predictions about Jupiter fundamentally incorrect? Or was something else maybe in play? One heavily embraced theory is that Jupiter, like Earth, exhibits all manner of atmospheric variations. And the probe just happened to plunge into more of a dry section. This idea is supported by multiple studies from the ground (using lots of -ometers) and, if anything, makes the argument that many more such probes should be dropped into the atmosphere of our most superlative-laden planet.

The entirety of Galileo's prime mission, having been in development for over twenty years by this point, would artfully loop the ship through Jupiter's complex system for at least ten unique orbits. It would do this using the gravity of the various bodies to adjust course, suckling only as needed from the precious propellant tanks. Altogether the party was meant to last nearly two years. "After each flyby," noted Phil Roberts, "the spacecraft's new orbit has to be measured by tracking from Earth. And then any errors in the path corrected with a small fuel burn."

Galileo began these closer-up, follow-on orbits, and new rounds of imaging ensued. These days, people take throwaway photos on brand-new smartphones at resolutions of thousands upon thousands of pixels. But the *Galileo* orbiter began discovering never-before-seen Jovian moons with a 1970s-vintage camera, shooting 800-by-800 pixels; such tech *still* signified a wondrous improvement over Voyager imaging. Upcoming picture sessions were anticipated to resolve surface features as small as 165 feet across. And study how the various surfaces changed over time. That might help answer some lingering questions. See, the aged Callisto, marble-like Europa, and cratered Ganymede had all shown, during Voyager, evidence of ice on their surfaces. With *Galileo*, higher-resolution images could be examined in context with the mechanics of each spacecraft flyby. The specifics of the gravitational deflections, measured at each moon, would provide clues to the distribution of mass within that moon. Pair that with surface photos to reinforce—or refute—the idea of a liquid salt-water ocean hiding underneath the surface of a moon. Maybe multiple ones!

Upcoming flybys of three high-priority moons would be at ranges of a hundred to a thousand times closer than Voyager. In general, the Galileo mission was expected to blow Voyager clean out of the water for uncovering Jupiter's secrets. "Just the fact of being in orbit was a huge, huge advantage," asserted Don Gurnett, who'd faced his own long road to this observational phase. Finally he'd be able to settle in and begin reaping the rewards. Gurnett had prepared, in part, by arming himself with paper. "For every one of the flybys, I got a notebook that's got our data in it," he explained. His collection would grow to exceed fifty notebooks.

Gurnett humorously categorized Iowa's Galileo instrument as "our most advanced design *that we were willing to take a chance on actually building*." And with much excitement, he launched into details, which can be formidable when the topic is plasma waves. Luckily, Gurnett—a valuable mix of scientist, engineer, and college professor—was adept at communicating in everyday terms. He likened Iowa's experiment to an old-school radio dial, where different stations are tuned-in by rotating a knob. With Voyager, Iowa's onboard hardware monitored only sixteen "stations," or frequencies. But for Galileo, the number had jumped to the hundreds. Tuning across them, in slow fashion, more broadly measures the spectrum of plasma wave concentrations. Resultant findings would be represented as *spectrograms*, visually depicting frequency versus time, and provide great insight into the enigmatic physics of the Jovian system.

Wouldn't it be great if those findings actually came back! Despite safe beginnings to the orbital tour, enough major hardware failures on this ship had already curtailed its science returns by approximately 30 percent. Now, compared to imagery, Iowa's spectrograms didn't require anywhere near as much data to represent. But neither were they featherlight. "We kind of realized we had a huge problem," admitted Don Gurnett. Debilitating limitations existed due to the broken high-gain antenna. Not to mention the handicapped tape recorder. As is always the case on even completely operational space missions, each bit and byte was practically a Wall Street negotiation.

The revised expectation constituted a real clogging of the arteries. "We had a factor of 1,000 decrease in the amount of information we could send back from *Galileo*." That was Bill Kurth talking. Despite the hassle of data return, he was still very much enjoying a research career at Iowa—and his collaborations with Gurnett on the mysteries of plasma waves. Amusingly, Kurth encapsulated his life's work within a potentially denigrating phrase. He called it wiggly-line science. "Guys get all excited about these wiggly lines. But they don't mean anything to the guy in the street! And, um, I think that's a real problem."

Sometimes he and his fellow plasma wave researchers try to explain it to people. Emphasis on *try*. "Waves? They're kind of hard to talk about. You can't see 'em, you can't touch 'em, you can't feel 'em. I mean, they're—they're just kind of a very abstract, uh, thing to talk about. And why would anybody want to study those things?" People who *do* study them acknowledged the struggle long ago. Kurth credited Fred Scarf for converting a few plasma wave readings into sound and playing them at Voyager press conferences. Genius move. Continued Kurth, "Probably didn't compete with images of Io. Or, you know, rings of Saturn, or something like that. But you turn something into sound and people can listen to it," he argued. "They can appreciate a little bit more than showing a plot or a graph."

Regarding Galileo, Kurth had lost no sleep over the functionality of their investigation. "I don't think there were any perceived risks for our instrument," he commented of the extra time it spent in storage following *Challenger*'s demise. Its core hardware was a small box of electronics nestled within the spun half and fed by antennas way out on the end of an extendable boom. "Any mechanical device is subject to issues, but that popped out like it was supposed to." Really, Kurth's greatest worry centered on two accessory antennas provided by French researchers. All those attempts to free the stuck high-gain antenna had resulted in their being exposed to incredibly low temperatures, which were never accounted for in the design. Later on during the mission, one French antenna would fail.

"And it may or may not," jabbed Kurth, "have had to do with the fact that they froze us to death!"

Obviously, he and Don Gurnett, having finally reached the promised land, wanted to maximize their returns in every way possible . . . despite this new paradigm of low-rate transmission. It demanded that everyone subsist on starvation rations. Discarding every last digital bit that could be omitted or interpolated around. Iowa could partially do that, at a minimal level, because sometimes they'd be recording vast tracts of nothing. "A lot of places are just noise," Gurnett suggested of the greater Jovian environment. Space is, after all, appropriately named.

Just ignoring the dead spots wouldn't amount to near enough of a reduction though. Nay, maximizing their harvest would surely hinge on data compression, which requires computer processing. "Now, as soon as I thought of that, I ran into a problem. I don't have a computer in my instrument," Gurnett conceded. The innards of Iowa's little metal box had been inflexibly hardwired, with no way to reprogram them. The device did one thing only. "And that is a result of my being still conservative. And I—I admit, I should have been, maybe, more pushing the edge of the art" when it came to the original instrument's design. How frustrating—he needed onboard resources he didn't have.

But wait a second! The camera system had them. Its own dedicated computer was already configured to compress data. (Albeit *image* data only.) Well, what else might work? Gurnett and Kurth talked through every component. The spacecraft itself possessed two main computers governing power, navigation, and communications. Despite high mission workloads, they occasionally went idle. Excellent! To boot, JPL engineers had farsightedly provisioned the ship with what amounted to a computer network. *Galileo* couldn't use Netscape Navigator to call up GeoCities. But it *could*, for the first time on any JPL spacecraft, move data around between the different subsystems.

Gurnett lit up. "Suddenly I realized that if we could—if we could get our frequency-time spectrogram sent to the imager people *as a picture*, they could process it to reduce redundancy by a big factor. And reduce our bit rate." One of the main spacecraft computers could totally pitch in to help. They could do this. They had time. Reprogram, send up changes, and get what they needed despite all the heartache and setbacks.

"You can redesign the whole darn thing in ways you never even thought of," rejoiced a triumphant Don Gurnett. "And that saved us!"

Curiously, the process of recounting this story also triggered in Gurnett a relevant pet peeve. "NASA tries to divide the world between physicists, scientists, and engineers," he accused (in a relaxed yet direct way). "They'll say, 'Okay, *that's* science, and that's to be run by the scientist.' The engineers over here are supposed to just do what the scientists say." Gurnett categorized himself as one who straddles these disparate worlds. "I can improve things sometimes by realizing what the *engineer* didn't think of, and the *scientist* didn't think of." Splice those realizations together into brand-new solutions.

Wasting no time, JPL software coders launched into what Bill Kurth termed a "superhuman effort of revisions." Galileo designers had, perhaps surprisingly, based its two computers on a more or less off-the-shelf processing chip from RCA called the 1802. In the words of one participant, this chip was "nobody's favorite choice." It ran slowly compared to other chips and was prone to damage from static discharge. But the 1802 boasted reliability and low power draw.

Its clock speed could be modulated—actually set to zero if need be—with no adverse effect on operation. Considering the era of project approval, this chip made sense. If you, in the late 1970s, operated a Montgomery Ward CyberVision 2001 Home Computer ("They're not cassette tapes; they're *Cybersettes!*") or perhaps RCA's dog-ugly Studio II video game system, then you used the same base chip that ran a billion-dollar spacecraft. JPL engineers had arranged six in two redundant "strings" with four devoted to *Galileo's* spun bus and two for the despun section. In software this was abstracted into six virtual computers all collaborating harmoniously to run the show. Three of them were considered "privileged," meaning they remained constantly active and would never change: one handling administrative tasks, a second devoted to fault protection, and the third for contingency procedures. When one needed to process a task, it did so immediately—regardless of what the rest of the ship might be up to. During high activity levels, both strings could operate simultaneously. If demanded by a specific task, both strings could also work together as one super string. If some piece failed, others could absorb its responsibilities. That made the design more reliable. And to Iowa's supreme benefit, flexible beyond its original intent.

As Bill Kurth explained, Iowa's revised strategy would prioritize data near the moons—with collection rates dialed up as high as permissible. But while transiting them, "we went down to an extremely low rate, and employed some very extreme data compression schemes in between." Not ideal but workable. They could live with it. Kurth likened the difference to creating a detailed street atlas versus a globe. "So it was kind of that level of science that we could do," he said of the lower-resolution readings. "You can draw trends as a function of large pieces of real estate as to what was going on." For example, "there's interesting things going on close to the planet, and close to the equator." Still very good science! Any detailed interactions of, say, particles *between* moons would be lost to the antenna failure.

Regardless, Kurth figured all the excitement was at the moons anyway. "Each of these moons is a world of itself," he grinned. "They all have their own story. None of them are boring."

Jupiter's neighborhood of moons could be structured into three distinct groups. Farthest from the host planet were the irregulars—miniature worlds tracing high, tilty orbits in a backward motion from Jupiter's. Possibly captured asteroids. A second group, closer in, likely included even more captured asteroids but orbiting in the same direction as all our planets. And then the pack closest to Jupiter included the aged Galilean moons first seen all those hundreds of years ago. Plus the tiny world of Amalthea. It signified the last Jovian moon to be discovered by eye—that is, by actual direct observation—as opposed to studying photographs or inferring its presence from gravitational wobbling. Mr. Galileo Galilei is to be excused for missing Amalthea because seeing it requires a twenty-four-inch telescope with at least 250-times magnification, during ideal nighttime conditions, and that Amalthea be farthest from Jupiter (based on an Earthly perspective). It's tough to see. Even if Galileo had located Amalthea, he would've had no way of knowing that one day this lopsided chunk of red will finally yield to gravity and spiral into Jupiter and be absorbed just as the probe would.

Back in JPL's TV studio, the well-dressed host of the Galileo broadcast turned to his current panel of guests. "Any final, closing remarks?"

Ed Stone had a few. A physicist by training, Stone had been a researcher at Caltech for decades already and served as a coinvestigator on Voyager. Although promoted to JPL's directorship in 1991, replacing Lew Allen, Stone had retained his position as Voyager's lead project scientist. If anyone deserved air time for commenting on a mission of planetary exploration, it was Ed Stone. Clad in a suit and tie of muted tones, he occupied the far-right seat of the TV studio desk. Stone maintained a lean physique and spoke from behind sensible eyeglasses that had seen a lot. "Well, I think *this*, as I said, is really what it's all about. I mean this is, really, what the space program brings," he evangelized to the host and camera. "It brings the opportunity to see things nobody's seen before. To learn *new* things. And to understand how the solar system, how the planets, how it all *works*." His professorial and fatherly tone of voice conveyed the simple amazement of it all.

"This is the beginning of a wonderful, two-year journey."

The culture of planetary encounters had sociologically changed. Back in Voyager times, encounters were few and far between, and most every scrap of information flowed directly from the earth's receiving dishes into JPL. There was no other place to get it. At least, not right away, which is when everybody wants it. So Pasadena hotels all filled to the brim, and the JPL campus turned into a convention center, with scientists and news crews arriving from everywhere in the world for a focused period of roughly a week at a time. The campus buzzed with activity, and coffee flowed nonstop. Experimenters such as Don Gurnett and Bill Kurth and Fred Scarf would hole up with other science teams in Building 264.

Press releases and news conferences happened every single day. People were so busy they sometimes slept on desks. Every so often a JPL runner would blitz through the science areas with a stack of oversize printed pages and hand them out like the morning newspaper. In the case of the plasma wave instrument, each delivery represented about half an hour's worth of readings. To get a sense of trends and the general big picture, everyone would tape their printouts together one after the other, using any open section of wall they could find. As Bill Kurth described it, "One or two of the floors of Building 264 were absolutely wallpapered with plots that came off of thermal printers." Experimenters would wander up and down, appraising the readings as they would an art exhibit—occasionally stopping to tap a finger on a page and say, "Well, I wonder what happened here?" Different teams would compare their plots to each other, judging whether any specific events lined up chronologically. As Kurth continued, "And of course, all the plots had different time scales. And so they were extremely difficult to handle!"

Most every instrument team had the ability to generate much more useful visuals back at their university or other home base. But that meant bouncing data onto reel-to-reel computer tapes and hustling out to the Burbank Airport to put the tape on a plane. In Iowa's case, someone from the Physics Department would then have to hightail it half an hour up I-380 to the Cedar Rapids airport and collect that tape. Drive another half hour back to Van Allen Hall. Find parking. Process the data, plot onto paper some custom way, then reverse the whole process to send the new plot to JPL.

"As you might imagine, you might do a plot and someone would decide, 'Well, you know, it would be really better if we could plot it *this* way instead of *that* way.' So then, somebody would have to do another plot," laughed Kurth. New rounds of printing and airport runs would ensue. Airplanes were literally flying pieces of paper cross-country. "So it was a very slow process." Fortunately that began dying out over the latter half of 1989 with *Voyager 2*'s Neptune encounter. According to Kurth, an officially unofficial policy allowed him and other experimenters to lug their own computers to JPL and connect them to the Lab's previously closed data network. "A much more efficient process!" he celebrated. It didn't end all the round-tripping of tapes and paper, but surely reduced it. From there, the use of dedicated phone lines emerged, and everyone was able to begin staying at their respective offices for the very first time in planetary encounters.

On Galileo, Iowa and the other teams went hybrid. Any event-heavy encounters continued to remain "a focus of the whole project," clarified Kurth, and beckoned people to JPL. "Everybody would go and be there at the same time. We did that for Galileo to a certain extent." Lab rules now officially *officially* permitted guest computers; Kurth and the others all brought their own, "so we could do our own processing." But visiting Pasadena every single time no longer made sense—if for no other reason than the heavy use of *Galileo*'s tape recorder often slowed inbound feeds to a sludgy trickle. "It literally took weeks in some cases for the data to come back," Kurth pointed out. Increasing reliance on the internet further reduced the value of convention-style encounters. Altogether this negated any credible reason to loiter at JPL (or in a Pasadena hotel room), waiting for something you could now have delivered back home.

When asked about the division of labor between Don Gurnett and himself, Bill Kurth explained that Iowa's incoming raw data was barely one step up from an unintelligible stream of gibberish. It had to be organized and consolidated and plotted onto graphs. It's been that way since Voyager started, and Gurnett always focused on the readings themselves versus how they came in. "He's not conversant with the software and so on," explained Kurth.

So when Voyager results first began splashing down into the receiving antennas on Earth, Gurnett had approached Kurth and said, "I expect you to do the plots and leave 'em on my desk every morning."

Decades later, not much had changed. "I've done that for the entire mission!" Kurth hooted. "That's just the way we work!" He was quick to respectfully note that Gurnett rightfully *should've* been sidestepping the mechanics of data processing, because that's not where his strengths are. "Don Gurnett is a much better physicist than I am." Kurth added that he still gets his personal moment out of it because the results come first to him. He is *the first* to make new observations about our solar system, and that means a lot.

"You're seeing measurements that absolutely nobody in the world has ever seen before."

Software changes for Iowa's data compression were uploaded to *Galileo* in March 1996. That May, the spacecraft made a close approach to Ganymede, then another in September. The largest of our solar system's moons, Ganymede may host a paper-thin atmosphere of oxygen and conditions suitable for the development of primitive life. Its saltwater ocean—unconfirmed yet likely—and rocky sea bottom constitute the major reasons people are thinking in that direction.

Callisto got its turn in the spotlight that November, with Europa up just over a month later. Readings from the ship's magnetometers suggested that Europa, underneath a miles-thick crust of ice, was almost certainly hiding its own ocean of *warm* saltwater. But follow-on readings would have to wait; *Galileo*'s next orbit included no close moon encounters at all. An impending solar conjunction, with the sun temporarily between Earth and Jupiter, would interrupt spacecraft observations and communication.

A similar interruption had also occurred in the past—specifically at the end of May 1610, when a tired-eyed Galileo Galilei suspended his own Jupiter observations due to the planet's being obscured by the rays of the sun. He used some of the resulting down time to envision a revised, updated edition of his landmark book—intending to wait until Jupiter reappeared so he could append a prolonged chronicle of observations. Come the end of that July, the planet began its reemergence in the morning sky.

But Galileo had already moved on, and no future edition of *Sidereus Nuncius* would ever reach the press. Instead, that same month of July 1610, he turned his attention to Saturn and logged yet another profound discovery. He first communicated it in a letter to the government of Tuscany.

> The star of Saturn is not a single one, but an arrangement of three that almost touch each other and never move or change with respect to each other; and they are placed on a line along the zodiac, the one in the middle being about three times larger than the other two on the sides; and they are situated in this form oOo.

How very pleased Mr. Galilei would no doubt have been to learn that in the very near future, an incredible machine—perhaps unfathomable to him at the time—would be leaving Earth to visit this "star of Saturn" and its intriguing arrangement of side stars.

23

Separation Anxiety

You're going to have a sense of loss. You need to be prepared for that.

—Cassini spacecraft manager Tom Gavin advising project workers of
what they'd be feeling after launch

Lynn Casani, according to her husband, could name any flower in the city
of Pasadena—and how to best take care of it. She excelled at taking care of
things. Including surprises. "There would be some out-of-town visitors, some
people from NASA Headquarters, or from one of the contractors that we were
workin' with." That's how John Casani began describing a repeat scenario.
And when it happened, he'd sometimes call the house as late as three o'clock
that same afternoon.

"Hey, Lynn, I got six or seven people that I'm bringin' over." Be it for cock-
tails or perhaps even a full dinner. "And that would be like with a two-hour or
a three-hour notice. And we got there; everything was ready. She *never* complained about that. She *loved* that. She *loved* entertaining, and meeting the
people that I worked with, and getting to know them."

Casani paused in his recollection. Lynn's been gone a long time already,
but her impact has never waned.

"She was very, *very*, I think, instrumental in the success that I had because
of that." He tried to put a finger on why such last-minute events were never
an issue. "My wife loved people. She was a people person. And I was a *thing*
person, you know?" Once, while at a neighborhood party, John watched as
Lynn bantered with someone whom John figured he was supposed to recognize. But didn't. The guy eventually sauntered off, and John asked, "Hey,
Lynn, who was that guy?"

"For Christ's sake, John, you've met him a dozen times!" she chided. "Why
the heck can't you remember his name?"

Wordlessly they gazed at one another for a few beats. And then, according
to John, "I could see sort of an epiphany came over her face."

Suggested Lynn, "Why don't you just think of him as a spacecraft component?" Because some days that seemed to be about the only damn thing he could remember!

Casani twinkled at the embarrassing memory. "And that's when I realized, you know, I thought in terms of *things*. I reacted to *things* and whether *things* worked well."

Of course, Lynn herself sometimes had visitors over and occasionally for much longer periods. Returning from work one day, John encountered his wife, who informed him, "I brought somebody home that I want you to meet."

That turned out to be not exactly the whole truth. "Well, she brought home a *family*." Specifically, a mother and three kids Lynn had noticed in the parking lot of the nearby Vons grocery store. Frantically was the mother trying to escape an abusive husband. Despite multiple boys filling the Casani house already, an extra bedroom remained—with the grateful family bunking in for several weeks while Lynn worked to relocate them with outstate relatives.

Indeed, for Lynn Casani, "taking care of other people was number one," according to John.

Once, Lynn brought somebody home for good. It started this way: "Mrs. Casani? We got a situation here and want to know if you could help."

On the phone was someone from the Loyola High School counseling office. Lynn had worked there long enough to befriend everybody in her radius as she typically did—including many students—but had since departed. What lingered at Loyola, however, was her reputation for assisting anyone in need.

Here, the need involved a teenage boy, Andrew—a junior at the school. He'd been living with a guest family—some friends of his deceased mother—but things weren't going well. Injecting Andrew into an existing family (with five daughters!) had upended the chemistry of the household to the point where nobody could take it anymore. "There was just too much friction," is how John later characterized the situation. Andrew's host family had just notified Loyola of the boy's pending eviction. He was going to be relocated to an apartment on his own. After learning the whole backstory about the friction, the school called Lynn.

Strictly speaking, Loyola's request went only as far as putting Andrew up for a while until longer-term arrangements could be made. But that night Lynn called a family meeting with John and all the boys. It began with her explaining the situation and pointing out how their house offered plenty of guest space. For any duration.

"But we're not gonna take him in for a year or two," she warned, "and then turn him loose or something. If we take this kid in, he's gonna become part of our family."

Lynn faced each boy in turn. "That means he's going to be *your* brother. He's going to be your brother *for life*." She turned to her husband. "This is gonna be *your* boy. *Your* son," Lynn cautioned. Emphatically. "Now. Do we still want to take him in?" A unanimous yes was the answer. And just like that in January 1996, the Casani family of South Orange Grove Boulevard in Pasadena, California, went from four boys to five in the loving blink of an eye.

Structural elements for the *Cassini* orbiter had arrived at JPL's Spacecraft Assembly Facility on November 20, 1995. A few days later, the main bus cable harness was ready for installation, courtesy of Clyde King and his team. King also oversaw responsibility for producing the diagrams used to actually build the cables. He was coming up fast on retirement, though, and deep in the process of mentoring colleague Mary Reaves on the specifics of taking over for him.

"There wasn't another job like this anyplace in the world," gloated Reaves of her years at JPL, doing what she did. "I used to tell the girls that worked with me that the only way for God to know what it felt like to tie a spacecraft harness was through our hands."

Mary Reaves is an uplifting example of where hard work and determination can take a person. Despite lacking even a high school diploma, she became a skilled aerospace technician at one of the most high-tech and reputable facilities that exists. The first woman to do this job at JPL. It happened because, in early 1975, her brother suggested it. He already worked there and reckoned she might fit in.

One of its nooks and crannies just *had* to be better than anyplace else she'd worked. There'd been a stint cleaning houses. Waitressing. Even go-go dancing. "By 1975 I had been married, divorced, and had three kids," she listed. Building cables for space machines represented an altogether different vocation. Her interview took place inside the Lab's cabling shop, Building 103, and according to Reaves she got the job because the interviewer appreciated a couple things. Not only was Reaves able to read and follow instructions, but also when it came to working on spacecraft, her compact stature made for a literal good fit. The offer of $6.52 an hour sealed her interest. But Reaves said her interest really piqued the moment she entered a work bay. "There in the middle of the shop was the mockup of the bus for Voyager. I took one look and knew what I was going to do for the rest of my life."

She earned her stripes on that machine, inhaling every last detail about wire types, gauges, connectors, soldering, potting, bundling, documenting. Individual connectors could have dozens of pins linking dozens of locations. "If you want 24 return wires going to the same contact, you have to know how to make that happen. And how to *show it happening* on a piece of paper that the technician building a harness can see."

She had to learn how to use a computer. Her mentors were Clyde King and the other old-timers from Building 18, and they had a launch deadline to contend with. "We worked through the night on problems with routing, and new wire or connectors, or potting materials, or how to hook up," she remembered. "Clyde was very careful to call out anything that he saw that he didn't like, or was suspicious of."

The finished wiring, taken as a whole, always resembled a product of nature. "By the time you got it all tested and mated, it looked like a brain. Now think, if only one wire went to the wrong place in your brain? What kind of a problem it could cause!"

For the *Voyager* launches, Reaves drove herself cross-country to the Cape. "Did you know there were not *any* bathrooms for women in any launch facility?" she claimed of the time. "The guys used to say they would watch the doors for me, but they didn't." Her time at the Cape lasted six months. Later, Reaves would learn that the original plan was to dismiss her after both *Voyagers* went. But people from the launch team showered her boss with compliments one after another. She was so great to work with. Never tired. Produced the neatest and cleanest bundles you ever saw. The hits kept coming—to the point where the logical response was to keep Reaves on board. Moving up the ranks. "I was getting a name of my own, and some respect." And now here she was, a seasoned team lead for *Cassini's* flight cabling, empowered to designate her own second in command. A professional. And intent on following their creation to the Cape to see it off safely.

"You can't have a bad day, or get sloppy with your work, because when the spacecraft launches, you can't fix a mistake!"

By April 1996 *Cassini's* subsystems and a few science instruments had been integrated. The remaining scientific hardware—instruments that'd been selected at the beginning of the decade—went on in early August. The University of Iowa's plasma wave gear was mounted in four distinct places on the spacecraft. Altogether it incorporated six antennas (with three coming from France), a Swedish plasma probe, a central data processor, and the ability to record hundreds of channels' worth of signals at the near-Saturnian environment. Presciently, it also included dedicated hardware for data compression!

Don Gurnett couldn't have been more proud. "My gosh, just a tremendous instrument, you know, compared to *Galileo*."

The trajectory for *Cassini*'s impending jaunt contained even more intricacies than VEEGA. Launching on a traditional, expendable booster (with a Centaur upper stage!), the vessel would trace something of a scenic route—going Venus–Venus–Earth–Jupiter–Saturn—taking nearly seven circuitous years to reach its destination. This added a billion miles to the flight path. But the collective gravity assist from all those intermediate planets equaled seventy-five tons of free rocket fuel, which justified the approach.

News articles of the period loved to play up *Cassini*'s impressive size and capabilities. How it represented the most sophisticated planetary exploration machine ever created. How audacious the outbound flight would be. How it better not hit Earth during the swing-by, or bad things would happen. How the amount of time for radio signals to zip from Earth to Saturn would take longer than an hour. How the *Huygens* lander would be dropping into an unknown abyss and might never actually land on Titan at all, depending. How the family-size amount of refined plutonium aboard the giant orbiter, there to generate electrical power, had environmental groups wailing in opposition. How lawmakers were getting badgered to intervene, with protests intending to "stop the launch." All of this and more commanded headlines.

One detail, however, escaped popular reporting of the time—exactly how *Huygens* would separate from its mother ship in advance of the Titan plunge. For all anyone could glean from magazine articles, you'd untie a length of string, and away it'd go. Whatever the actual mechanism, it had to support the weight of *Huygens* while securely mounted to *Cassini*. It had to survive the trauma of launch without disengaging or otherwise activating. Once in space, its components would have to wait patiently, in a state of extreme reliability, throughout seven soaking years of crazy temperatures. We're talking a range

37. (top) A trailblazer in her field who helped shatter gender barriers, cabling tech Mary Reaves reaches up to complete a final adjustment on *Cassini* prelaunch. "A degree didn't matter. How hard you were willing to work—and being there daily or nightly or both—was what mattered." Courtesy NASA/JPL-Caltech.

38. Don Gurnett poses with elements of the flight spare *Cassini* plasma wave instrument in 2006. To his side is the storage and deployment system for the three electric antennas. Flattened and rolled for launch like party-favor noisemakers, each antenna would unfurl in space to over thirty feet apiece, aligned like the corner of a cube. Gurnett holds a length of antenna, which contains small holes to reduce the chances of warping in space from temperature swings. Resting on the table is the instrument's central data processor. Courtesy the University of Iowa Department of Physics and Astronomy.

of minus 148 degrees Fahrenheit, at the low end of the scale, up to 482 degrees at the high. Periodically, *Huygens* would need to be awakened for electronic health checks, like a nurse coming through in the night. Approaching Titan, it'd have to separate at a specific velocity. And angle. And spin rate. At a precisely commanded time. It would have to do this cleanly, flawlessly, with no chance of bumping the mother ship. If something didn't work perfectly—if the moving parts hung up even slightly, if the ship popped away off-kilter—Earth could lose the only funded opportunity to visit the surface of Titan.

Being an engineering-type situation, this procedure called for a number. A standard for guiding the engineers. A stupendously high one. In accordance, the probability of successful probe separation was specified by mission rules to be at least 0.996 out of 1.0. The *Huygens* prime contractor, Aérospatiale of France, naturally reasoned that such a precise and demanding value should be produced in a region of the world accustomed to delivering stratospheric precision. The job went to Switzerland.

At the outskirts of Zürich near the Seebach city border, just south of the A1 motorway and a tennis club, the matter of the *Huygens* separation mechanisms came into the hands of Udo Herlach and his brainy associates at the modest engineering firm Contraves. The company had built various equipment and instruments for scientific and commercial satellites. As well as fabricated payload fairings for the French Ariane launch rocket. Now gifted unto the team was a fifty-pound weight budget for solving two problems at Titan: First, safely liberate *Huygens* from *Cassini*. Then, once inside Titan's atmosphere, cleanly discard the lander's protective front heat shield and back cover.

Anyone entering Contraves's campus had to badge in past a secured perimeter. Ahead lay a couple of small-scale office buildings, walkways, and parking. Other largely nondescript structures housed the exciting bits: clean rooms for assembly, a vacuum chamber, and machining facilities offering the latest gear available. Altogether they composed sort of the ultimate workshop. Pretty much anything could be made at this place with submillimeter accuracy.

Nearly five hundred people worked there. And of the eight to ten people assigned to *Huygens*'s separation, one of them, mechanical engineer Platon Tatalias, would occasionally get teased by his colleagues for being "too Swiss," as they put it. Even though Tatalias was actually Greek, had earned his master's in London, and moved to Switzerland only because few engineering jobs were to be found in Greece at that time. Besides, his wife was Swiss, so residency came easily. He tried to explain the joke of being too Swiss. "In the late seventies, when I came to Zürich, the mentality in general was *conservative*," Tatalias began. "As a young graduate from more open London, in combination to Mediterranean *easy way*, I had to learn the *more restrictive* way."

A pivotal happening early in his career helped crystallize this "more restrictive" mindset. Only one week into Tatalias's time at Contraves, a senior manager requested his opinion on a design.

Tatalias wisecracked, "I hope it will work."

The manager exploded. "We send a 200 million satellite. And no mechanic can go after it for any repairs. And *you say you* HOPE!"

Tatalias could only gulp. "That lesson stayed with me my whole working life." Quickly he'd buttoned himself down more firmly. Checking every last detail in any set of plans. Calling for more testing when appropriate. Reminding others that the basic laws of physics were always right. It was to the point where native Swiss colleagues thought he was almost *too* buttoned-down—or too Swiss, as they put it.

Right at the start of the Huygens job, Tatalias was still in the midst of wrapping up other work on a gadgety thing for deploying a satellite. He moved over after the Huygens contract was signed and already underway. In group meetings, everyone freely vacillated between English and the Swiss-German dialect, and to keep with the flow, you had to know both. (All official reports were written in English exclusively.) A preliminary design from ESA and Aérospatiale did exist. But preliminary nonetheless, and nowhere near ready to build. Contraves's prime deliverable was called the Spin Eject Device. Front and center sat that hard limit on weight—which would become something of a flash point—along with the basic geometry of *Huygens*, its operating conditions, separation parameters, and trajectory needs. Also a specified budget for power. Although the core team remained small, Tatalias stressed that its members had the resources of the entire company behind them, as well as specialized third-party firms also laboring on the greater Huygens effort.

From the meetings and reports and informal chitchat, a workable approach began to coalesce. Spacecraft components, like gas-powered car engines, largely bolt together. A bolt is about the simplest fastener there is: easy to configure, readily made, and perfect for repeated assembly/disassembly (unlike, say, rivets). The properties of metal bolts are well understood and consistent. It's easy to predict how metallic alloys will age or fatigue. When they will break. Understanding all that, the act of physically separating the Titan lander from its mother ship—and then from its shielding—would, conceptually, be as easy as freeing every bolt holding them together. And then giving the right amount of push!

Ideas flowed throughout the campus, with an invisible yet powerful undercurrent of energy and momentum driving progress. As Platon Tatalias made his way to meetings or presentations, the walls greeted him with framed pictures of previous space missions where Contraves had played a role. He could pause at any one of the multiple snack bars located on campus that were run by an outside company and stocked with all sorts of coffee and nibblies. Lunchtime usually meant visiting the sole cafeteria.

The team reviewed some known structural details. A simple lattice of tubular struts would extend from one side of *Cassini* like miniature scaffolding. And converge on three equidistant hardpoints. Then *Huygens*, shaped more or less like a truck wheel, would fit its flat side against them—one hardpoint every 120 degrees. So three bolts represented the minimal number of links between mother ship and lander. Each bolt would interlock with a common aerospace industry fastener known as a pyrotechnic nut. They're not nuts in the traditional sense of threaded, hex-shaped bits from the hardware store; instead, regard them as a locking receptacle that the end of the bolt fits into.

Next, engineers broke down the separation into its discrete actions, going step by step. To release these bolts at the magic moment, *Cassini* would send an electrical signal to a specialized device mounted at one end of the pyro nut. This "NASA standard initiator" is about the size of a C battery and contains a miniature cup of explosive material with a single wire running through it. The electrical signal heats the wire, igniting a small wad of zirconium fuel that goes *boom*. Combusted fuel turns into a hot, pressurized gas that races through the far end of the initiator with enough force to separate the locking mechanism of the pyro nut.

With the divorce having been finalized, the next action for *Huygens* would be fleeing the scene. Physically departing the side of *Cassini*. No question: Do that with springs. Mechanically simple and requiring no electricity, springs could sit compressed between the two machines and push *Huygens* away once the bolts freed. The Contraves team, after calculating the amount of force required for this motion, could thus specify what to make the springs out of, how compressed they'd need to be, and so on. A complicating detail here was how seven years of tight compression would age the springs and change their expansion properties.

Huygens would then be in the process of leaving its ride. Over the three weeks between departure and Titan entry, the unpowered lander would need to dependably coast in a specific direction. In space that's done by imparting a spin motion to the ship. It wouldn't take much—about seven and a half rotations per minute. Straightforwardly achievable by mounting the lander on short lengths of angled track to set it going. Similarly angle the springs and, voilà, you got yourself some spin.

Finally, with the departure barely underway, any electrical contacts between the two machines would need to be separated as well. It called for male-female connectors that are press-fit together while in use and pull apart under a specified amount of force. But once the connectors were apart, their live contacts would be exposed to risk. Draftspersons at Contraves sketched in aluminum flaps, under tension, to flip into place and cover each end of the now-unmated connectors—thus protecting against debris or similar risks of accidentally shorting out the contacts.

The entirety of these separation actions would occur in less than half a second and be described in ESA literature as "a strenuous event for Cassini."

Having arrived inside Titan's atmosphere, *Huygens* would then need to shed its front heat shield and back cover. Again, bolts could serve as fasteners. But the wraparound nature of the covers, and such cramped confines, really precluded the use of pyro nuts. The team instead went with frangible bolts—strong enough to hold the covers in place yet specifically meant for splitting in two. Using bolt cutters, of course! The business end of one is a strong metal blade driven right through the center of the bolt, snapping it like a carrot.

The action of driving each cutting blade would begin with another NASA standard initiator—a "first fire," as described by industry publications. But that isn't enough to force a blade through a bolt. Its energy output must be stepped up. Gas from the initiator blasts out its far end and into the leading face of a NASA standard detonator. The two are threaded together like a faucet and hose bib. Incoming gas carries enough force to detonate a small amount of weak explosive at the leading face of the detonator. It generates a larger shock wave, firing a main output charge and creating a humongous shock wave. *This* sends a small metal disc—small enough to rest on your fingertip—down a short metal tube at more than seven thousand miles an hour. Carrying enough energy to indeed force a blade through a bolt, separating covers from lander.

Now, nobody should go around snapping bolts like that without some way to prevent the remnants from drifting into places they shouldn't. A special bolt-catching receptacle accompanied each bolt. *Huygens's* back cover would be lightweight enough to just fall away on its own. But in the case of the front heat shield, more springs—liberated by the snapped bolts—would push it away like a distraught maiden rejecting her suitor.

Within their capable Zürich facilities, Platon Tatalias and his colleagues organized the scheme of bolts, pyro nuts, initiators, springs, guide track, disconnects, detonators, bolt cutters, and bolt catchers into three unified assemblies called nodes. Functionally, each did the exact same thing. But the geometry of the *Cassini* struts mathematically dictated a unique shape for each node. Some of their components would be either stock or provided by others. From a contractor's catalog, Contraves chose a model of pyro nut with built-in fittings for two standard initiators. This provided redundancy—with both initiators receiving simultaneous yet independent firing signals. As Tatalias commented, "Pyrotechnics have long, extensive, expensive, and painful development histories before they can be declared as qualified. This is done by few specialized companies only." Getting them in hand would be as simple as issuing a purchase order and signing upon delivery. The bolt cutters themselves, as another example, came from Dassault—a venerable French manufacturer of both civilian and military aircraft.

Why the two different types of fasteners? Didn't that complicate things? As explained by Tatalias, the interface between mother ship and probe had to carry high structural loads—but separate with minimal vibration, so as to not alter the trajectory into Titan. And for that, pyro nuts work great. The exact flight nuts could even be used in ground testing, operated with an air compressor instead of the special (and one-use-only) initiators. "An equivalent bolt cutter," continued Tatalias, "would require a large amount of pyro charge, which would result in increased vibrations."

Discarding the front shield and back cover didn't have to occur under the same delicate circumstances. "Vibration considerations are secondary, since *Huygens* is already on target inside the Titan atmosphere." Both could simply be jettisoned. And the summative loads weren't nearly as high, permitting the team to specify thinner bolts to help reduce overall node mass.

When choosing springs, the team decided to go with stainless steel models. Their properties are defined and understood—to a point. "A pre-stressed spring may be subjected to creeping when it is exposed to variable temperatures." Similar to what would happen over the course of this mission. Contraves therefore needed to calculate and engineer, for the exact precise moment *Huygens* left *Cassini*, exactly and precisely with how much force that should occur. The spec sheets from the spring supplier contained only so much information. Therefore, "an aging procedure, at high temperatures, was developed and applied to the springs." Partly this involved overcompressing and then vibrating spring samples.

Developing what they did involved one round of mini challenges after another. Endless puzzles. But according to Platon Tatalias, his team's greatest engineering challenge was mathematically modeling, with high accuracy, the trajectory of *Huygens* once it completely left the tracks and was on its own.

Machining custom parts from aerospace-grade metals is hugely expensive. So the initial prototyping of Contraves's nodes was done via stereolithography—a forerunner of today's 3D printing. For all the capabilities they boasted in-house, stereolithography wasn't one of them, and the work had to be fulfilled by an outside contractor. This still-used but costly process focuses a laser into a vat of liquid resin. The pinpoint laser beam solidifies the resin and slowly builds up a part, layer by layer. An initial prototype node handed to Platon Tatalias consisted of translucent yellow plastic. What his team was creating could be held for the very first time; he turned it over in his hands as he would a Rubik's Cube, beholding it from every angle. On one side, designers had added a ball-and-socket interface joint. The front heat shield, as big as it was, had been predicted to bend and deform slightly during launch and cruise, so this joint allowed the shield to float just a little.

Constantly present was a battle with weight. And as node development progressed, Tatalias had to admit that this battle now included the weight below his neck. See, in the time since graduating university, he'd gained a bunch of unwelcome extra pounds. To help restore balance, and combat the ever-present stress, he started exercising with a colleague. "Jogging after work helps get rid of the tension accumulated during the day." A few streets east would take them into a park. And going only a few streets west had them at the edge of a cluster of farm fields. These outings became a necessary part of every week. A respite.

39. The cheese between: one of the Contraves separation nodes. At left, angled downward, hangs the enclosure for the *Cassini* separation spring, which stayed behind. Wide base plate departs with *Huygens*. A squarish box on top holds the bolts and bolt cutters for the front and back entry shields. Dual springs atop push away the front shield. © European Space Agency.

Of their initially dictated mass budget, Contraves requested—and had already been granted—an increase of just over 10 percent. Still the nodes kept coming in heavier than allowed. The team worried that ESA managers would force them to literally shave each extra gram from the structure walls. "Which would be a time consuming and expensive effort," complained Tatalias. "Thought to be risky, and may lead to strength problems. Everybody had a mass budget and most companies exceeded it!"

Prior to manufacturing the first test round of metal hardware, ESA conducted a giant review of how much every Huygens subsystem weighed. It smacked of getting called on the carpet at fat camp. Painstakingly, the Contraves team listed out every third-party component it had no control over: bolts, washers, cabling. Plus the initiators and detonators. Continued Tatalias, "We added the dimensions of the nodes, and brackets and thicknesses, with the resulting mass calculated." And in terms of weight margins, well, there were none. The amount of margin was zero. ESA reps heard that, and their faces went sour.

Taking the floor, Tatalias pointed out two hopefully obvious points: Not only were all the third-party components decided and qualified and therefore "sacrosanct," as he put it, but also the Contraves nodes, as of their latest iteration, "were porous like a Swiss Emmental cheese." Before the group he unfurled a table of node specs. "These are the wall thicknesses and their corresponding structural safety margins," he informed the group, which had begun desperately scanning the figures in vain hope of finding margin somewhere, anywhere, please. But fabricators had shaved down every last millimeter as if it was their wedding day.

"I do not know where we can reduce them and save mass."

The meeting ended, everybody went home, and later on from ESA came forth the grumbly edict that one single kilogram had been benevolently gifted to Contraves's weight budget. This represented something of a divine reprieve for the exhausted team and had Tatalias and others wondering why they couldn't have just been given that extra kilogram to begin with.

Initial samples of flight-grade metal nodes began emerging from fabrication. Into precisely machined holes went fittings and wiring and frangible bolts. The bladed bolt cutters had been made redundant—two blades per, screaming in from opposite sides, with either blade capable of doing the job. Elsewhere, metal parts in contact with each other received special coatings to guard against fusing together in space—a common issue known as cold welding.

Functional node examples left the workshop benches, and testing began. Checking structural rigidity. Electrical continuity. To simulate launch and transit stresses, the assemblies were shaken on a specialized table and then exposed to hot and cold temperature extremes. So far so good. If a Contraves node failed in flight, it wouldn't be due to insufficient testing.

Past a certain point, the individual nodes had to be evaluated as an organic unit because they might well behave differently when strung together. Between them and the Cassini struts would hang a rigid ring of carbon fiber, unifying struts and nodes into one load-bearing system. All future tests would employ this arrangement to match flight conditions as closely as possible. Aérospatiale pressed for brute-force kinds of tests. One saw the installation of six hydraulic rams to mercilessly shove and twist the rig in multiple directions, sometimes all at once. Now command the separation. Again it passed, just as each month of the year was rapidly doing. On a continual basis, Platon Tatalias found himself reorganizing his workout bag from warm-weather jogging clothes to cold-weather ones and then back again.

In mid-April 1995 a complete, high-fidelity, flight-like Huygens test article went to Esrange Space Center—an established facility for sounding rockets and scientific balloon flights. Located way the hell up in northern Sweden, near the old mining town of Kiruna, it offered some forty-four-hundred-plus square miles of emptiness within which a high-altitude, end-to-end drop-test of the probe and shields could occur. This test would focus on the parachute sequence and probe descent dynamics while allowing various discardable elements such as covers and spent chutes to separate and fall uncontrollably without fear of damaging private property. For a real-world evaluation of *Huygens's* descent characteristics, really no other site in Europe fit the bill.

The Huygens test article most resembled its counterpart on the exterior. Size and shape were all consistent, though not identical, and included the flight-spec ring of stabilizing aerodynamic vanes around its circumference. The inside, however, wasn't even close as no actual science instruments were aboard. Rather, the space was consumed by batteries and motion sensors, plus radios for data and commands. Also aboard were two film cameras: one pointing upward and the other down. On May 14, 1995, with the test unit fastened to a support gondola and large-volume research balloon, they let 'er go, and everything floated up to just over twenty-three miles of altitude.

On command, *ksss* went the separation from the balloon. Although pyro nuts and a spinning ejection weren't part of the day's itinerary, bolts cracked and covers tumbled away and chutes popped in succession—all exactly as intended—and eighteen lively minutes following balloon separation, the thing was back on the ground. Still operational! Retrieving the various discards of descent took two days. Film shot from inside the test article confirmed how both front and back covers had tumbled away without brushing the lander—a persistent concern now dispelled. Every parachute deployment had occurred precisely on cue; all inflated fully with no trouble.

Despite the outside possibility of failure during the balloon drop, Platon Tatalias said he began feeling assured they were going to come out okay given all the rigorous advance testing. ESA ultimately accepted the Spin Eject Device for flight and away went Contraves's baby to the JPL assembly bay where *Cassini* awaited.

40. *Cassini* in an assembly bay prior to receiving its final insulation blankets. Hanging on it is *Huygens* (*left*), inside the heat shield. Gurnett's equidistant plasma wave antennas are mounted just below the top dish and on *Cassini*'s body. Courtesy NASA/JPL-Caltech.

"No one left the Huygens project due to stress. It was a fulfilling project with full support of the management," praised Tatalias—going on to celebrate, for the nth time, the steadfast backing of the greater ESA community.

"Perhaps the only thing to do differently was to challenge the mass budget from the very start!"

All subcontractors delivered their respective hardware, and the large Huygens acceptance review took place in late March 1997. As final preparations commenced for an October launch, veteran JPL engineer Tom Gavin made the rounds of the project staff. Cassini's roster now numbered about seven hundred people. Gavin wanted to have a few quiet words with some of them, low key, one-on-one. Specifically, he targeted the younger members.

"You're about to experience a feeling of separation," he cautioned them. "You've been working now for five or six years with all of these people. You're a part of this great Cassini team here at JPL. We're going to launch it, and then all of this is going to go away." Indeed, returning to his office after launching *Cassini*, even the office wasn't Gavin's anymore. He'd been unceremoniously evicted. The lion's share of cubic footage was now piled full of boxes from Cassini's *incoming* manager of flight operations. Gavin stood there for a moment while sizing up the encroachment.

Presently, the new occupant materialized in the doorway, saying, "Hi, welcome back! When can you be out of here?"

Platon Tatalias reported no such Gavin-esque, separative feelings. Although very proud of his team's deliverables, a Huygens node was "something we had just in our hands" and nothing more. Contraves's work had finished. Just enough time remained for them to file the node drawings and empty the trash cans and get another quick jog in before sitting down to enjoy a sister moment of engineering success—the *Pathfinder* landing of America's first rover on Mars.

24

A Man for Others

He doesn't look like a teddy bear. He looks like a bulldog. Right?
He's gonna take charge. And make everything happen.

—Former JPL engineer Mike O'Neal discussing John Casani

"I wasn't really too keen on retiring."

John Casani admitted that openly. The calendar read 1998. His current title read chief engineer—a new position, created specifically for him only a few years prior, reporting straight to the JPL director on core engineering standards and practices. More than forty years in, and only two away from welcoming a new millennium, Casani wanted to extend his run at JPL. Wherever there might be a need.

Others were bailing. Even Kane had already taken an attractive option to retire. His younger brother! One year prior.

John's financial assistant kept railing on him: "You're too old to be working. You ought to retire."

Aged sixty-five by that point, Casani parsed any implication that came to mind. "And I said, 'Well, why not?'" He spoke of feeling satisfied over past accomplishments. That's a bit intangible, and maybe too generic, considering his life in a world of precision. So how did the feeling manifest itself?

"I get my satisfaction out of knowing that what we built did what it was supposed to." Whatever imaginary quota dwelled in his brain had been fulfilled. John Casani was ready to be done building.

It'd been a hell of a career. Interestingly, all these decades later, Casani dismissed his time at the Rome Air Development Center as part of his career per se. To him, he'd had only one career, ever—at the Jet Propulsion Laboratory. A place where he'd brought machines and missions and mind storms into creation. Starting with missiles. "You know, a guided missile only has to work for three or four minutes," he pointed out. Longevity was never a priority. The Explorer Earth-orbiting satellites of the late 1950s, and follow-on *Pioneer* deep-space probes, were expected to last only a matter of weeks, tops.

Lifespans quickly lengthened. Come the early 1960s, Casani had been on Mariner's design team—a new pedigree of spacecraft intended for our nearest planets. Where Mars was concerned, that meant up to nine months of outbound travel time. "Nine months!" he wailed. Just to get there! "Holy mackerel, what the hell are we doing?!" He likened probe electronics in the 1960s to television technology of the day. Real touchy stuff. It worked most of the time. TV sets had knobs for archaic adjustments including "vertical hold"— about as relevant today as star-69 on your phone. But if the vertical hold on your TV went out, the picture would roll like a bowling ball. And when manipulating that knob brought no relief, it was time to call the TV repairman. He'd show up with a giant case of electronic components, which themselves could be just as finicky as whatever was already inside the TV. But at some point a magic combination of parts would be discovered, hopefully in time for *Bonanza*.

According to Casani, the parts crammed inside just one early *Mariner* equaled that of approximately two hundred TV sets. And when reporters (or Congress) wondered why the next *Mariner* struggled so much during integration tests, Casani used that as a point of comparison. "Just imagine lining up 200 television sets against the wall," he once rejoined, "turning them all on, getting 'em tuned up, and then not having to touch them for nine months."

From Mariner, spacecraft grew in complexity and endurance. JPL began to reliably make their Vikings and Voyagers last years and function beyond expectations. Prime missions received extensions. Casani's lowest point during Galileo was the *Challenger* disaster. No question. "Those were the darkest days for me," he acknowledged. "But I never gave up hope." *Galileo* had gone on to orbit Jupiter while outlasting the radiation, the antenna failure, the budget, the politics. Another flagship orbiter had just recently embarked for Saturn. And a rover had even survived landing on Mars and sent back little movies of itself trundling around. Times had certainly changed.

By 1998 John Casani thought his work was done. That turned out to be not-not-not the case.

The retirement party was one worthy of Caligula. It went down on a weekend night at the Pasadena Civic Center and hundreds of well-wishers showed.

41. Here's John Casani performing a favorite party trick during his suspenders phase. "I think I got into the habit of wearing that when I started gaining weight" during the transition between Voyager and Galileo. "My time was getting a little bit more difficult to manage," he continued. "What I lacked in smarts and capability—I compensated that with persistence, so then I was spending more and more time at JPL." Casani's regimen of swimming and handball subsequently vanished. Courtesy the Casani Family.

Plenty came from NASA Headquarters. An emcee presided. John Casani made a grand entrance with a smallish cardboard boat suspended around his middle; he and a buddy had invested in a real one together. "My wife thought that owning a boat in San Diego was the perfect draw for getting me out of Pasadena," John later claimed. His buddy, Louie, was the very same one who'd called all those years ago about moving from New York to California. The party rocked into the night with skits and singing and gifts galore, and many took turns at a microphone to pay respects.

But the following Monday, newly retired John Casani's phone unexpectedly rang at home. JPL director Ed Stone needed help. Quick as that, retirement ended for what Casani termed "a little bit of a crisis" over an upcoming shuttle flight involving JPL. The winged orbiter would be carrying an advanced, JPL-provided, dee-luxe radar system to map Earth's surface at high resolution, generating sexy data on things like terrain elevation.

Similar to what flew on *Magellan*, the shuttle-borne system incorporated dual radar units trained on the same piece of ground: one perpendicular to it, the other offset at an angle. One of these units occupied most of the shuttle's payload bay. The second, as big as an office desk, had to sit on the end of a boom arm jutting nearly two hundred feet sideways into open space. That's a pretty long lever by anyone's standards. Despite all the engineering behind this arrangement, plus control thrusters at the end of the boom, some at the Johnson Space Center reckoned it might not be completely safe. So Casani joined a team of JPL experts who huddled at JSC with another team from the astronaut office—in conjunction with even more teams from Langley and Marshall.

"So we spent a week down there and managed to get that all quieted down." A grueling technical review ultimately cleared the radar rig to fly, boom and all, and the mission went great. "But now I was back as an interim employee," noted Casani. That meant being on JPL's payroll for six whole months. No benefits. Most of his working papers had already been boxed up and trucked to the Lab's archives. Someone found him an office to sit in. Casani didn't have a whole lot scheduled except waiting for the clock to run out.

That is, until *Mars Climate Orbiter* went dark in late September 1999. The spacecraft, essentially a Martian weather satellite, was also intended to catalogue the planet's vaguely understood distribution of ice and water. Just prior to its entering Martian orbit, the thing broke radio contact, and nobody heard a peep from it again.

Within a month, JPL convened two failure review boards as part of an overall effort involving multiple boards at multiple facilities. Casani got asked to lead JPL's board. He did so, conceding that retirement had essentially failed for the time being, and dug into the details of *Climate Orbiter*'s mission operations—a failure of another kind. Slowly, over the course of three months, the various participants collectively peeled away more and more layers of the onion. And finally their fingertips brushed across the bedrock of the problem.

It centered on a software routine aboard the ship called Small Forces. Relating to thruster performance. Its output fed a different, higher-level software routine that expected the provided data to be metric. But Small Forces mistakenly used English measurement units, without converting. The values were different enough to matter.

"There had been some navigation inconsistencies noticed during the mission," sighed Casani "*well before* the failure at Mars orbit insertion." The outbound trip required nine months. And during that spring and summer of 1999, naggy discrepancies had continually arisen among the various methods of tracking the spacecraft. Nobody resolved them. Project workers were understaffed and inadequately trained. All the while, minuscule errors were accumulating in the ship's trajectory, shifting *Climate Orbiter* lower and lower to Mars relative to where it should've been. But the person who noticed these inconsistencies hadn't followed the established process for addressing concerns that pop up during flight operations. It's a simple process, yet an important one, and begins with submitting a form called Incident Surprise Anomaly. "It's only one page long, but it's a formal report," explained Casani. "Once it's submitted, it gets tracked," much like today's customer service tickets. During the course of any mission, hundreds of ISAs may be produced. "Once you write an ISA, it becomes a permanent record in the system. It gets reviewed. Its ongoing status gets reviewed. Its *closure* gets reviewed." Repeat meetings occur. People have to sign off on the proposed solutions. It's a managed situation.

Casani offered more detail about what happened. "In the case of the *Orbiter,* the person who noticed this problem didn't use the ISA form. He wrote an email message to the person that he thought could solve the problem. *That* person got the email message, and he looked at it, and worked on it for a while. Then his boss gave him something else to do that this individual judged to be of higher priority than working on the problem outlined in the email." When considering the navigational anomaly as a whole, this wasn't the only factor. But it stood as a deciding one. Upon reaching Mars and positioning itself for orbit, the spacecraft was roughly 106 miles lower than intended, and down went *Mars Climate Orbiter*—either shredded in the atmosphere or bouncing off it into space forever.

"Here is a case where the guy who noticed the problem thought he was doing the right thing. He wanted to get the problem taken care of quickly. He sent the email out." And completely sidestepped the incident reporting system. Sending an email is like writing a letter; there are few guarantees. People don't always check for it, and some don't always read it all the way through. Sometimes they prioritize based on quick glances, fail to address the substance of the original message, forget what they wrote after replying, or completely put off any response until making time to compose something well thought out. New mail arrives and pushes the older stuff down, out of sight. Email is fraught with the potential for sloppiness and incompleteness. It lacks reliability as a management system because nothing about it is mandatory. Had the spacecraft tracking concern instead been communicated via the ISA, "it could not have been forgotten. It would have become a permanent part of the engineering documentation."

In no way was Casani trying to tamp down on the value of email. He described it as an incredible communications tool. But one that, according to him, is "just too easy to blanket the world with," leading—ironically—to breakdowns in communication. Back in the ol' days, group memos went out via Ditto machine. This now-archaic technology involved writing or typing on special two-ply "master" paper containing its own embedded ink. Subsequently the master would be fixed onto a simple rotating drum that generated copies by mechanically squishing the master page against blank sheets. A painstaking process, it required care in preparation and was good for about thirty copies or so, until the embedded ink in the master ran out. So everyone had to write economically, limit the distribution list, set aside enough time to see it all through, and in general behave rather differently than they do in the present day. When writing an email, people can mindlessly clack away to their heart's content and then zing it out to hundreds of recipients with nothing more than a mouse click.

"That was my gripe about email," Casani summarized. "I was seeing a deterioration in our ability to document technical problems. Or to handle issues in a way that could be examined and commented on—reviewed easily—by other people." So the lesson is one of larger significance than email etiquette or even the downfall of *Mars Climate Orbiter*. "We need to reinforce the distinction between the need for *rapid communication* and the need for *engineering documentation*, which creates products that can be peer-reviewed. And that leave an audit trail for engineering follow-up and closeout."

In recapping the failure review board's findings, Casani further stressed that a major factor was the sea change in working conditions. "At the time, we were operating under extreme pressure from HQ to do things *faster, better, cheaper*," he reminded—an environment ordained by NASA administrator Dan Goldin. "We didn't really know *how* to do that, but we did know that real savings would require that we do it with fewer people." Staff cutbacks occurred left and right, and management tried to press on.

"In the end, we had very good people. The problem was, as stated by the chair of the other failure review team, that we just didn't have enough of them."

Casani closed out his paperwork, and only weeks later it happened again. Another Mars-bound spacecraft, *Polar Lander*, was in the process of its terminal descent to the Martian surface when it terminally died. Radio contact disappeared and never resumed. In something of a minor insult, *Polar Lander* was originally to have depended on *Climate Orbiter* as its communications relay. Oof. Both were now toast.

Shock waves rippled through America's space agencies. These two Mars missions weren't grad student thesis projects; they carried a combined price tag of $236 million, which didn't even include booster rockets. John Casani, still around and still on the clock, joined *Polar Lander*'s review board. Far less information was available on this one. Everything right through the beginning of terminal descent had been going tickety-boo. Regardless, the failure analysis team ultimately agreed on a likely scenario. With the lander on final approach, its legs deployed as expected—but induced a mechanical shuddering that made the onboard software think it'd touched down already. The craft lacked a way to check altimeter readings alongside the landing sequence and verify actual touchdown. Its landing thrusters switched off while still airborne. *Polar Lander* crashed from confusion.

Again, Casani wrapped his work on a failure board. Ed Stone resurrected JPL's Flight Projects Office, with its attendant vigor for such proven disciplines as systems engineering and mission assurance. The Lab had erred in mothballing the very structure that'd garnered so much success. Casani served as one of the key people restoring things to the old standard. "And by that time, my six months was up!" he laughed. "So I left!" Over a year had passed since his original retirement party. Although explicitly saying "I left," Casani actually didn't. He stayed on at JPL, reporting directly to Ed Stone as "special assistant to the director" and then to Charles Elachi after the directorship changed hands in 2001.

"It was kind of a hokey title. But it was meant to be broad, and allow me to do basically anything anybody wanted me to do," Casani explained. His boys were all grown and out of the house, and time ran slower. With more balance in life, he and Lynn traveled some. He picked away at restoring a Mustang—amazingly, the same one bought new for his first wife, Marie, all those years before. At JPL, people would occasionally drift by his third-floor office to ask if they'd be getting back the retirement gifts that had been given to him. "Look," he told every single one of them, "you're not getting it back." To him a gift was a gift. But when he finally retired for real, he assured them, "you should not feel in any way obligated to give me *another* gift."

The Casani family milestones continued accruing. Adopted son Andrew graduated college, and even his biological father attended the ceremony because

Lynn had seen to it that the two developed a basic relationship. During a large reception afterward, Andrew made a point of bringing friends around. He introduced them to Lynn, saying, "This is my mom." Then moved to John with, "And this is my dad." As Andrew progressed around the table, it was, "And this is my natural father, and then that's his wife." According to John, Andrew did this multiple times over the course of the event.

"He always referred to Lynn as his mom. And he always refers to me as his dad, or called me Pop," smiled the very proud John Richard Casani. "And still does."

Only one majorly bad thing happened, in 2004: "During that time, Lynn came down with cancer." Their boy Drew had been a star football player and went on to work for the National Football League as a scout. He relinquished the pro ball scene to come home and be with his mom. Dutifully John attended to Lynn—chauffeuring her to appointments and addressing every need—but needed for himself the continuity of going into JPL every day. "I chose to continue working," he said, with minuscule yet detectable regret in his voice. "I would have been better off spending that time with her." But Lynn, according to John, had insisted that he operate however he needed to and that *his* mental state mattered as much as anyone's. Neither had any qualms about it at the time. She let go in June 2008 at the same hospital where her sons had first entered the world. Family, friends, and John were all by her side.

"I had the most wonderful wife and family relationship you can imagine."

He continued at JPL in similar capacities, lending his opinion to future mission technologies and concepts. A month after Lynn's death, project workers on Cassini welcomed a twenty-seven-month extension dubbed the Equinox Mission, to prolong operations through most of 2010. Don Gurnett, Bill Kurth, and other Cassini scientists had much to look forward to: supplemental ring observations, icy moon studies, and wads of attention on Titan. From a distance, Casani followed the proceedings but without nearly as much emotional investment. "My job is generally done once the damn ship is launched," he reflected. He said he never got really excited about the data coming in and probably wouldn't understand it even if he did.

"The thing that I get the biggest charge out of, you know, is just—and I've used this expression—*just getting shit done*. Making things happen. You know, setting out for whatever you're doing, and finishing it. I get a *great* deal of satisfaction out of that."

When it came to project management, Casani volunteered that one of his operational maxims came from Herbert Bayard Swope, an early twentieth-century American journalist and three-time winner of the Pulitzer Prize. Swope was once quoted as saying, "I can't give you a sure-fire formula for success. But I can give you a formula for failure: try to please everybody all the time."

A perfect application of this philosophy was Casani's decision—judged by some as the most controversial of his entire career—to delete *Cassini*'s boom arms for the science instruments. Jilted mission scientists had erupted in unified outrage. And so, with a thoughtful reprocessing of Swope, Casani promulgated what he termed an optimum approach for managing project costs: "An *optimum approach* is one that makes everyone equally unhappy."

Once those boom arms went away, multiple types of data could no longer be taken at the same time. "This added a lot of unplanned operational complexity to the mission, which would mean a lot more work and cost for the mission operations team. Making *them* very unhappy," he continued. "But not a deal breaker!" It facilitated leaving Earth on budget, and that's what Casani was on the hook for. Everyone would still go to Saturn. Fast forward a couple decades, with *Cassini* discovering amazements one after the other, and John Casani could take comfort in that such hard decisions enabled the mission to happen at all.

"I never thought of myself as a good manager," he admitted in a surprisingly straightforward fashion. Afternoon sun glinted off his mostly bald head. "I think of myself more, if I could use the term, a *leader*. And *then* a manager. And in some ways, I think you can learn to manage, if you're interested in it. But I think it's hard to learn to *lead*. I think that's more, part of your DNA."

Just look at Galileo, where being a project manager meant backing a team beaten down by congressional funding threats, shuttle development issues, celestial mechanics, and ongoing launch delays. "An important part of any project manager's job is to sell a project. Not just to get the project off the ground, but to keep the project *alive* when surmountable obstacles arise. That selling may require creative thinking to frame the project in a way that makes its value more apparent to project sponsors." In other words, to live "in a world with no corners." Some say that everyone, no matter their job title or career description, must engage in marketing. Casani's experiences support that.

Sometimes he just needed to preserve basic hope. "I knew my team would eventually find a way to get *Galileo* launched, and I knew what the spacecraft could deliver. But it wasn't an easy sell. When I went in front of senior NASA management, I made an opportunity cost argument to them. I pointed out that for the increment of funding we still needed, they could, in essence, buy an entire mission. The sunk cost didn't count because they couldn't recover that. It was water under the bridge. So what was the opportunity cost of that additional increment that we would need to finish? Could they buy something of more value for that same amount of money?"

Despite his already-cramped schedule, Casani made the rounds on Capitol Hill, pressing the flesh to help shore up support so *Galileo* could go the distance. "The project manager doesn't do that anymore," he clarified. "Headquarters does. But even at the time, I got to do things not usually done because a lot of people had written our project off." Casani stressed the importance of modeling positive attitudes and leading from the front. "A project team takes its lead from the project manager. When managers make clear their own commitment to, and belief in, their projects, they empower their teams to overcome problems that crop up."

Casani leaned toward his restaurant guest to address the topic of his upward rise through JPL's ranks. "You asked me the question, 'What did they see in me?' You know, maybe *that's* what they saw in me. You know, the ability to get people to work together, all pulling in the same direction." Sometimes he did this with gimmicks—such as the prelaunch parties or, for heaven's sake, that whole crazy thing with the goat. But other times he simply listened carefully to all involved personalities, without taking sides, and worked the situation to "the least unacceptable solution." Naysayers, who maybe thought Casani to be unrecoverably stuck, would sometimes hear him say, "Well, we haven't concluded that there isn't a solution."

Managing large technological projects is not a drop-in kind of job. It requires a basic assertiveness—a confidence in ability—coupled with a day-to-day fluidity in managing oneself and the greater task at hand. Coupled with maximal communication at all times. Early in Casani's career, he worked underneath a group supervisor who took on a project manager job for one of the Mariner flights. This person was a seasoned engineer. Had been at the Lab for years. But the guy had a problem with his new managerial role: "He was fretting all the time."

They were talking once. "John," began the other man, "I—I really don't know about this project."

"What's the matter?"

"There's three or four hundred people working on this program. I don't know five of 'em," whined the guy. "People are making decisions every day that's going to affect the outcome of this project. In terms of whether it's gonna be a success or a failure. And I don't know who they are or what they're doing." Undercurrents of doubt were coursing through him, forming their own deltas into more worry and uncertainty.

According to Casani, the guy went so far as to not see the value of a project manager at all. Despite performing that exact job on a $30 million project! To Casani, he described an imaginary scene of hundreds of ants occupying a log that's floating downstream. Every ant has a paddle and contributes—except for the project manager, who merely watches where the log is heading, in virtual detachment from the proceedings. The scene just kept looping in his head.

Casani thought, *Jesus, this poor guy.* "He just couldn't get comfortable with the people. And that he had any connection with them, or any control over really what was happening." The guy's crippling fatalism hindered any managerial advancement up JPL's ladder. Despite bringing the project in under budget!

"So, uh, it sort of broke him up badly."

After a measured sip of wine, Casani gestured to his guest. "You have to get comfortable with this. You have to make sure that everybody knows what the objectives are and what the order of priority is," he insisted. "What things *matter* in a rank order." To a degree, it *is* as the Mariner guy said: Clicking away underneath project managers are hundreds of people making decisions one after another in the absence of direct visibility by upper management. But by presenting clear leadership, and by communicating the vision, could a project manager walk the entire team forward. And do so with the basic assurance that every invisible decision is occurring in an environment of clarity.

Confidence. Communication. Clarity. These sorts of qualities were baked into Casani's whole self. Qualities that had him again trying retirement in 2011—the same year that America's space shuttle flew its last mission. And out at Saturn, despite miscellaneous thruster problems, *Cassini* continued to perform. Its mission had been extended yet again, potentially until 2017, depending on how things continued to work. About 20 percent of its maneuvering propellant remained in the tanks. It could totally keep going.

Unlike John Casani's, the retirement of the space shuttle proved truly permanent. The orbiters went to museums, and the United States entered an uncertain period of having no human space launch platform of its own. On July 1, 2016, Mike Watkins took over as JPL's new director from Charles Elachi. Subsequently, Casani was invited to speak at a formal-ish Lab event recognizing a longtime coworker. And later on during the event, Watkins casually strolled up. The two knew each other already, as Watkins used to be a project scientist at the Lab.

"You know, John, you really know a lot about this place."

Casani admitted that he supposed he did.

"I think I could use your help," continued JPL's newest director. "I'd like you to come back."

Watkins had specific reasons behind his request. Years later, Casani maintained that the Caltech board of directors, who hired Watkins, harbored unease over how the Lab was being run. "And he told me that the board of directors was really concerned about the risk posture of the Laboratory. From a point of view that, uh, every one of our engineers and specialists and everything was fully committed to a project." The board wanted recommendations for properly managing this highly educated, specialized, and relatively priceless workforce. Watkins already had someone heading up the overall effort but needed seasoned people with recognized, time-tested, and farsighted perspectives. Enter John Casani, who grabbed three other retired JPLers "that were retired also, but experienced, and well-known and respected at JPL. And so the four of us went to work on that for a couple months."

With the better part of the day already spent, but sunlight only vaguely beginning to fade, Casani's guest at the Italian restaurant brought up a topic for which the mood now seemed appropriate. Something hopefully straightforward to discuss, meriting perhaps a couple brief sentences in response. But a question that could only be presented after the necessarily long process of securing trust.

"I had a question for you about your Catholic upbringing. Obviously, you know, very strong, big, close, Italian Catholic family," he prefaced. "Seems like that sort of carried over into your own parenthood. That your boys went to parochial schools and whatnot. And I was curious about whether you would consider yourself a religious man, and were there any ways in which your religious beliefs sort of helped you get through tough periods at work?"

At that, Mr. John Richard Casani—manager of international, high-tech, billion-dollar projects; veteran of innumerable podium speeches and press

42. No stranger to technology, interruptions, or multitasking, John Casani fields
interview questions—and occasional text messages—in 2016 at the Italian restaurant
in Pasadena, shortly before going to work for JPL director Mike Watkins.
Bread pudding in foreground. Photo by author.

conferences and live TV interviews; recipient of at least one honorary PhD;
survivor of repeat congressional hot seats, not to mention defuser of law enforce-
ment raids—remained stone quiet for a noticeably awkward period. The guest
repeatedly stopped himself from breaking the silence.

Finally Casani offered, "Now, *that's* an interesting question."

He wasn't sure where to go from there. The habit of attending Sunday services had stuck with him into adulthood. Well, for a while, anyway. "Went to church every Sunday. Even after I first got out here to JPL. But then pretty soon, going to church got to be not something that was particularly interesting, or important, or, uh, that I felt any need or compulsion to do." Dyed-in-the-wool Catholics may react to that with disappointment.

But Casani then came up with something more to offer than mere attendance stats. He retreated into childhood, resurrecting memories of his stalwart father, Jack. "I would say *he* was a pretty religious guy, probably more religious than any of us," is how the full response began. Such beliefs, surely coupled with resentment for the candy business, are what propelled the educational plan set by Jack and Julia Casani for their children—Jesuit schools. No question, little Johnny and his siblings would be educated in the progressive ways of the Jesuits.

"When it came time to go to high school, I went to a *Jesuit* high school," the story continued. "It was never a matter of discussion. I—I didn't even know where I was going. And the same thing when it came to go to college." Casani said it all seemed to happen on autopilot. "I just sort of did what my father wanted. There was no argument." If you were Jack Casani's kid and asked him why, the predictable response came back: There is no why.

From this educational continuum, what would ultimately coalesce in John Casani, and persistently stick with him, was a guiding foundation of core Jesuit philosophies. Consistently, he referred to it as "my training."

The Jesuits, to be clear, are a religious order. They sprang from the Catholic faith and are not a self-contained religious organization. They came about from the enlightenment of one Ignatius of Loyola—a Spanish priest who, alongside half a dozen other like-minded men, founded the group in the 1500s. While certainly devoted to Christianity and the pope and general Catholic beliefs, the Jesuits distinguished themselves by emphasizing such core tenets as integrity, courage, love, and generosity. Education. Becoming a true citizen of the world. One pillar of Jesuit tradition involves *discernment*—basically, the discipline of making an informed decision. It starts by investing enough time to truly understand the issue at hand. Sound familiar?

Rooted deeply in John Casani was another fundamental aspect of the Jesuit tradition: "They talk to you about being a man for others." He'd been hearing that all his life.

Repeatedly, Jack Casani practiced what he preached by taking his family to Philadelphia's orphanages. Places where at-risk children needed serious help locating meaningful pathways in life. "He would take us there, and we would usually give them clothing." And of course, candy by the bagful. Their chief goal being to provide for those less fortunate—kids who'd been dealt a bad hand in life, who were struggling in circumstances beyond their control and maybe just needing someone to listen. To empathize. Overly protective parents might fear their children would be traumatized by witnessing such deprivation. Clearly, though, Jack Casani knew he was a privileged man and never wanted his children to take anything for granted. In them he wanted to embed a lifelong mindset of thankfulness, positivity, and continually giving back. To always be in a state of helping. For the Casani family patriarch, this attitude represented the base layer of good citizenry. Take most every cue from the Jesuits.

Young John could've just gone through the motions. He said he did, at least for a while, but his training began to sink in. It guided him on how to accept failure and take responsibility when unfortunate things happen. When they don't come out right. Because this *will* happen. He failed to appreciate a liberal arts education. Took responsibility by rerouting his path through college, via electrical engineering. He failed at marriage. Took responsibility by remaining an active and involved father, to the point where his son chose to live with him after divorce. And in the long run, John Casani proved that he could indeed be quite successful at marriage.

Certainly, nothing in this paradigm ever came down to luck. Casani cemented his opinion on luck in the 1960s at JPL; a specific event during Mariner development stood out in his mind. The spacecraft's maneuvering thrusters had been the subject of much concern, and a key test was about to go down in the Lab's giant vacuum chamber. As Casani relayed the story, "So we fired the engine up, and—and sure enough, in a couple of seconds, it was clear that the system held. It was stable. We didn't need to worry about it."

Afterward, he was standing around outside the chamber with twenty to thirty other people, and somebody teased, "Well, Casani, you're pretty damn lucky."

In the midst of processing that remark, a superior turned to Casani and put a hand on his shoulder. "John, being lucky is pretty close to being correct."

At the restaurant table, Casani had been looking slightly off axis and away during his telling of the Mariner story. Now he refocused on his companion. "And, you know, I carried that away with me sort of as a principle. I mean, it was just embedded in my brain." He thrust his hands out in front of him as nonverbal punctuation. "There's no such thing as good luck or bad luck. It all depends on whether you've thought through all the things that could go wrong and have taken some reasonable measure to prevent them from happening in the first place." If that process doesn't occur, then what wasn't thought through will surely become the pointy end of the situation—and probably in front of respected colleagues whose opinions are of great value. Casani referred to this as "engineering by exposure."

The Jesuit-minded parents, teachers, and coaches of John Casani also had something to say about fighting. (Ah, not in a physical sense.) "They talk about argumentation. Which is, you know, maybe you could call it debating or something like that," he began of the lessons. "But the idea is that no matter what the argument is, you ought to be able to defend it. Or attack it equally well." Essentially, no matter which side of the argument a person might stand on, true cognizance of *both* sides absolutely must exist for the issue to be resolved properly and fully. And the process of making your case needs to be done in an atmosphere of respect—not by freaking out or shouting people down or hurling insults. "You should be able to argue it dispassionately," Casani pleaded. Seek out faulty reasoning. Never go ad hominem. "You attack the idea but not the person."

Consider this philosophy in the environment of creating space missions. It applies almost perfectly. Endless design reviews see people offering make-or-break concepts (such as, oh, dual spin), and eventual mission success depends on each detail surviving a reasoned, thorough analysis. "You have to expose it to critical thinking by other people," Casani reinforced. "You're always on the one side of defending what you've done. But if you were on the *other* side, as a reviewer, your job is to attack what's been done. And people that are in this line of business frequently find themselves on *both sides* of an argument."

In sum, you have to endeavor to understand the situation from every angle. Spend at least as much time finding flaws in your own ideas as you do in others'. "See, there's no way I can know *all* the problems," he stated. That sounds obvious but perhaps isn't. "Maybe you can find most of 'em. Maybe you *can* find all of 'em. Probably not, but that doesn't mean you shouldn't look damn hard. Right?" Then, setting your ego aside, unroll those findings in front of your peers and submit to their mercy. Drop all attitude. Show them you want what's best for the project—for others—and not necessarily what's best for your own career or some arbitrarily significant percentage of correctness. Show them you're not afraid to be wrong. "If we *do* see what could happen, there's likely something that we could do to mitigate it!" After a pause, Casani suggested that one of the most impressive things anyone can say in a design review is, "We thought of that already, and here's our approach."

His volume dropped, lest he disturb the neighboring table. "You do everything that needs to be done, correctly, and it'll work. *It's not a question of luck.* If you fuck something up? You forget something that's important, or you don't do something right? That's not bad luck. *You fucked up.*"

Collectively, revealed Casani, these philosophies from his background were everything at JPL. And the legacy of John Casani's work there is not, as he professed, gone or in need of being thrown out. His experiences contributed massively to two guiding JPL reference standards: *Design Principles* and *Flight Project Practices.* Born from the reconstituted Flight Projects Office. The Lab even began a class for prospective project managers in which the roles and responsibilities are presented to candidates, thus equipping them to decide whether the position is one that fits.

"Casani has a very systematic approach in examining issues or problems." Such was the insistent praise by longtime JPL engineer Tom Gavin, who worked alongside John for decades. "When you had to present a problem and the potential solution, Casani would very quickly work the discussion to the boundary of your understanding of the issue. He always worked it *with* you so that you were discovering the soft spots in your solution. It was always a constructive learning experience with Casani."

"He helped lay the foundation of how we approach project management and build spacecraft—essentially how our nation would explore space." So offered Charles Elachi, Magellan scientist and eventual JPL director. "And when today we send these rovers and other spacecraft on missions across the solar system, we are really standing on the shoulders of giants like John Casani."

And why? Because he lived to be a man for others? Holy cow, that explained pretty much everything in the guy's life. The calm perseverance. Listening and compromising. Communicating. Avoiding corners. Helping. Adopting. Recognizing this master key for what it was, Casani's guest bolted upright and practically knocked over the bread-dipping sauce on their table. "If it's not shepherding something like *Galileo* through all of its tribulations or, for heaven's sake, John, taking this boy, Andrew, into your home and having such a rapport with him to the point where he's calling you *Pop*? I mean, right there, *a man for others*! It's, like, when were you *not* there for others?"

"Well, thank you," Casani responded, maintaining eye contact. "My initial reaction is that I think my wife exhibited that to a greater extent than I did. But maybe that's, uh, you know, that's unnecessarily self-demeaning." He sighed, unmistakably preoccupied by a distant longing for the companion who could no longer be with him.

But then John Casani finally answered the original question after all. "Okay. Uh, I uh, I don't think I was that religious."

And from there, he proposed the next topic—dessert. Maybe bread pudding?

25

All Things Go

This is the kind of people we are. People who are attracted
to these first of a kind, one of a kind projects. We love to
solve problems. That's how we value ourselves.

—JPL software engineer Dan Erickson

Galileo's long-deferred prime mission ended on schedule in December 1997, two years to the day after first reaching Jupiter. But that didn't mean curtains just yet. After eleven orbits and many more discoveries, the dual-spin machine—albeit damaged and imperfect—still had miles in the tires and garnered a two-year extension. Doubling its orbit time! But good news almost always is tempered by a downside, and here it involved money. As in, less of it. The project would have to survive on $15 million a year, which meant drastically reducing staffing. Fewer people doing less work but disproportionally so.

Less money also constricted *Galileo*'s scientific pursuits. Only two days of data on each flyby would be collected instead of seven. Not to mention fewer readings of Jupiter's magnetic fields. Otherwise, full speed ahead! Where the ship went next, and what it did, largely came down to agreeing on which new discoveries warranted further investigation. Well, one in particular topped everyone's list—Europa. Had to be. Its thick ice crust was doubtlessly hiding a liquid ocean underneath. Made of saltwater. Encircling the entire moon. Maybe even breaking through the surface in places. It could be home to life of some kind. Readings at Ganymede and Callisto suggested the presence of subsurface water there also, though the data seemed nowhere near as convincing. A series of follow-on observations, directed at Europa's crust and delicate atmosphere, could reinforce (or refute) what they thought they knew.

Bill O'Neil was in his office when Bob Mitchell dropped by to ask something. What were O'Neil's plans for the extended mission?

O'Neil told him, "I don't really have any."

Pause.

"Well, Jim Erickson and I are kind of wanting to move up." In casual tones had Bob Mitchell dropped a rather strong hint, which O'Neil immediately recognized. And welcomed.

"That's fine with me. I can step down and stay in an advisory role until I find something else." After more than seven tough years at the helm, Bill O'Neil was "*ready* for something else," as he put it. Ready to stop being Galileo's project manager (PM). And Bob Mitchell was ready to start.

"I can cover your salary for a few months until you do," promised the incoming man. And in the space of this one-minute informality occurred a changing of the guard.

Years later, O'Neil would reflect on his personal investment in Galileo. "I just *lived* to get this thing done," he charged, with a dash of emotion. "Whatever it took!" After vacating the project office, he accepted a Lab gig pursuing technology for future Mars missions. Within months it transitioned into a new, full-blown PM job for a Mars sample return program. With fresh money having come through, NASA intended to mount an assault on Mars—partnering with France's space agency on multiple launches in 2003 and 2005. (Not so much a partnership as an *interdependence*, it would emerge.) All to obtain a handful of dirt from the next planet over.

Bob Mitchell's involvement with Galileo stretched back eighteen years. The relationship had started in mission design and trajectories, enveloping his supervision of Roger Diehl throughout the miracle of VEEGA. By the end of 1988, all trajectory work was done for the '89 launch, so Mitchell left Galileo to supervise the entirety of the Mission Design Section. That lasted only a few years until a critical Galileo manager died, and Bill O'Neil pleaded for Bob to come back. Mitchell did so, having occupied most recently the role of mission manager, until news came through of the Europa extension. Prompting that visit to his colleague's office.

As Bob Mitchell continued, "My first and biggest challenge was cutting the team size down to fit within the budget that NASA HQ had allocated for this mission phase." Under him, the operations team as a whole slimmed to about a fifth of its original roster, in a process that was not exactly pain free. "Many long-time team members had to go. Not necessarily terminated from the Lab, but at least had to find another job *not* on Galileo. I wanted to keep the best performers for obvious reasons."

"Mission manager" sounds similar to "project manager." Typical bureaucratic duplicity? Well, while the latter stood at the absolute top of the pyramid—juggling administrative particulars such as budgets and personnel—a mission manager focused more on day-to-day operations. That wasn't set in concrete, however, as project managers certainly dabble in the doings at lower levels. According to Mitchell, "The split of responsibilities between the two probably depended more on their own expertise than on what may have been written in role statements." A more technically oriented PM such as Bill O'Neil, for example, often reviewed software changes before they went up to the spacecraft.

Seven consecutive Europa flybys took them through early 1999. Passing much closer to this moon than during the prime mission, *Galileo* imaged the icy Europan surface at three-times higher resolution to document its never-ending process of heaving, cracking, and renewal. Not one single crater appears on Europa's surface because change is always afoot. Repeat observations proved it so.

And what about the probable ocean deep underneath all that ice? Salty oceans near Jupiter have a tell: They generate magnetic fields. And *Galileo* had already detected as much in Europa's vicinity during the prime mission. But closer examination had been prioritized to determine whether they were coming from the moon itself or Jupiter or what. These latest flybys brought the situation into focus.

As Europa orbits Jupiter, it passes through the giant planet's magnetic field. This motion drives measurable and consistent patterns of electrical activity inside Europa, further inducing new magnetic fields within what can only be a saltwater ocean. If the interior was solid, you'd get different results. *Galileo*'s measurements determined the approximate thicknesses of both ice *and* ocean. The spacecraft's instruments even allowed a whack at how much salt the ocean contains—a key indicator of Europa's habitability. All from something as benign sounding as magnetic fields!

Amid the Europa loops, in the summer of 1998, Bob Mitchell got handed a bit of a shocker—a high-level request to move over to Cassini as *its* new project manager. What, already? He'd been running Galileo only a matter of months. The Jupiter ship was chuffing along well enough, getting good data and all, and yeah, there'd been some itchy personnel issues. But nothing crazy. As part of the staff reductions, Mitchell had explicitly asked people for two-year project obligations—the length of the Europa extension. It wasn't meant to be so much a hard-and-fast rule as a statement of intention. "We didn't want

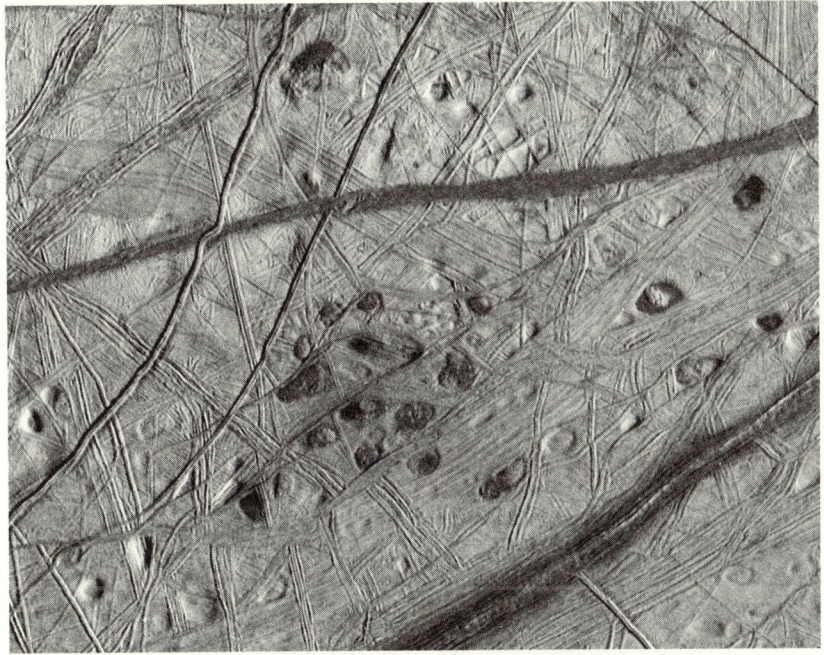

43. An amazing close-up of Europa's surface, courtesy of *Galileo*. Easily visible are innumerable cracks and channels. Dark acne-like spots, known as lenticulae, are thought to be the product of heated, subterranean material welling up from below. Courtesy NASA/JPL/University of Arizona/University of Colorado.

people to take a Galileo position as a parking place until something better came along." That's what the Cassini offer smacked of. Shouldn't he himself be honoring the same terms?

"Initially I said that I couldn't take that job because of the commitment I had made to the Galileo Team. But Gael Squibb, my boss at the time, told me that he wanted me on Cassini and that he would take care of my two-year Galileo problem." A sudden management change such as this might seem disruptive. But Bob Mitchell clarified that this sort of thing often happened when the phase of a mission changed. "John Casani was the original PM for Cassini when the task was to sell the project to NASA HQ. Secure funding, negotiate with international partners, get through all of the politics that are inevitably present, and make it all happen. Casani was a master at this," explained Mitchell. "The next task, once the mission is approved and funded, is to flesh out the design, build enough flight software to launch and fly the craft, build or contract for the hardware subsystems, assemble, test, and launch." *That* phase called for its own specialized manager. "Next is navigating the spacecraft, building sequences for both engineering and science observations, writing and maintaining the flight software, operating the spacecraft, and herding the cats—scientists—who want to do mutually exclusive things with their instruments. This lined up quite well with my 33 years of experience at JPL at the time!"

Mitchell packed his office and moved and unpacked. "It all worked out, but I did deservedly take some heat from the troops for it." Assuming Mitchell's place on Galileo was Jim Erickson—originally the sequence team chief from back when the high-gain antenna first failed to open. He'd subsequently worked up the Galileo ranks, in various managerial roles, to serving as Bob Mitchell's deputy. Erickson therefore was no stranger to the management level.

Smoothly Erickson assumed the chair that Bob Mitchell had occupied for only six months yet had a sizable impact on. "I would say that my greatest contribution to the Galileo extended mission was getting the transition done," Mitchell suggested. "And clearly identifying to the team, especially the scientists, what we could do and what was off the table due to funding cuts. The scientists obviously didn't like the new limitations. But they were mostly very understanding and supportive."

Even after winging past Europa as many times as *Galileo* did, months of extension remained. The ship next buzzed Callisto four times in a row. This big moon—nearly the size of Mercury—currently holds the title of most heavily cratered object in our solar system. Its ancient battle damage indicates a lack of seismic activity. Underneath this cratered and icy surface, however, some 155 miles down, an ocean likely lurks. At least, that's what could be discerned from *Galileo*'s magnetometers. The data would need to be further dissected and discussed. Alternatively, Callisto's interior *may* consist of ice mixed with rock and metal; no ocean after all. A future descendent of *Galileo* will be needed to settle these remaining questions.

Plenty of the spacecraft's dwindling attention also focused on Jupiter proper, with efforts to better understand the distribution of water in its atmosphere. *Galileo* flew near the planet multiple times, collecting details on the circulation of the wind. It snapped cloud photos. It witnessed raging Jovian thunderstorms many times larger than the largest ever seen on Earth. By proxy *Galileo* functioned as the eyes and fingertips of those who had longed for this

44. James Van Allen (*left*) with former JPL director and Voyager project scientist Ed Stone in October 2004. These two spent decades working together. The picture was taken in Iowa City during a celebratory "Van Allen Day" retrospective. Courtesy University of Iowa James A. Van Allen Collection.

chance beginning over thirty years ago—when James Van Allen first rallied for Jupiter with, "The time is ripe to get on to this."

Then, having used Callisto to lower its orbit and with only a few months remaining in its extension, the ship visited a hellish home of more volcanoes than anywhere else in the whole entire solar system—the weirdly perturbed moon of Io. It'd been intentionally left till the end because of risk. Too close to Jupiter, too much radiation. The original shot at Io had been missed during that uncertain period late in 1995, right as *Galileo* had first entered orbit. Some reckoned they'd never get so close again. But now it was happening!

In early October 1999, as *Galileo* approached its long-delayed first run at Io, Jupiter's ever-present radiation bit down hard. High-energy electrons bombarding the spacecraft cruelly flipped a single bit on one of its memory chips. Recognizing the anomaly, onboard software quarantined the orbiter in a "safe" mode, which protectively switched off all nonessential functions. Unfortunately including, um, the operation of science instruments. It happened in the wee early hours on a Sunday morning, and closest Io approach was only nineteen hours away. Urgently the ship needed fresh instructions for how to overcome its condition and follow through on observations. But most of the specialty personnel had been dropped because of funding! Whatever new software went up would take thirty-three minutes to arrive. They made it, though, returning the spacecraft to full functionality only two hours before closest approach. Clocks in Pasadena said 8:00 p.m.—a stressful Sunday indeed. The effort paid off when *Galileo* imaged hot lava pulsing through Io's surface—lava that couldn't have been more than a few minutes old.

Seven weeks later, the ship completed a tight loop through the backstretch and gunned for Io again and came even closer to losing it. Naturally, celestial mechanics had ordained that closest approach occur at the inopportune time of 8:40 p.m. on Thanksgiving Day. Some team members on duty that night picked at lunchbox turkey dinners. Others wolfed down coffee in medically unsound amounts. Fair *Galileo* embraced the spirit of bad timing by again "safing" itself with only four hours to spare. A previously unknown software gotcha had suddenly arisen, conflicting the recent October patch with other software present in the spacecraft since launch. Updated instructions righted *Galileo*'s brain only three minutes ahead of closest Io approach.

"I have very little memory of those four hours," one team member would later comment. But in return, the world got to see jaw-dropping images of a mammoth volcanic eruption. Its twelve-mile-long lava flow earned the name "Curtain of Fire."

"It really builds a camaraderie," is how JPL's Eilene Theilig described the outcome of such nail-biting experiences with her colleagues. It'd been dangerously close this time. It could've devolved into a real underwear-changing moment. Alas, victory. "This has been one of the most special times in their lives. And may never be matched again."

A planetary geologist specializing in lava flows, Theilig rejoiced in discovery as much as she did in overcoming technical challenges. Geological enthusiasm had brought her to the Lab in the 1980s. A side interest in operations is what delivered her to Galileo specifically. "I joined the mission seven months before launch, in March 1989. I was then a sequence integration engineer, which means that I helped put together, design, plan, then build the command sequences for the spacecraft." Come 1997 she rose to be team chief for sequencing and was on duty the night of the Thanksgiving scare. "The fact that everybody knew just what to do and was right on top of it was an exciting evening. And then to see an image that we got back, which was our first image of an active volcano. I mean, we caught the Curtain of Fire!"

The concept of a *command sequence* is utilized across all types of space missions reliant on automation—that is to say, a mission where the operations are happening so far away that nothing can be controlled in real time. Like a latchkey kid, the ship is temporarily being left alone due to circumstances, unsupervised, and requiring explicit direction on what to do.

"We build commands for the spacecraft that cover anywhere between three or four days up to several months," articulated Theilig. "Everything the spacecraft does in that time is built into this one sequencing command. So it's a continuous process of building these." Having completed the prime mission, JPL's sequencing workflow had changed—partially due to a change in the timescale of events. "In this extended mission, we have one sequence, or 'load' of commands for the encounter period, and one for the cruise period—in which we play back all the high-rate data recorded during the encounter."

To the surprise of some, the two-year extension concluded in November 1999 with their Jupiter machine *still* operational and more or less healthy, albeit not in perfect showroom condition. Approval materialized for a *second* farewell tour, to the tune of $9 million, with the idea of trying to answer even newer questions raised during Galileo's previous Europa-centric phase. People called this third round the Millennium Mission because of its official commencement in January 2000. Unlike Judas Priest, though, no more farewell tours would be forthcoming—largely because the low fuel light had figuratively lit up. Its tanks were nearly empty.

"*Galileo* has already succeeded beyond expectations. And we have the opportunity to learn still more in coming months," mused Eilene Theilig. "But it is sad to see the end of the road up ahead."

Indeed, proceeding with the millennium phase begged a painful question: Just exactly how to douse the lights? To bring operations to a close? Look, the orbiter would have to be destroyed somehow, and for that a precedent did actually exist. Back in 1994, the majestic *Magellan* Venus orbiter—having mapped 98 percent of the planet's surface and generally accomplishing way more than anyone reckoned—had been deliberately flown into Venus to end its mission. The first time such a thing happened.

But what to do here with *Galileo*? Easiest would be simply letting go of its controls for the delicate harmonies of gravity to usurp. But that risked the ship's accidentally striking a habitable moon like Europa. "The issue was, the spacecraft was not very clean," emphasized PM Jim Erickson in summarizing their dilemma. No reason had existed to sterilize the machine prior to launch. Continued Erickson, "Instead of leaving it in orbit for at least fifty years, the odds on hitting one of the Galilean satellites were just too great. They weren't *very large* odds. But still." And then a follow-up: During *Galileo*'s concluding loops toward destiny, how much good science could be done along the way? If better science meant taking extra time, well, how much risk would that entail?

As might be expected, the Space Studies Board and COMPLEX entered this situation to comb through the tously scenarios of cause and effect. They met in late March 2000 at the Beckman Center—a serene and pristine facility of the National Academy of Sciences located on the northwest periphery of UC Irvine's campus. Just south of San Diego Creek. Still Galileo's mustachioed project scientist, Torrence Johnson drove an hour down from Pasadena to brief the committee on the various options at hand. Subsequently COMPLEX chatted up various other scientists working the mission, did some outside brushing up, and in late June submitted a dense, utilitarian, six-page opinion. It contained little fat.

"Despite *Galileo*'s general spaceworthiness, it is unrealistic to assume that it will remain both controllable and scientifically useful for the indefinite future." So began COMPLEX's report, which gently reminded readers of the 1967 United Nations treaty on avoiding planetary contamination. "*Galileo* has a relatively high probability of eventually colliding with one of Jupiter's satellites unless some action is taken to achieve an alternative result. Thus, *Galileo* must be disposed of in a controlled fashion." Oh, the agony of throwing it away after so much had been endured for so long.

One option was to gravity assist the ship out of Jovian orbit and into faraway loops about the sun, lasting pretty much forever. That could be cool and all—with a sort of transcendental, *it never actually died* kind of woo-woo afterlife vibe—but carried a catch: "the very small, but nonzero, chance of eventual impact with Earth," asserted COMPLEX. Even such a remote possibility as this one meant formally reviewing the implications of *Galileo* striking Earth, and that review wouldn't be cheap. COMPLEX estimated some $7 million for the analysis, or more than the current annual budget of the whole project. This went into the discard pile. Instead, they'd crash it somewhere in the Jovian system.

"Personally, I like the impact idea," endorsed Eilene Theilig. "I like the definitive conclusion."

Well, Europa was right out—not even up for discussion, owing to the near certainty of its subsurface ocean. "Although any terrestrial organisms on *Galileo* have now been exposed to the vacuum of space and irradiated along with the spacecraft, it is impossible to be certain that none have survived," argued COMPLEX's report. "Nor is it possible to be certain that all surviving organisms will perish upon impact with Europa."

Two other moons also fell from contention. "Ganymede's biological potential cannot be shown to be zero." It was nixed. Same with crater face: "Callisto displays magnetic characteristics indicative of a global ocean of salt water," the text established. Now, an "indicated" ocean surely didn't carry the same weight as "strong indirect evidence" for a Europan ocean (as the report categorized it). But with better options remaining, the hammer also fell against Callisto.

Using Io could work. It'd be like hurling the One Ring into Mount Doom. Jupiter seemed obvious, too, and had precedent of its own. The probe, for heaven's sake, had been deliberately sent in with no advance sterilization whatsoever. Nobody had ventured a millisecond of concern. In deliberating the possibility of a shattered *Galileo* releasing live organisms into the Jovian atmosphere, COMPLEX concluded that "the chances of survival are essentially nil." If stowaway germs were aboard, folks, they surely wouldn't last long.

"Thus, COMPLEX concludes that collision with either Io or Jupiter is the most appropriate planetary-protection strategy for the disposal of *Galileo*." In his characteristic warm and fatherly tones, Torrence Johnson spoke for the Lab in responding that, all things being equal, they'd certainly have an easier time hitting Jupiter. And COMPLEX concurred.

Target resolved, the discussion pivoted to what kinds of science might be done with the trailing orbits of this mission. Everything had to end not near but *at* Jupiter; plots of trajectory options had *Galileo* sailing into the swirly clouds between late 2002 and late '03, depending. If the mission went with a high-risk/high-reward option, but *Galileo* turned unpredictably labile, what was the real likelihood of it hitting Europa? Maneuvering propellant was super low. Any positional tweaking drained the tanks just a touch more. One ninety-degree twist could slurp nearly seven pounds by the time the ship reacquired its original position.

And what about the rest of the machine? After years in a high-radiation environment, *Galileo* had degraded. Just look at its camera—outputting lots of overexposed, saturated images. Switching it off and back on helped sometimes. As did reloading software. Problems also abounded with the system *Galileo* used to manage its orientation in space. The low-gain antenna had to be pointed within four degrees of Earth, and to maintain position, the craft tracked multiple stars. But radiation flashing through the star scanner occasionally confused the ship in terms of what actually was a star and what wasn't. A thousand flashes *a second* could ping in areas of highest radiation. Occasionally the star scanner would crash outright, reboot, then crash again. Generally this could be mitigated by training the scanner on a particularly bright star—one more intense than any radiation flashes. Operations were becoming a series of workarounds one after the other.

A year into the Millennium Mission, in January 2001, Eilene Theilig took over as project manager for a gaggle of Galileans now down to less than sixty people. The health and operation of their onboard tape recorder stood front and center as the main stress point. Yeah, plenty of lightweight science readings could be sent in real time. But images and other heavy data still counted on the deteriorating reliability of the single and distinctly radiation-soaked tape recorder. With no advance warning, the moving length of tape could hang up on any number of heads or capstans. One preventative routine wound the tape halfway down its length and back again, exercising the working parts. It usually did the trick of keeping things moving. Someone realized how, when doing that, they always stopped the tape in the same location every time, and wouldn't it make sense to not do that? The process was reworked.

For a briefly exclusive period in February 2001, geriatric *Galileo* even had company. Passing through the Jovian system was none other than noble *Cassini*—on its way to Saturn and what promised to be an equally exciting regimen of discoveries. Both spacecraft were able to examine the same Jovian features at the exact same time and from unique vantage points—a situation never before occurring at the outer planets. (Both *Voyagers* journeyed through approximately two years apart.) In particular, *Galileo* and *Cassini* both observed a giant eruption on Io that sent material over 230 miles above the moon's surface, blanketing terrain as far as 435 miles from its center.

Galileo looped back for a tight Callisto flyby in May 2001. Its concluding observations loomed. These final, dangerous, radiation-bathing trajectories could be sanely undertaken only by a ship on its very last legs, and the juicy target was Io. Closer inspections of this netherworld had been prioritized by COMPLEX in its report.

Skimming Io added a year to the mission, during which time any number of things could go sideways on the ship. High radiation during encounters triggered all sorts of sweaty-palm glitching, with short circuits across the spin bearing assembly, for example, prompting the ship to reset itself. "We saw it in the gyroscopes and the attitude control system," Eilene Theilig elaborated of the random knockdowns. Crap could happen multiple times during a single encounter. And more trouble struck in early August 2001 when *Galileo* came in low and tight above Io—barely 120 miles up. Soaring overhead, nearly all imaging was lost due to a camera malfunction. But what pictures the ship did manage to return home were jaw-dropping. Displayed on JPL monitors were surprises in succession: Lava lakes. Lava *curtains*. Fifty-thousand-foot-tall mountains collapsing under their own weight. How could mountains even *form* on such a redheaded oddity like Io?

Then there's what the spacecraft *sampled*. Quite inadvertently, *Galileo* flew through the top of an active volcanic plume blasting three hundred miles high. It took the first such direct measurement. No more than five minutes elapsed between eruption and fly-through. "Totally unexpected," marveled one person from the University of Iowa team. "We've had wonderful images and other remote sensing of the volcanoes on Io before. But we've never caught the hot breath from one of them until now."

Just over two months later, *Galileo*'s next dalliance began with an overflight of Jupiter's turbulent north polar vortex. The ship logged five unique hours of observations on this one feature alone. It also examined the probe's entry site—by chance a relatively dry and cloudless area. Io then came in range with *Galileo* still on the moon's dark side; a comprehensive mapping of surface temperatures ensued. This pass darted over Io's south pole and drew even closer

than previously—within 112 miles of the zany surface. All six fields and particles instruments continuously gulped in data for twelve straight days. The
camera also functioned well, harvesting shots of lava channels and prominent volcanoes.

And then it happened. While playing back recorded images, the tape stuck
again. After resting it for three days, they had another go and lucked out; all
pictures were recovered. Tape-management routines always involved two complete playback sessions just in case the first spit up any errors. Full playback
could take weeks upon weeks, and Io's last flyby awaited only a few months
downstream, shortly after the New Year. Please, they begged of the recorder,
no more sticking. If these successive flybys demonstrated anything, it was how
quickly Io's features could change, and everyone wanted to see.

January 2002 began. Orbit number 33 for *Galileo*. Its maneuvering propellant was about gone as the final Io pass began. This one came in over the
southern hemisphere, again on the dark side, moving sunward. Venturing to
within sixty-three miles of Io's unstable and indigestive terrain, it marked
the tightest approach to any Galilean satellite for the whole mission. Just
over sixteen total days of science collection occurred and included readings
from Jupiter as well.

The pass wasn't without drama. Some twenty-eight minutes ahead of closest Io approach, *Galileo*'s lower, despun bus unexpectedly reset and went inop-
erative. It'd happened so often in the past that a custom software patch was
already in place. But the patch failed to recognize this particular manifesta-
tion of the reset. All incoming data meant for the tape recorder was lost to
the ether. Luckily, the intellectual horsepower of the project techs brought
the ship fully back online in time to acquire a passionately desired image of
moon Amalthea, enabling cartographers to refine their understanding of its
exact position in space.

Galileo then snapped its very last image—a color one of Europa, taken from
the side facing Jupiter. The camera was done forever. Tape playback started in
late January and included various recordings not yet returned.

January's Io swing-by had supplied the mission's final gravity assist. Not
one more time would the ship need such a thing or even the most pitiful of
tweaks from its thrusters. It was permanently on course and set. In September
of the following year, operational or not, *Galileo* would be eternally embraced
by the planet Jupiter.

Months went by, the summer of 2002 giving way to autumn, with one final visit to occur before plunging. The COMPLEX report had specifically echoed a core scientific interest in flying closely past Amalthea—one of Jupiter's innermost satellites. "Amalthea may be a fragment of an object that formed closer to Jupiter than the Galilean satellites," the report had noted. Its recommendation was to monitor radio transmissions from *Galileo* as it passed the moon. Requiring no functional science instruments—just the radio—this would help estimate Amalthea's mass and might enable definitive conclusions to be drawn about the formation of the Jovian system in general.

Closest Amalthea approach happened on November 5 while only forty-four thousand miles above Jupiter's psychedelic cloud tops. It pushed *Galileo*'s total radiation exposure to approximately four times the amount it'd been originally designed to suffer. "In that close, we actually did take substantial damage," commented Eilene Theilig. And wouldn't you know it? After encounter, the tape recorder wouldn't play back. Oh god, the recorder again. A key man on the troubleshoot began examining a nearby spare with a depleted, game 7 kind of look and asked himself, "What's going to die because of radiation in a machine that was working perfectly before we flew through it?"

He took it around to others on Lab, soliciting opinions. Their collective insights boiled down to a strong belief that radiation had affected a mechanism that senses the position of the tape path motors. Specifically, this mechanism contained an array of tiny LED lights that shine on position detectors. After encounter, the LEDs were outputting maybe 20 percent. Evil radiation seemed to have rearranged atoms inside the crystal lattices of the LEDs themselves, dimming the lights enough to halt recorder operation. At the moment it wouldn't run for longer than a second at a time.

Oddly enough though, pushing electricity through the LEDs for extended periods would force their atoms back into place. This condition was fixable. Cautiously at first, the procedure gradually lengthened to the point where LED output improved to about 60 percent, thus enabling the recorder to operate for up to an hour at a stretch. Long enough to play back all the revealing data from Amalthea's flyby.

Celebrated Eilene Theilig, "It is one of *the* most astounding recoveries, I think, the mission has made."

What data came off that tape contributed greatly to our understanding of Jupiter's rings. Turns out the planet has multiple ones. Their boundaries are actually defined by the orbits of Amalthea plus three other close-in moons. Eons' worth of impacts by meteorites and comets upon all four of these moons kick up waves of dust and debris that are no match for Jupiter's gravity in stripping it all away. (Amalthea isn't really dense to begin with—resembling more a pile of rubble.) The loose material becomes trapped 'round Jupiter and begins circling it forever and—abracadabra!—rings.

Even the thickness of each ring corresponds to the orbital inclination of each moon. Closest in, Jupiter's wispy "halo" ring comprises mostly dust. Farther out is the sharply defined "main" ring. Then beyond it are two faint "gossamer" rings.

Arcing away from Amalthea, *Galileo* began rounding the bend. The final orbit of its $1.39 billion mission—*the* mission of some eight hundred souls in all—would trace an elongated, ten-month loop out and away from Jupiter and then back in for good.

Galileo's final project manager, Claudia Alexander, assumed control only a month before the end. "By that time we were all pretty attached to the spacecraft," she remembered. "You felt you had almost a living thing that was palpably returning all the love we'd lavished on it. Some of my colleagues were very emotional about saying goodbye."

What a survivor. Legislative cannonballing. Delays. Tragedies of engineering oversight and lost human lives. *Galileo* overcame it all, journeying some 2.8 billion miles while making everyone who touched it proud. After all these many years, putting one's role into perspective could be tough. "The thing that I connected to was the dedication and commitment of the people who built it." That was John Casani, speaking from his then-current position as assistant JPL director.

"A lot of people were sorry to see it go," he went on. "You know, I didn't think of it that way. It was out of fuel, and there was nothing much more we could do with it. It was going to die one way or another." As with all milestone events, Casani felt the impending send-off to be worth gussying up for, and he had selected a festive rose-colored dress shirt with patterned tie. The ship deserved a tie.

Durable *Galileo* took the plunge on a Sunday morning. In Pasadena the calendar read September 21, 2003. NASA TV, a thing on satellite and cable since the early 1980s, went live to JPL's von Kármán Auditorium as a standing-room-only crowd welcomed a discussion panel of four previous Galileo project managers. Introduced first was Casani, who reacted to the crowd's applause with a millisecond of visible disbelief. Next to him sat Dick Spehalski, who'd taken over after Casani for a couple years before tossing to Bill O'Neil. Alongside Spehalski sat Jim Erickson and Eilene Theilig. Up high behind them, a line of three projector screens showcased video from von Kármán and data displays from the spacecraft. Off to one side of the stage, near Theilig, a large *Galileo* model had been positioned atop a black-skirted table. Nice accent. And beyond that, an oversize rear-projection TV counted down the ship's remaining minutes in bland sans serif numerals.

"We're going to start with John," said the host. "How did this mission get started?"

"Whoa," blurted Casani, who then laughed as the impossibility of summarizing such a thing settled upon him.

Half a billion miles from the auditorium and TV cameras and hype, *Galileo* began its own live transmission—this one consisting of science data trickling from its instruments. Between the ship and Jupiter lay six hundred thousand radioactive miles that were closing fast. Radiation had yet to trigger any glitches.

The NASA broadcast alternated between discussion panels in von Kármán and stand-ups with Claudia Alexander from the control room. "We are *still* collecting science data," she cheered to the amiable host alongside her. "It's been a great day so far." Fittingly Alexander had chosen all black for this funerary finale and clearly had been anticipating the worst. "I am absolutely thrilled about that. I was quite pessimistic that we were gonna make it this far, still transmitting science data."

Another panel kicked off in von Kármán, with science peeps this time, including Torrence Johnson. Seated next to him, the panel's chair asked about *Galileo*'s most surprising discovery.

"Well, I think it was the, the evidence for oceans on these icy moons," Johnson instantly responded. Adorning his neck was the latest Torrence Johnson hallmark of a string tie. "We had been oscillating back and forth in the science community, over the previous decade, as to whether it was possible to sustain liquid water oceans. And *Galileo*'s data has definitely tipped the scales over toward, ah, believing that we really *do* have these subsurface oceans on these satellites."

Tracing a lazy curve toward obliteration, *Galileo*'s radiation levels had now grown so high that the onboard star scanner could no longer function.

Unconcerning though because the software had already been told to expect no more stars. The ship's magnetometers captured their final readings. Less than 89,000 miles away, gradually falling in, and the strength of Jupiter's magnetic field at this proximity would've completely overwhelmed the magnetometers anyway. Picking up speed. Forty-five minutes later, the ship had closed to within 35,700 miles. No undo. No abort.

"Personally, I feel a little sad. I, um, really had the time of my life on Galileo." In a chair one over from Torrence Johnson on the von Kármán stage, Rosaly Lopes traced her own interest in the mission back to 1979 and a chance witnessing of a volcanic eruption in Sicily. It motivated her pursuit of a PhD in planetary science because, in her words, "I was hooked on volcanoes!" Education complete, Lopes joined Galileo in 1989 with an obvious focus on Io. The level of volcanic activity present there utterly dwarfs what is found on Earth. Across Io's surface, upward of a hundred actively erupting volcanoes can be raging forth all at once. Even lava changes over time, and the temperatures of Io's eruptions—as Lopes pointed out to the audience—are so hot as to indicate a lava type that hasn't erupted on Earth for at least two billion years. So in a very real way was a modern spacecraft examining the ancient earth.

"I am a little sad to say goodbye to an old friend," she eulogized, as the crowd listened politely. "But I know that there are new missions and new worlds to explore. So we carry on."

With only a few minutes to go, less than six thousand miles above the Jovian cloud tops, *Galileo* offered its very last transmission and then passed into the darkness of Jupiter's shadow from which it would never emerge. The cloud tops met their visitor. And then—for the first and only time in its entire existence—the majestic *Galileo* orbiter caught a breeze, experiencing just a mote of what its companion probe did.

"We are now less than a minute away from the end of mission," said JPL's on-camera host. In the Lab's mission control room, there was no longer anything to control, and many at the consoles were no longer seated before them. They were standing up, aiming compact cameras and home-video camcorders, and committing the scene to recordable media as lens caps on little black strings dangled below their hands. People counted down the final seconds as if a launch were occurring versus the exact opposite.

After collectively chanting, "Two," then, "One," nobody said, "Zero." Instead, the Lab fell completely silent for two whole seconds as into Jupiter went their *Galileo* at 108,000 miles an hour.

Forces began acting on the ship. Tearing away the long boom arm holding its magnetometers and Don Gurnett's carefully crafted plasma wave antennas. The fragile mesh of never-unfurled, high-gain antenna caught the atmosphere and disappeared, leaving one to wonder if it might have finally opened all the way for even an instant. Hugh von Delden's blankets and sun shields were no match for the conditions now being imposed. Vanished. The maneuvering thrusters and pipework went. The plutonium power generators. Metal panels. Clyde King's and Mary Reaves's wiring bundles. The tape recorder. The camera. The bus. Circuit boards and computer chips. Every last bit obliterated.

And in these concluding moments of death, the members of this upstanding community of the Galileo mission to Jupiter could find solace in how they had remained unbroken. They did it. They did it for Jupiter. They did it for understanding. And they did it for Mr. Galileo Galilei himself.

"*Galileo* is gone," saluted the on-camera host. She said it respectfully, the way a hospice volunteer would.

After which the applause began.

Fifty-two minutes and eighteen seconds later, the ship's final whispers reached the Goldstone Observatory's receiving dish of the Deep Space Network. Their beloved creation had ceased to be.

"The damn machine was just a machine," said John Casani.

26

An Endless Pursuit of Wonder

To the outside world, a spacecraft experiment may be a gadget.
A metal box with complicated insides. But in the NASA world, it is a
collection of bright individuals whose notions and emotions must
be harmonized before anything can be built.

—Charles Kennel, physicist, Voyager coinvestigator, and former
NASA associate administrator

Sixteen months after *Galileo*'s poignant farewell, more excitement burbled from the outer solar system as a 701-pound golden disc approached Titan. Zürich transplant and "too Swiss" engineer Platon Tatalias initiated his personal celebrations by flying up to Frankfurt, then taking a car forty minutes south to Darmstadt. He got in around midday—in time to attend a dinner organized by the European Space Agency. Tatalias knew plenty of folks there but not plenty of others, and the scene only reinforced how an immense diversity of talent is required for visiting the types of worlds that are barely visible from Earth. Three weeks had elapsed since *Huygens* separated from mother ship *Cassini*.

Next morning, Tatalias made his way to ESA's European Space Operations Centre. It butts up against the Waldfriedhof, a serenely forested nature preserve and cemetery. The main control floor of any space mission facility is no place for even well-disciplined visitors to loiter. And with the jobs of people such as Tatalias long since complete, he found a spot in a nearby observation room alongside other nonessential personnel. Wing Ip joined the festivities in the center's main press room.

The day's excitement traced a progressive crescendo of multinational engineering and integration. Within the hurtling disc, a trio of redundant countdown timers hit zero and prompted a chain of batteries to begin powering up what lay nestled at the center of things—an aluminum amalgamation that was *Huygens* itself. Computers, sensors, instruments—all rousted after years of peaceful slumber. Electronics warming up. Shaking off the chill. Nearly four and a half hours later, the arrangement reached the uppermost whispers of Titan's chubby atmosphere. Date: January 14, 2005. Speed: just over 13,400 miles an hour. Angle: precisely 65.4 degrees. Altitude: almost exactly 789 miles above a surface heretofore unseen and undefined. Humanity had come a-knockin'.

Barreling straight in, the leading face of its nearly nine-foot-diameter disc supernaturally glowing, all that burning and pain suffered initially by gold-colored foil, then by glued-on tiles of silica fiber and resin that slowly eroded and gave themselves to the heat. Some 12 g's of force converged at peak load as *down, down, down* it plummeted with nobody knowing whether *Huygens* would strike a ridge or a lake or even if its batteries would last long enough to reach anything at all waiting below.

Within four and a half minutes, the craft slowed to 693 miles an hour on friction alone. Then, still very much supersonic and 96 miles high, deceleration sensors cued an explosive mortar to fire. Bolts severed and freed the back cover of the heat shield as the mortar sent a pilot chute bursting upward through a tear-away roof panel and—one-Mississippi, two-Mississippi—the pilot chute sliced away for replacement by a larger-diameter main. That pilot chute had waited through seven years of cold boredom for two and a half seconds of action and in a blink did its job perfectly. Then went goodbye as it whistled away in the haze. Thirty seconds ticked. *Snick* went redundant blades. *Crack* went mounting bolts as they sheared in two, and away went the forward heat shield. And *on* went the radio transmitters as various intake ports on *Huygens* popped their covers like Christmas crackers.

I'm alive, cooed the probe.

Platon Tatalias reminisced about the scene. "We were listening to events data as they came in, while on the screen an animation was shown. First excitement was when it was announced that the beacon signal from *Huygens* was received." That came around 11:40 a.m., as reported by America's Green Bank Radio Telescope in West Virginia. People cheered. It meant the probe's back cover really was off and the transmitters definitively up and running. Major milestone. *Huygens* had begun slinging data upward to *Cassini* as the massive orbiter trundled overhead at some thirty-seven thousand miles distant.

The "beacon signal," as Tatalias called it, represented early notice of probe vitality. See, radio signals from *Huygens*, by design, were never intended for

45. During *Huygens*'s assembly, its back cover is lowered into place.
Visible are the front cover's protective heat-blocking tiles prior to being
concealed underneath gold foil. © European Space Agency.

receipt directly on Earth. Its transmitters weren't powerful enough. So the probe would instead radio all its findings to *Cassini*'s onboard recorders for eventual transmission home much later on. The big orbiter couldn't simultaneously receive *and* relay because its antenna dish could point either at *Huygens* or at Earth but not at both. The situation meant hours of waiting for even basic confirmation that the probe was operational. Okay, maybe that's an inconsequential amount of time considering the years of its outbound flight. But who wouldn't want to know as soon as possible? In response, Green Bank stepped up. During the critical moments after Titan entry, the telescope's gaping, hundred-meter dish would enjoy a direct line of sight to *Huygens*. And likely be able to snatch its "carrier" signal—a constant base tone—though not any big data like pictures. But even minor Doppler shifts in the carrier tone *could* inform them of jarring events such as parachute deployments. And hopefully a touchdown!

Inside the atmosphere now. *Huygens* on chute. Events proceeding as scheduled. Dual twiglike sensor booms next swung out and snapped into place. Instruments inhaled whiffs of Titanian atmosphere and ducted the samples to hand-built analyzers mounted deep within. A lens cover discarded itself so the camera behind it could see out. What that camera saw was dark, fiery-colored haze and nothing more. Everywhere.

As only one face of the project, a French planetary scientist named Jean-Pierre Lebreton spoke to reporters. He'd been with ESA since 1978, joining right after the completion of his PhD work, and had functioned as the Huygens mission manager since its inception. With the probe finally dropping, rampant speculation abounded as to how things might play out.

"During the project development phase," Lebreton later recalled, "three minutes of science from the surface, after landing, remained as a realistic goal. Although survival on a totally unknown surface was never guaranteed!" He stressed that any postlanding run time was always viewed as a *goal*, versus some defined requirement, with *Huygens* capable of lasting beyond three landed minutes. It probably had enough battery life; they'd strategically planned this descent.

All good, but what about hitting a lake? As clarified by Lebreton, "Strictly speaking, it was not designed by requirement to float." If ESA had mandated such a deliverable, he said, validating it would've been difficult. Exactly what liquids would it need to have been test-floated in? And for how long? If it sank just a little, would it still have had to work? Up to what depth? Luck-ily, an analysis conducted back in development showed *Huygens* to be rea-sonably watertight almost by accident. "Hence," Lebreton continued, "there were hopes that it would float—for a least a few minutes—had it landed in liquid." Answers would come soon enough.

Huygens descended smoothly under its main chute. Just over ninety miles to touchdown. Still nothing for sky but opaque orange. Every time news had come through of a successful parachute deployment, Wing Ip heard cham-pagne corks popping. "It turned out to be the celebrations of the contractors," the people who'd built the parachute systems, he explained. "For them, the mission is already a resounding success."

"Don't underestimate the will to celebrate," added Platon Tatalias. "*And* the thirst of the engineers!"

Fifteen minutes after deploying, the main chute gave way—superseded by a final and compact "descent" chute responsible for briskly escorting the works down to whatever kind of landing awaited. Dual-redundant altime-ters kicked on and fed stats to the instruments. Less than 200,000 feet to go. Ideally, the machine would complete its descent while some fractional per-centage of battery power still remained and, by golly, do so before *Cassini* dis-appeared from radio contact.

The initial hour under its descent chute went kind of roughly. Lots of heavy turbulence. Thirty-six tiny vanes mounted around the periphery of *Huygens's* shell made it rotate. Such motion aimed to provide the best operational con-ditions for multiple onboard instruments. But the spin rate began to slow. It then stopped altogether, only to begin spinning in the opposite direction. This unexpected behavior complicated data analysis later on.

Presently, winking shapes in the thick orange background hinted at more to be seen. Coming down smoothly now, 190,000 feet. Smog clearing. Below it, *Huygens* glimpsed eroded mountain ranges and ridges—mesmerizing features hundreds of feet tall. There were mountains! Down to 72,000 feet, dropping, strong chute, dark streaks in the mountain ridges. Exaggerated cracks and fractures wandering up the sides of one towering range. Settling in, 16,000. River deltas now. Scarps and shards. Carved channels in rock—that meant flowing liquid. Maybe not in present day but definitely at some point. Nearly there and coming down easy. Less than 2,300 off the ground. A thousand.

Some two hours and ten minutes after the descent chute first went out, a

46. At the landing event, Jean-Pierre Lebreton shows off the full-scale *Huygens* SM2 model. This is the actual unit test-dropped in Sweden. Said Lebreton of the overall *Huygens* concept, "It was designed as a straight atmospheric probe," relaying measurements of the environment in case it never landed. © European Space Agency.

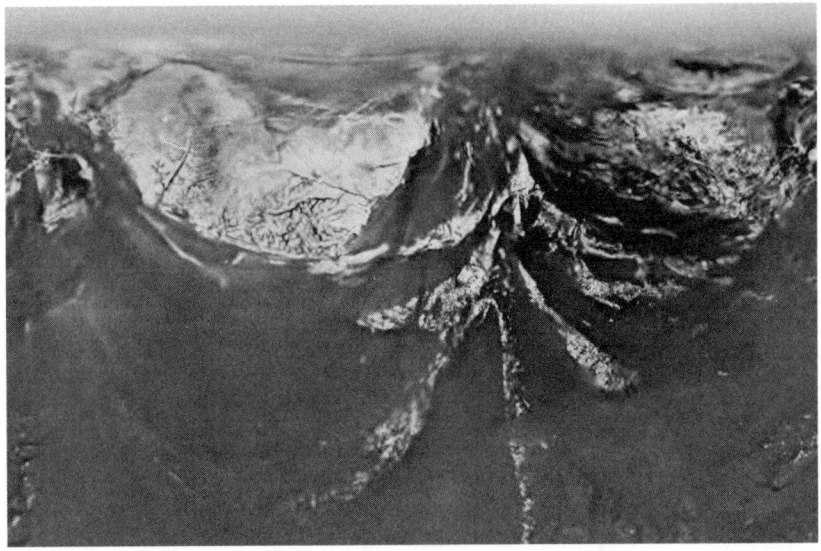

47. Below Titan's haze yet still high above ground, *Huygens* snapped images such as this one rapidly enough to create a movie of its descent. Note the river channels carving through surface features. © European Space Agency.

lamp on the side of *Huygens* spit to life and began illuminating Titan's imme-diate surroundings. Final approach. Data pulsed upward to *Cassini*, still over-head. Two minutes and five seconds later, *Huygens* thumped pay dirt at just over eleven miles an hour and executed a poetic sideways baseball slide wor-thy of Carl Crawford. Its motion stopped.

Underneath *Huygens* an instrumented, protruding metal wand had inten-tionally served as the lander's first point of contact. It suggested a complex surface consistency—similar to a hard crust with softer material underneath. Well, shoot, how could something that weird be communicated to the press so that nonscientists might understand? Members of the project team brain-stormed analogies, and somebody came up with a great one.

TITAN TEAM CLAIMS JUST DESSERTS
AS PROBE HITS MOON OF CRÈME BRÛLÉE

Alison Abbott, Darmstadt

The high-risk mission had functioned better than its designers had dared to expect—and they quickly reported that the moon looked even more interesting than they had hoped.

"My worst nightmare had been that Titan would turn out to be a bland object, a simple icy surface with no structure or variety," says John Zarnecki, principal investigator on the Surface Science Package at the Open University in Milton Keynes, UK. "But it is fascinating."

Zarnecki coined a description of Titan's soft but crusted surface that looks likely to stick: *crème brûlée*.

Huygens sat there, in one piece and completely functional, on a world never before seen by humans. Let alone visited! Temperature: 291 below zero Fahr-enheit. Pressure: almost 50 percent higher than Earth's at sea level. Atmo-sphere: mostly nitrogen and methane. Titan, as would be discovered, operates a methane cycle akin to Earth's water cycle. A drop of liquid fell past the cam-era lens, beyond which sat a number of happy little ice pebbles. Their general roundness indicated a history of tumbling in liquid.

Layered against the cold with internal foam blankets, batteries giving their all, the visitor purred through its preprogrammed tasks, chucking every result upward on the radio link as soon as it became available. The jolt of landing had produced a dramatic jump in the amount of detectable methane—as if it sur-

48. What *Huygens* saw: the view from its final touchdown spot. Titan's horizon disappears into the mist. Its surface, darker than expected, shows evidence of erosion. The prominent, rounded pebble below center is estimated to be 1.5 inches across. © European Space Agency.

rounded the probe. Methane was *everywhere*, Titan's surface damp with it. The relative warmth of *Huygens* even boiled liquid methane pooled underneath its metal shell. And let's not forget the argon! Its presence, both up high and on the surface, meant geologic activity. The probe also found ethane. Carbon dioxide. Benzene. In sum, a rich inventory of complex organic compounds with the potential of sustaining life.

This side of Titan still had 131 hours till sunset, although there was no way the batteries would last long enough for *Huygens* to see it. Even so, the little nipper validated every prediction that this particular moon was indeed a place worth visiting. Titan had so much on offer that it would come to define and consume people's entire careers. Planetary scientist Tobias Owen mused, "Titan has seemed to be some kind of perverse machine that's been put into orbit around Saturn by a superior race—as a kind of intelligence test for earthlings, to see if they can unravel what's going on out there."

Four and a half hours after *Huygens* initially went in, *Cassini* dipped below the horizon of the landing site and severed the radio link. All in all, more than 220 minutes of probe data reached the mother ship, of which 72 minutes' worth came from the surface. *Cassini* then crisply rotated to beam home what had been collected.

Offered JPL's Charles Elachi: "In my mind, the two biggest surprises were, first, that it has lakes with rivers coming into them. All made of hydrocarbons." Some were gigantic, approximating the Great Lakes in the United States. (The largest of which, Lake Superior, boasts a nearly 2,800-mile shoreline and holds almost three thousand cubic miles of water.)

"It looks like Earth to some extent!" Elachi went on, giddy. "The other big surprise was the sand dunes. We did not expect that there would be fields of sand dunes." For anyone who thinks that every kid grows up and no longer wants to play in the sand, check out Charles Elachi and his headful of wonder. "It's like this with every scientific exploration. You answer a certain question, and it raises new questions. What are the sand dunes made of? Do they change? What is the liquid made of? How deep are the different lakes?" His questions kept coming.

"We're starting to think about the next mission for Titan."

Since Elachi made his comments, that next mission has been designed and approved and will see a descendant of *Huygens* take to the skies. Titan's thick atmosphere is ideal for a drone-style, multi-rotor, airborne science platform repeatedly day-tripping across the moon's surface in pursuit of intrigue. Called Dragonfly, this NASA mission aims to spend over three years studying Titan's atmosphere and geology. Among other investigations, the machinery will analyze actual surface samples in its belly, looking for biologically relevant compounds, and study Titan's weather. Dragonfly's hardware mock-up is the size of a snowmobile. We will be returning to this fascinating moon in style.

Cassini, having relayed all the *Huygens* data, obediently picked back up with its four-year prime mission and some seventy-two orbits still to come. The huge machine had been performing brilliantly. None of this migrainous, *Galileo*-esque pinballing from one technical snafu to another. The fixed antenna dish atop *Cassini* didn't need to unfold. The solid-state data recorders inside its body would never experience tape sticking or fogged LEDs or any number of klaxon-sounding chest grabbers that its Jovian colleague had suffered. Also, don't discount the impact of a healthier environment: Saturn's radiation belts are wimpy-small compared to Jupiter's, posing far less of a risk.

About the only thing *Cassini* didn't have was those boom arms, giving the science teams something to bellyache about until their dying days. Broad disappointment was understandable in part because their deletion resulted not from engineering failure or inadequate development time but simply a lack

49. *Cassini* entered Saturn's orbit in late June/early July 2004; that naturally meant a party. A bearded Bill Kurth (*center*) nurses a Corona while speaking with John Casani. Just behind Kurth's shoulder is Charley Kohlhase, *Cassini*'s science and mission design manager. Courtesy NASA/JPL-Caltech.

of funding. Not enough money had been available to the project. Iowa's Bill Kurth made a valid argument about the price of complicated coordination: "Deciding who gets to point where, and when, occupied most of the mission planning efforts. And probably cost *much* more than the scan platform would have cost. My opinion."

He wasn't alone, of course. Everyone made compromises. The plasma spectrometer team, for example, had negotiated the mounting of a dedicated little platform on the orbiter's body. It enabled the rotation of their plasma sensors up to 160 degrees and clawed back part of the measurements they would've originally had with the turntable. "Being able to scan the entire sky is critical to plasma observations," cried one of their reports, and the surrogate platform still couldn't do it. Long after *Cassini* had launched and was on its way, telemetry would come back suggesting the spectrometer's rotating parts had snagged on an insulation blanket. Mission managers thus further limited the degree of rotation this instrument was allowed.

Regardless, success abounded. Each orbit saw *Cassini* tweak its trajectory, via deliberate gravity assists, to produce a dynamically variable tour of a system every bit as diverse as Jupiter's. The last time planet Earth visited Saturn was during the Voyager days. And no question, both *Voyager* flybys revealed much. But an orbiter with a four-year residence would continually build upon those findings. Start with Iowa's Don Gurnett: "We didn't detect any whistlers at Saturn with Voyager for some reason. I'm not sure. We just weren't in the right place." For him, *Cassini* would ultimately deliver on the promise of discovery at Saturn. "There's no thunderstorms, sometimes, for long periods of months. And then a thunderstorm will build up in the atmosphere, and then we hear a lot of lightning. And sometimes whistler signals!"

Every scientific discipline celebrated its own wonderment. Take Saturn's hexagon. Partially imaged by the *Voyagers*, then comprehensively by *Cassini*, is a mammoth jet stream feature bearing six curiously well-defined sides. It rages at Saturn's north pole and would never occur on Earth because of our disparate, rumply surface of oceans and deserts and mountains. Not to mention Earth's relatively thin atmosphere that fluctuates in temperature. But at Saturn, the lack of terrain—plus a deep, deep atmosphere with a hundred times less sunlight—allows the winds to just go, baby go, and the hexagon formed. It measures more than twice the width of our home planet. Wind speeds can hit three hundred miles an hour, equaling that of the fastest-ever tornado recorded on Earth. Smaller vortices garnish the hexagon like a plated gourmet entrée. At its center exists a massive void, similar to the eye of a hurricane, that is fifty times larger than any given Earth hurricane. Within that void is an emptiness plunging down as far as *Cassini* could see.

And what of the planet's rings? Consisting mostly of water ice, they proved to be not only beautiful but homey, containing moons that impart beautiful gravitational ripples in the ring material. During its travels, *Cassini* found six previously undetected (and now verified) moons. The presence of more is forever conceivable as people never stop analyzing mission data.

The spacecraft repeatedly flew past Titan, snapping images in visible light and with radar. (Mission rules actually required the orbiter to continually reencounter Titan. Reason: a baseline dependency on the large moon's gravity to help align *Cassini* for upcoming orbits.) Those splendorous sand dunes, which commanded so much of Charles Elachi's attention, were originally spotted by *Cassini*'s radar. Current thinking holds that the dunes, while nearly identical in appearance to dunes on Earth, are not at all the same thing. Earth's sand comes from rocks that have been weathered over lengthy timescales by wind and water. But dunes on Titan seem to be made of complex organic molecules forming way high in the atmosphere before falling to the ground. How this "sand" forms as it does is still being debated.

Perhaps *Cassini*'s greatest discovery was at Enceladus. At first glance, it's a gleaming-white yet otherwise unremarkable-looking moon of Saturn. Going into the mission, its general lack of craters had some positing whether volcanic activity might be the culprit—by continually resurfacing the place. Definitely worth a close look. Another point for volcanoes: material sheeting off Enceladus seemed to be feeding one of Saturn's outermost rings.

Ultimately the secret of this moon was revealed not by pictures but in some unexpected data from Cassini's magnetometer team. Why, they questioned, did Saturn's expansive magnetic field take the long way around Enceladus? As if the moon had a force field in place around itself? What, some kind of protective bubble?

Follow-up imaging gave it away. No force field. No volcanoes. Instead, plumes of icy particulates were jetting off the moon's south pole—a shockingly *warm* south pole. What's ejected is responsible for populating an entire ring around Saturn, though much flutters back onto the moon's surface in a never-ending snowfall. This discovery alone cracked the riddle of why Enceladus is the brightest and most reflective moon in our solar system.

The plumes spew from a distinctive grouping of four fissures aligned at the moon's south pole. Researchers call them tiger stripes because, well, that's what they look like. *Cassini* purposely flew low through the plumes and managed a good lick. It verified the existence of something on Enceladus that no other moon in our solar system is known to have: Spraying out from its stripes, at eight hundred miles an hour, are hospitable geysers of salty water vapor, methane, and other rich organics. All ejected from a humongous underground ocean that's continually warmed by harmonic tugs of Saturnian gravity. We are talking temperatures of 190 degrees Fahrenheit or greater. About double that of your average hot tub. This constitutes hydrothermal activity. It was completely unexpected at Enceladus.

This moon has warmth. It has water. And food in the form of methane. It is quite capable of supporting life.

Much appreciation for telling us, *Cassini*!

Famed author Arthur C. Clarke addressed the project's troops in mid 2007. He spoke via prerecorded video, shown on screens to those mustered in von Kármán Auditorium. "I want to thank everyone associated with this mission and the overall project. It may lack the glamour of manned spaceflight, but science projects are tremendously important for our understanding of the solar system." The occasion honored *Cassini*'s impending close flyby of the moon Iapetus—a Clarke favorite. In the video he wore a plain gray button-down with a mandarin collar and read from notes. Behind Clarke towered shelves full of books such as his famous *2001: A Space Odyssey*. "Who knows?" he continued. "One day, our survival on Earth may depend on what we discover out there." He smiled and wished the group a successful flyby.

Like *Galileo*, the flagship Saturn orbiter couldn't last forever. Over years of flight, its reserves of maneuvering propellant gradually waned, triggering discussions on how to safely destroy *Cassini* without contaminating Enceladus or Titan or other potential habitats. "It's room temperature inside the spacecraft. A hearty microbe could easily have hitched a ride along," suggested one project member.

"It is going to be a dramatic end," waxed Charles Elachi in 2015. "We are planning, on purpose, to have the spacecraft enter Saturn's atmosphere and burn up before it completely depletes its fuel." That occurred on September 15, 2017, in another emotional countdown that JPL termed the "Grand Finale." Starting from its original proposal by Wing Ip, Huygens took six years to be approved. Add eleven months for a congressional thumbs-up to the Cassini mission proper. Plus eight more years until launch. Finally, tack on almost seven more long ones just to reach Saturn. That's enough time for a day-old newborn baby to grow, attend school, and graduate college.

"In our business, you have to be patient," advised Charles Elachi. "It takes a long time, particularly for the outer planets."

27

Grab On

Oh my goodness, you didn't say bread pudding came with ice cream!

—Casani to Nancy the waitress at the Italian restaurant

Imagine for a moment that history is a rope in motion, running horizontally through time. Braided and thick, it defines the past and helps contextualize the present and is always, always moving. Whether you grab on or not comes down to personal choice.

Some hold onto that rope for dear life. They seek to understand today by posing questions about yesterday: Why are spacecraft so small nowadays? Why no more one-shot, "flagship" missions to the outer planets? Why are we trying to do things faster or better or cheaper? Every answer is an individual strand in that moving rope of history. Just grab on. You'll find the responses interwoven with those for such ponderables as, *Why did the Immovable Ladder become immovable?* Or, *Who was D. B. Cooper?* Or, *Why did the dinosaurs die?*

If we cease our efforts and stop making sense of today—which is easy to do by disregarding history—then that rope will get away from us. It'll fly from our hands, and we'll stand there in solemn ignorance of when we determined that planets and moons could be toured with gravity. Or what space music really was. Or how oceans and geysers thrive in so many places. Or how we even rose to the challenge of exploration in the first place.

And of course, we will no longer experience teachable stories of how people harmoniously intermeshed with common goals to invent ingenious machines that departed our home planet for the great beyond. Where might that rope have pulled us? What if we'd held fast just a bit longer? Imagine the bewilderment of someone who never grabbed on at all—someone who was never pulled backward through blurry phases of time to such historic waypoints. What they won't know is *how we know* what is out there and with *what* we made the determinations and *who* created those *whats*.

Instead, they will stare blankly upward at the firmament. Unaware of the astounding adventures and distant discoveries that all took place within the expanse of our solar system, in regions barely visible from here "on the good Earth." All courtesy of driven humans who envisioned their machines from pure curiosity, overcame repeated setbacks via ingenuity, and operated those creations with love and nurturing throughout their journeys.

Machines, you say? What machines? There were machines in the solar system that told us all these wild things?

And quite possibly ask themselves, Well, how did they get there?

Cassini's encore was neither a flagship Uranus orbiter nor one for Neptune but instead a diminutive bundle of instrumentation and hope called *New Horizons* that departed for Pluto in January 2006. It merely flew past, in July 2015, but still: A spacecraft from Earth had finally visited that occasional planet lurking near the tail end of our solar system.

Aboard *New Horizons* were no instruments from the University of Iowa. That didn't upset Bill Kurth one bit as his hands were already full with Cassini. And long before Kurth hit his stride on the outer-planet flagships, the University of Iowa fronted a modest research effort called Hawkeye. It had none of the pomp and circumstance of any piloted spaceflight but did launch a barebones Earth orbiter in 1974 to study our planet's magnetic field. James Van Allen headed up the team. "He sat us all down in a room one time," remembered Kurth, "and he said, 'You know, there's gonna be reporters. And this is gonna be in the news, at least here in Iowa.' And—and he said, 'Someone is bound to ask you, "What's this mean for the farmer in Iowa?"' He says, 'I don't want anybody to try to make a case that a farmer in Iowa is gonna benefit directly by what we're doing with Hawkeye.'"

According to Kurth, Van Allen then flipped into evangelist mode, preaching, "The reason we're flying *Hawkeye* is that we're learning about the universe. We're adding to human knowledge." It made for an inspiring sermon. "Trying to understand the world that we live in." Kurth tied that back to present day. "I think that's the most important thing that we do with plasma waves, is we learn about the universe we live in." Practical or not, useful or not to the everyday Joe, who can argue with learning?

Kurth recalled being in the midst of Cassini proceedings when he had a tangential debate with his peers. "I remember having a discussion one day about what the greatest mission of all time was. And you know, there's obviously some people that say, 'Well, Cassini is,' and Cassini is a great mission.

It's really superlative in just about every way. But Voyager takes the cake. And people said, 'But, well, why do you say that?' I said, 'Well, if you didn't have Voyager, we would not have had Cassini.'"

The point Kurth succinctly makes is that these missions build upon each other out of the natural evolution of our learning. "You'd have to know some things about where you're going—to have the right instruments, to make the right measurements, to ask the right questions. And Voyager was really the pathfinder to go past each of the giant planets and tell us *what we didn't know.*"

Despite everything we now *do* know about plasma waves and magnetic fields, Kurth—who remains in his position at Iowa and continues sifting data from the three flagship missions—has the same complaint as so many others. He laments the struggle for adequate budgetary support that has been a fixture of his career going back to its very beginning. "We wrote the textbooks on Jupiter and Saturn," went the contours of his argument, even though the lion's share of NASA's budget is fed to human flight. The solar system always gets the leftovers.

Bill, so what's the secret to effective public engagement?

"Pictures, pictures, pictures!" His response was immediate. Granted, nobody can even see plasma waves, and that's part of the issue. So keep sending back amazing shots of Saturn's rings and hexagon. The tiger stripes of Enceladus. Jupiter's crashing color bands of chemicals. Or the ribbons of lava on Io. Stir up some wonder. Fascinate people with the solar system and they'll demand more, and the money will come. In the meantime, Kurth—alongside the cascading generations of space scientists flowing through Iowa—will continue to chase his own unending wonderment.

By the time *New Horizons* launched, Phil Roberts had been gone from JPL for decades already. He left due to the 1980s' decline in funding for planetary missions. Roberts jumped ship to private industry, which offered much in the way of benefits—but also plenty of secrecy, as his government-funded work dealt in laser weapons. "When people asked me what I did for a living, there was little I could say," he reminisced. "As a result, I mainly told them of my previous work at JPL instead." He stayed on until 2008, took actual retirement, and then began volunteering with his wife, teaching phonics to kindergartners.

"I like kids, and the one-on-one tutoring was enjoyable."

Another wholly satisfying part of his life remains the groundbreaking contributions made to the efficiency of space mission development. Phil Roberts had come along at a critical time in the history of computer use and saw the obvious—automation. His efforts turned out to be more robust and hard wearing than expected. "Twenty years after I left, if I met someone from JPL, they knew of the 'Roberts Routines'!"

Many visitors to Oregon's Evergreen Museum no doubt consider the Spruce Goose to be the facility's major attraction. But during his own visit, Hugh von Delden took extended pause at the Sergeant missile launcher. There, in a *museum* of all places, sat his very own design, his brainchild, his work—which came to be during JPL's nascent, missile-heavy period just before the space age began.

Von Delden wasn't his management's first choice. "I was in the design section, and they had a lot of the companies bidding on it. And I said, 'Give me a chance.'" Conveying the memory, a pleading excitement in his voice came through as strong as it must've been back in the mid-1950s. "I drew up a design for a launcher, and they liked it, and it was built that way." Now, there sat his work on display!

"That was fun to see."

He thinks fondly of the JPL days, his oddball transition from missile work to thermal control, and reckons he went to Friday morning status meetings for twenty years straight. "During my whole tenure at the Lab, every day I enjoyed going to work. And every day was doing something that had never been done before. And there wasn't any books or literature I could read to help me out with what I was doing. It just wasn't there. We were the ones producing it." He spoke as if high on vivid satisfaction. "Every day we were working on something that was gonna make history, and that—that really kinda kept me going."

What an amazing turn of events for a guy who just needed a job and just happened to find a calling. "It sometimes surprises me too. And I don't know how I had the talent to do this."

One historymaker that ultimately sidestepped Hugh von Delden's talent was *Huygens*. After its operations ended, the Contraves company threw a party at its offices, and Platon Tatalias naturally attended. "It was done in our cafeteria and all employees were invited." Victoriously they ate, drank, and gave speeches. As a group they watched videos about this unparalleled mission to land on Titan—a mission dependent on their expertise. "Looking back on

my working life in industry," Tatalias began, "the team spirit and cooperation was incredible." It's a comment that those at the front lines of the space
business frequently utter.

Tatalias offered advice for anyone interested in joining up. What must exist
at a fundamental level is inquisitiveness regarding how things work. "Try to
find a job that fits your character." His recommendation was to avoid "a pure
engineering office," as he put it, where the focus is on one specific discipline
such as calculating loads or producing shop drawings. Pay attention to the
attitudes of the managers as well; at Contraves they were superb colleagues.
"The management support was exceptional or unconditional," which fostered
a healthy environment.

Don't underestimate the satisfaction of delivering a complete product to a
customer—whether it's a rocket engine or a zero-g Coke can or a way to break
apart two spacecraft in flight. And be courageous enough to stand behind your
ideas. "Do not be afraid of responsibility for solutions that you may suggest."

In 1987, for discovering Galileo's VEEGA trajectory, Roger Diehl received one
of the highest-level NASA accolades—the Exceptional Engineering Achieve-
ment Medal. "I felt very honored," he disclosed with sincerity. Over the years,
Diehl had received other NASA awards—awesome for sure but relating to various group efforts that weren't in the same class as being singled out for VEEGA.
"All the years I had spent in college and graduate school and, you know, ten
years in JPL, it all sort of culminated in being able to make an important contribution. And so that just gave me a lot of self-satisfaction."

Diehl retired in 2014 after thirty-nine total years at JPL. He and Denise
moved to the Pacific Northwest largely because of the excellent wines available there. No, really! "Within a half hour of our house there's probably about
two hundred wineries," he gushed of the area. "My wife and I, we *love* Oregon pinot. We're in the perfect spot for that." Despite being retired for years
already, Roger keeps his hands in the mix by volunteering. "The local high
school where I mentor has an engineering and science program, which is *way*
beyond what I experienced in high school."

Well, what's your take on the depth of talent in this next generation?

"I think we are in good hands for the future."

Certain high school teachers would not have associated Diehl's statement
with one specific former student of theirs. But people such as Bill O'Neil can
outgrow their mischievous ways. When *Galileo* crashed into Jupiter, O'Neil

50. In 2021 Roger Diehl showed off his custom license plate. "Nobody asked me about the license plate at the JPL parking lot, but several times gas station attendants asked me what VEEGA stood for!" The certificates behind him are for various Galileo project achievements. Photo by Denise Diehl, courtesy Roger Diehl.

wasn't at the festivities on JPL's campus. He was in Europe at a space conference and out for the evening with an old NASA chum. "At the moment of impact I was having dinner with Wes Huntress. We toasted the event."

O'Neil spent his entire career at the Jet Propulsion Laboratory. "It was a wonderful place to be!" he crooned. "I never, never considered going somewhere else. Even though I had some people approach me." What could be more worth doing than the unprecedented? "For me, the real key was *first*.

Everything I worked on there was never done before. And there weren't many places you could go and do something like this."

As to whether the multiple Galileo reconfigurations ever took the wind out of his sails, O'Neil responded with a glimpse into what the team cohesiveness had been. "So many of us were so absolutely dedicated to getting this done," he replied, with more than a whiff of emotion. "We would just never quit." O'Neil called to attention that perfect rallying cry, *Failure is not an option*. "We were not going to walk away" from the challenges of Galileo, pledged O'Neil. Despite every delay and redesign, JPL would always have been able to fulfill the core mission. "How we were going to achieve the objectives pretty much stayed intact throughout all of these—is it *five* major reprogrammings? One, two, three, four, five . . . yeah. Yeah. So what changed, was how we could *implement* it."

The work took an obvious toll on family. "In retrospect, I can say I wish I would have done more for them," he divulged of his children and spouse. "But at the time, I was where I wanted to be and doing what I wanted to do. It was a passion. You know, it was a—I called it at the time, it was a magnificent obsession."

And with that ever-present passion in his voice, Bill O'Neil finished with this: "Imagine being able to wind up and say, 'Nobody else. Nobody else did this first. *I* did this first.'"

Another who did so much for the first time, Clyde King, retired from JPL at the end of 1998 after thirty-five years. He lived to experience the space age from Sputnik 1 to the *Curiosity* Mars rover, with everything in between, and died in 2016 at eighty years of age. Here was a man who set standards. Part of his obituary read, "Clyde is remembered as having an incredible work ethic, character, and integrity."

Certainly that's the impression he left among those closest to him. Explained son James, "After he died, and the family went through his belongings, we discovered that he had received a NASA service medal in the months prior to his retirement." Indeed, during an annual awards event in June 1998, the elder King had been one of only twenty-two that year to receive NASA's Exceptional Service Medal. This particular award is given for (big breath) "significant, sustained performance characterized by unusual initiative or creative ability that clearly demonstrates substantial improvements or contributions in engineering, aeronautics, space flight, administration, support, or space-related endeavors that contribute to the NASA mission."

But at the time, the King family had no idea. "None of the kids were aware of it," James indicated. "It made me a little sad to find out after the fact. But it also goes to tell about how subdued he was about his accomplishments. He just wanted people to do their jobs well."

"Things that were hard for other people came easy to Dad," added Seaton King. "He had no interest in doing anything except what he did, which was to build stuff that had never been built before."

Clyde King not only worked on all four of the flagship outer-planet space-craft but also did it while simultaneously mentoring the next generation. Including Mary Reaves. She told a story about Hugh von Delden driving along a road one day and noticing some wiring bundles off to the side. Hugh thought, *Mary wouldn't have done it like that.*

Reaves cherished the memory. "I wanted a perfect wiring harness!"

Her own retirement came at age sixty-two after thirty colorful years at the Lab. "I loved my job. I grew up at JPL and was so proud. Coupled with the vigorous work there was fun, and feeling of belonging to a family." Reaves has much to be proud of: Her handiwork went to the end of the solar system. She codesigned Mars rover wiring. Her name's on a patent. Reaves and only one other tech were the ones who mated the plutonium power supplies to both *Galileo* and *Cassini.*

And she was the first woman to do what she did at JPL. "Today, there are many fine female technicians," she applauded. "We are proud of what we accomplished."

One of the people who depended on her meticulous perfectionism was Don Gurnett, who simply loved reminiscing about his own unusual career. "My wife usually kicks me in the shins if I talk too much," he joked. "That's what a professor does, you know. They teach classes and talk about all these strange things."

After sixty years at the University of Iowa, Gurnett finally retired in late 2019—just in time for the worldwide COVID-19 pandemic. He spent more than a year unable to visit even his own office to fully clean out. But the unplanned hiatus gave him time to reflect.

"My life has been—in terms of science—has been following and trying to explain interesting things, without particularly asking, 'Is this really important to people?' A lot of scientists are like that, I think." His comments echo Bill Kurth's because pure interest is a fundamental tenet of science itself. Follow your heart, o curious one, and worry little about the overall usefulness of what's being studied. That part will come.

Like many, Gurnett can only grimace at the now-primitive circumstances of space work during those way-back times of yore. "To take a spacecraft out to a *farm*?" he gasped. Today they're tested in the strictest conditions of cleanliness—and so much more thoroughly. With a corresponding increase in the number of delays! "Look back at what we did on the Injun flights," he spotlighted. "We conceived, designed, and flew *in less than a year* a spacecraft with, you know, twelve instruments on it!"

Gurnett lamented today's increased difficulties in getting aboard a mission—such as the layers of committees picking apart every last sentence of your proposal and diode of your hardware and bullet point of your résumé. "You have to be able to write and propose good science. Maybe even *top* science. But you also have to be able to build the instrument as you propose it. And you have to have a track record for *delivering* the instruments you propose." He called it a game.

Although Gurnett hung up the lab coat, he continued interacting with Iowa's Department of Physics and Astronomy—partly because he's one of those for whom the work never ends. But partly because of a disturbing trend: The number of Iowa professors in space science was decreasing. At one point, according to Gurnett, all the department had left were him and one other guy.

"I feel very strongly about the continuation of space physics here." So strongly that he and wife Marie donated his childhood farm to the university. Yes, that same farm he grew up on and talked his dad into cutting the power to so an experiment could run. The university's sale of the property funded the 2015 creation of the Donald A. and Marie B. Gurnett Chair of Physics at the University of Iowa. Its website calls this position "an essential tool in recruiting and retaining world-class faculty," with the end goal being a healthy and ongoing space science program at Iowa.

The retirement of Donald Alfred Gurnett failed to cap his own wonderment. From the twin *Voyagers*, aboard which Iowa's plasma wave instruments are *still* operating as of early 2025, streams a constant bounty of data—including the density of the interstellar plasma. "And for some reason, the density has been gradually increasing. It's increased a factor of 2 already, and now we're out there at 155 AUs." (Rather outdated and sadly anthropocentric, an astronomical unit is 93 million miles—the approximate distance between the earth and the sun.)

"So what's happening is, I think, the solar system is plowing through the interstellar medium. And one possible explanation, it's kind of 'piling up' the interstellar plasma ahead of the heliosphere, kind of like a snowplow piles up the snow ahead." Gurnett is wondering if and when "the pileup" will peak as the *Voyagers* reach a new zone of the interstellar medium that's unaffected by the sun's influence.

Gurnett further talked about a desire to "complete the Voyagers," as he put it. This would be a huge milestone for him. One of those shining career moments. "I'm getting old," he prefaced in mid-2020, "but I want to live long enough to see the last bit of data coming out of Voyager," before their power supplies finally expire. Sadly, Gurnett's death in January 2022 precluded his wish.

Centaur engineers Larry Ross and Joe Nieberding still lived in the Cleveland area as late as 2018, separated by no more than ten minutes in the car. "Larry and I have worked together for fifty years," mentioned Nieberding at one point, in a noticeably casual way. As if everyone works together for that long.

"What a trial," deadpanned Ross, in response to his friend. They'd actually sat together to discuss their Centaur days because they were *always* together.

Nieberding decried the elimination of shuttle-launched Centaur boosters. They would have been *advantaged* Centaurs, beginning their operational lives from all the way up in Earth orbit. The strategy would've unlocked an exclusive category of flight options otherwise impossible with the booster inventory of the time. "That's why we were so devastated when it was cancelled," mourned Nieberding of Shuttle/Centaur. "The country had lost an opportunity for a game-changing new capability." Mature and powerful and reliable, Centaur upper stages remain in use on expendable rockets more than six decades after introduction.

Larry Ross recalled a Centaur anniversary party, years downstream of *Galileo*'s demise, where John Casani gave a talk. Later on that evening, Casani approached Ross in praise: "I've used something you taught me in a number of projects, and it works beautifully."

Ross blurted out, "What in the hell is that?" His inner voice phrased it differently: *What could I possibly teach John Casani?*

And Casani told him, "You had that quarterly review with all these people. You called it 'Senior Manager's Panel,' and I loved it. And I've done it ever since." Ross couldn't believe the compliment.

Yes, John Casani has a way of positively impacting everyone he encounters. Even his associate at the Italian restaurant. Both were still seated there, although not for much longer, as Casani pushed back the empty plate that until moments before had held not crème brûlée but a luscious bread pudding with ice cream. Then in a practiced, finishing-school sort of motion, he daubed at his mouth with a napkin. It functioned as a lead-in to his saying that he'd never be exactly sure what Jack James saw in him that led to a job offer at JPL. But of course, he's thankful it happened. "I always did well," Casani suggested, in a monumentally understated way. "I was able to get done whatever they wanted me to do."

Quickly—and without prompting—Casani brought into focus all the committed, hardworking individuals who labored with him on his decades-long run of space projects. "I worked with a lot of good people," began the accolades. "And it was because I had the *opportunity* to work for good people that good things happened. And so, you know, I probably got unfairly credited for more that I should've." He referenced not only the incomparable nature of their efforts but also the fine madness of managing those efforts. "They were mostly jobs that had never been done before, you know. So nobody really knew quite how to *direct me*, you know? They were sort of willin' to maybe stand back a little bit and see how he does." Casani went into the character of a generic supervisor: "If he's a screw-up, we'll let him go!" And then, in perfect dad style, laughed at his own impersonation.

Well into his nineties, Casani still lives in Pasadena—on the very same property where all those wild New Year's parties went down. "I don't have any more friends my age," he lamented in late 2024. But aging has not loosened the man's schedule. His smartphone buzzes at unpredictable intervals with messages or calls from friends and loved ones wanting to coordinate gatherings. Although his shirt pockets no longer carry index cards with million-dollar tasks scribbled on them, Casani still enjoys dirtying his hands on occasion. The Mustang restoration, to call out a low point, took multiple decades but did actually finish.

Adversity tested him. Casani struggled through the 2008 loss of Lynn and the 2012 death of younger brother Kane. Two knee-replacement surgeries have sapped John's general stamina. "My father would probably chalk it off to the vicissitudes of old age," he suggested; memories of Jack are always close by even today. Five sons blessed John's own rewarding fatherhood. Their vocations run the gamut from entrepreneur to football coach to, yes, working in the space industry.

Speaking of, JPL occasionally still calls up John Casani for contract work— long after his formally clocking out for the *I-mean-it-this-is-the-last-last* time.

"Why do you suppose they're calling *you*?" enquired his companion at the table. "I mean, they could call anybody in the world, right? But they're calling *you*?"

"Well," he responded, "I don't know." As if it never occurred to him. The words came out staccato. His train of thought then led to a declaration: Casani felt strongly that the most important accomplishment in his whole entire JPL career was enabling the creation of messages to the cosmos. Each *Voyager* spacecraft carries a set of them, encoded onto identical gold-coated copper phonograph records. With deliberate overtness, one is mounted conspicuously on the exterior of each ship, stylus and needle tucked behind, then enclosed by a protective cover engraved with pictograms intended to be a user manual. Considering the vastness of space, these records will last essentially forever. Should alien beings one day recover either *Voyager*, and successfully decipher the pictogram instructions for playback, greetings from humanity would ensue as the records contain salutations in fifty-four languages. Plus the sounds of whales. Volcanoes, earthquakes, thunder. Frogs. Music is on them also, including Bach and "Johnny B. Goode" from Chuck Berry. Even images are encoded: trees, dolphins, buildings, families.

John Casani didn't create these things; the records were produced by a small group working with famed astronomer Carl Sagan. But Casani was the one who, as Voyager project manager, recognized in these craft the possibility of a sort of ultimate thought experiment, contacted Sagan, and found enough holes in the prelaunch schedule to make it happen.

"I don't know, maybe I should have, but I didn't ask anyone for permission, and nobody stopped me." He wanted a message, Sagan's team delivered in spades, and off they went.

Casani himself now had a question. It pertained to the title of a book being written that would feature him prominently. "You were saying, *Born to Explore*?"

His guest confirmed that yes indeed, that was very likely going to be the title.

"Well, I am *not*." And in otherwise pleasing tones, John Richard Casani voiced a prior commitment to meet with someone else that same evening. It was time to go. He stood up, bade his visitor goodbye, then smoothly threaded his lean self through the scattered outdoor tables of the Italian restaurant— past ordinary citizens seemingly unaware of who just went on by. Nobody looked up as he disappeared 'round the corner en route to his next engagement.

His guest at the table could only ponder, *There goes John Casani, having set the world on fire*, and waved for the check.

***

John Richard Casani did not live to see the publication of this book.
He left us on June 19, 2025, at the age of ninety-two. Surrounded by family.
May his mighty legacy and outstanding accomplishments endure.

Sources

Books

Baker, David. *Space Shuttle*. New York: Crown Publishers, 1979.

Barbieri, C., Jürgen H. Rahe, Torrence V. Johnson, and Anita M. Sohus, eds. *The Three Galileos: The Man, the Spacecraft, the Telescope*. Boston: Kluwer Academic Publishers, 1997.

Barry, Robert C. "Eight Microprocessor-Based Instrument Data Systems in the Galileo Orbiter Spacecraft." In *Aerospace Applications of Microprocessors*, 7–15. NASA Conference Publication 2158. Washington DC: NASA, 1980. Preprint for a workshop held in Greenbelt MD, November 3–4, 1980.

Bauer, J. L. "An Electrically Conductive Thermal Control Surface for Spacecraft Encountering Low-Earth Orbit Atomic Oxygen." In *Proceedings of the NASA Workshop on Atomic Oxygen Effects, November 10–11, 1986*, edited by David E. Brinza, 156. Washington DC: NASA and Jet Propulsion Laboratory, June 1, 1987.

Bille, Matt, and Erica Lishock. *The First Space Race: Launching the World's First Satellites*. College Station: Texas A&M University Press, 2004.

Burgess, Eric. *By Jupiter: Odysseys to a Giant*. New York: Columbia University Press, 1982.

Butrica, Andrew J. *To See the Unseen: A History of Planetary Radar Astronomy*. NASA SP-4218. Washington DC: NASA History Office, 1996.

Carter, Jimmy. *White House Diary*. New York: Farrar, Straus and Giroux, 2010.

Chaikin, Andrew. *A Man on the Moon: The Voyages of the Apollo Astronauts*. New York: Penguin Books, 1994.

Chertok, Boris. *Rockets and People*. Vol. 2, *Creating a Rocket Industry*. NASA History Series. Washington DC: NASA History Office, 2006.

Conway, Erik. *Exploration and Engineering: The Jet Propulsion Laboratory and the Quest for Mars*. Baltimore: Johns Hopkins University Press, 2015.

Dawson, Virginia P., and Mark D. Bowles. *Taming Liquid Hydrogen: The Centaur Upper Stage Rocket, 1958–2002*. NASA History Series. NASA SP-2004-4230. Washington DC: NASA History Office, 2004.

Deese, Samuel G. "Application of Microprocessors to Interplanetary Spacecraft Data Systems." In *Aerospace Applications of Microprocessors*, 29–35. NASA Conference Publication 2158. Washington DC: NASA, 1980. https://ntrs.nasa.gov /api/citations/19810003141/downloads/19810003141.pdf.

Dessler, A. J., ed. *Physics of the Jovian Magnetosphere*. London: Cambridge University Press, 1983.

Ezell, Edward Clinton, and Linda Neuman Ezell. *On Mars: Exploration of the Red Planet, 1958–1978*. Mineola NY: Dover Publications, 2009.

Fimmel, Richard O., James Van Allen, and Eric Burgess. *Pioneer: First to Jupiter, Saturn, and Beyond*. NASA SP-446. Washington DC: NASA, 1980.

Galilei, Galileo. *Sidereus Nuncius or The Sidereal Messenger*. Translated and with commentary by Albert Van Helden. Chicago: University of Chicago Press, 1989.

Gawdiak, Ihor Y., Ramón J. Miro, and Sam Stueland. *Astronautics and Aeronautics, 1986–1990: A Chronology*. NASA SP-4027. Washington DC: NASA, 1997.

Griffin, Michael D., and James R. French. *Space Vehicle Design*. 2nd ed. Reston VA: American Institute of Aeronautics and Astronautics, 2004.

Heppenheimer, T. A. *Development of the Shuttle, 1972–1981*. Washington DC: Smithsonian Institution Press, 2002.

———. *The Space Shuttle Decision: NASA's Search for a Reusable Space Vehicle*. NASA SP-4221. Washington DC: NASA History Office, 1999.

Herlach, Udo R., P. Tatalias, B. Schmid, and D. Mussett. "Huygens Separation Mechanisms." In *Proceedings of the Sixth European Space Mechanisms & Tribology Symposium*, edited by W. R. Burke, 111–17. ESA SP-374. Paris: European Space Agency, August 1995.

Hubbard, Scott. *Exploring Mars*. Tucson: University of Arizona Press, 2011.

Ip, Wing, Daniel Gautier, and Tobias Owen. "The Genesis of Cassini-Huygens." In *Proceedings of the International Conference "Titan—from Discovery to Encounter,"* edited by Karen Fletcher, 211–27. Noordwijk, Netherlands: ESA Publications Division, 2004.

Ip, Wing-Huen, Tobias Owen, and Daniel Gautier. "Prologue 1: The Genesis of Cassini-Huygens." In *Titan: Interior, Surface, Atmosphere, and Space Environment*, edited by Ingo Müller-Wodarg, Caitlin A. Griffith, Emmanuel Lellouch, and Thomas E. Cravens, 10–21. London: Cambridge University Press, 2014. https://rapid.lib.ncu.edu.tw/bookboard/topic/2018_10/PDF/50.pdf.

Johnson, Michael R. "The Galileo High Gain Antenna Deployment Anomaly." NASA Technical Report N94-33319. In *Proceedings of the 28th Aerospace Mechanisms Symposium*, 359–77. Washington DC: NASA Lewis Research Center, 1994. https://ntrs.nasa.gov/api/citations/19940028785/downloads /19940028785.pdf.

Johnson, Michael R., and Greg C. Levanas. "The Galileo Tape Recorder Rewind Anomaly." In *Proceedings of the 31st Aerospace Mechanisms Symposium*, 231–48. NASA Conference Publication 3350. https://ntrs.nasa.gov/citations/19970021613.

Johnson, Nicholas L. *The Soviet Year in Space: 1986.* 6th ed. Colorado Springs: Teledyne Brown Engineering, 1987.

Jones, J. C., and F. Giovagnoli. "The Huygens Probe System Design." In Wilson, *Huygens*, 25–45.

Langford, Jerome J. *Galileo, Science, and the Church.* Ann Arbor: University of Michigan Press, 1971.

Logsdon, John M., Linda J. Lear, Jannelle Warren-Findley, Ray A. Williamson, and Dwayne A. Day, eds. *Exploring the Unknown: Selected Documents in the History of the U.S. Civil Space Program.* Vol. 1, *Organizing for Exploration.* NASA SP-4407. Washington DC: NASA History Division, 1995.

Logsdon, John M., Ray A. Williamson, Roger D. Launius, Russell J. Acker, Stephen J. Garber, and Jonathan L. Friedman, eds. *Exploring the Unknown: Selected Documents in the History of the U.S. Civil Space Program.* Vol. 4, *Accessing Space.* NASA SP-4407. Washington DC: NASA History Division, 1999.

Lorenz, Ralph, and Jacqueline Mitton. *Titan Unveiled: Saturn's Mysterious Moon Explored.* Princeton NJ: Princeton University Press, 2008.

Marov, Mikhail, with David H. Grinspoon. *The Planet Venus.* New Haven CT: Yale University Press, 1998.

McNab, David, and James Younger. *The Planets.* New Haven CT: Yale University Press, 1999.

Meltzer, Michael. *The Cassini-Huygens Visit to Saturn: An Historic Mission to the Ringed Planet.* Chichester, UK: Praxis Publishing, 2015.

———. *Mission to Jupiter: A History of the Galileo Project.* NASA History Series. NASA SP-2007-4231. Washington DC: NASA, 2007.

Morrison, David. *Voyages to Saturn.* NASA SP-451. Washington DC: NASA, 1982.

Murray, Bruce. *Journey into Space: The First Thirty Years of Space Exploration.* New York: W. W. Norton, 1989.

National Research Council. *Space Research: Directions for the Future.* Washington DC: National Academies Press, 1966.

Neufeld, Michael. *Von Braun: Dreamer of Space, Engineer of War.* New York: Alfred A. Knopf, in association with the National Air and Space Museum, Smithsonian Institution, 2007.

Neuman, James C. "Solar Thermal Vacuum Tests of Magellan Spacecraft." Report N91-19127. In *16th Space Simulation Conference: Confirming Spaceworthiness into the Next Millennium*, edited by Joseph L. Stecher III, 1–20. NASA CP-3096. Washington DC: NASA, 1991.

Ryder, Graham, ed. *Proceedings of the 18th Lunar and Planetary Science Conference.* New York: Cambridge University Press and the Lunar and Planetary Institute, 1988.

Sagan, Carl. *Murmurs of Earth: The Voyager Interstellar Record.* With F. D. Drake, Ann Druyan, Timothy Ferris, Jon Lomberg, and Linda Salzman Sagan. New York: Random House, 1978.

Stockman, David A. *The Triumph of Politics: How the Reagan Revolution Failed.* New York: Harper & Row, 1986.

Swift, David. *Voyager Tales: Personal Views of the Grand Tour.* Reston VA: American Institute of Aeronautics and Astronautics, 1997.

Thomas, Jack S. "A Command & Data Subsystem for Deep Space Exploration Based on the RCA 1802 Microprocessor in a Distributed Configuration." In *Aerospace Applications of Microprocessors*, 17–20. NASA Conference Publication 2158. Washington DC: NASA, 1980.

Tomayko, James E. *Computers in Spaceflight: The NASA Experience.* Washington DC: NASA, March 1988.

Van Allen, James A. *Origins of Magnetospheric Physics.* Iowa City: University of Iowa Press, 2004.

Von Kármán, Theodore, with Lee Edson. *The Wind and Beyond: Theodore von Kármán, Pioneer in Aviation and Pathfinder in Space.* Boston: Little, Brown, 1967.

Weissman, Paul, and Marcia Neugebauer. "The Comet Rendezvous Asteroid Flyby Mission: A Status Report." In *Asteroids, Comets, Meteors 1991: Proceedings of the International Conference Held at Northern Arizona University, Flagstaff, June 24–28, 1991*, edited by A. W. Harris and E. Boswell, 629–32. Houston: Lunar and Planetary Institute, Houston, 1992.

Wilson, A., ed. *Huygens: Science, Payload and Mission.* SP-1177. Noordwijk, Netherlands: ESA Publications Division, August 1997.

Zak, Anatoly. *Russia in Space: The Past Explained, the Future Explored.* Burlington ON: Apogee Prime, a Division of Griffin Media, 2013.

Periodicals and Online Articles

Abbott, Alison. "Titan Team Claims Just Deserts as Probe Hits Moon of Crème Brûlée." *Nature* 433 (January 20, 2005): 181.

Abramson, Rudy. "NASA Cancels Jupiter, Sun Rocket: Modified Centaur Fails Safety Criteria for Shuttle Missions." *Los Angeles Times*, June 20, 1986. http://articles.latimes.com/1986-06-20/news/mn-11406_1_space-shuttle-program.

Allcock, G. McK. "A Study of the Audio-Frequency Radio Phenomenon Known as 'Dawn Chorus.'" January 2, 1957. https://www.publish.csiro.au/ph/pdf/PH570286.

Anderson, Ian. "A Wing and a Prayer." *New Scientist*, December 10, 1987, 32–33.

Anderson, Porter. "Eilene Theilig: Faith, by Jupiter." CNN, December 11, 2001. https://www.cnn.com/2001/CAREER/jobenvy/07/16/eilene.theilig/index.html.

Australian Associated Press. "Agnew Proposes Man on Mars." *Canberra Times*, July 18, 1969, 4.

Bagenal, Fran. "The Magnetosphere of Jupiter: Coupling the Equator to the Poles." *Journal of Atmospheric and Solar-Terrestrial Physics* 69, no. 3 (March 2007): 387–402.

Barber, T. J., F. Krug, and K. P. Renner. "Final Galileo Propulsion System In-Flight Characterization." American Institute of Aeronautics and Astronautics, 1997.

Barcena, Jorge, and Maria Asun Mendizabal. "Advanced Materials for High Temperature Application (D3.10)." Tecnalia, SML-TEC-AMHTA-CO/3010, Issue 1B, January 31, 2018, 48.

Basu, Ramanuj. "Cassini Begins Its Final Act: A Conversation with Charles Elachi." Caltech News, October 18, 2015. https://www.caltech.edu/about/news/cassini-begins-its-final-act-conversation-charles-elachi-48611.

"Beatles Wallpaper Arrives in America (February 21, 1964)." *News of the Odd*, accessed April 11, 2018. https://web.archive.org/web/20160405222040/http://www.newsoftheodd.com/article1004.html.

Beckman, John C., and Ellis D. Miner. "Jovian System Science Issues and Implications for a Mariner Jupiter Orbiter Mission." Paper presented at the American Institute of Aeronautics and Astronautics and the American Geophysical Union, Conference on the Exploration of the Outer Planets, St. Louis, Missouri, September 17–19, 1975. https://arc.aiaa.org/doi/abs/10.2514/6.1975-1141.

Berger, Eric. "A Cold War Mystery: Why Did Jimmy Carter Save the Space Shuttle?" *ArsTechnica*, July 14, 2016. https://arstechnica.com/science/2016/07/a-cold-war-mystery-why-did-jimmy-carter-save-the-space-shuttle/.

Blagdon, L. J. "Galileo Lithium Battery." From the 1979 Goddard Space Flight Center Battery Workshop, April 1, 1980. https://ntrs.nasa.gov/api/citations/19800012343/downloads/19800012343.pdf.

Bluth, John. "Von Karman, Malina Laid the Groundwork for the Future JPL." *Jet Propulsion Laboratory Universe* 24, no. 14 (July 15, 1994). https://web.archive.org/web/20010117040500/https://www2.jpl.nasa.gov/files/universe/un940715.txt.

Brown, Stuart F. "Man Fixes Tape Recorder." *Fortune*, June 23, 2003. https://money
 .cnn.com/magazines/fortune/fortune_archive/2003/06/23/344584/index.htm.

Canby, Thomas Y. "A Generation after Sputnik: Are the Soviets ahead in Space?"
 National Geographic Magazine 10, no. 4 (October 1986): 420–58.

Casani, John. "Engineering Memos." *ASK Magazine*, June 1, 2004. https://appel
 .nasa.gov/2004/06/01/engineering-memos/.

———. "A New Spin." *ASK Magazine*, June 1, 2004. https://appel.nasa.gov/2004
 /06/01/a-new-spin/.

Chang, Kurng Y., and Dennis L. Kern. "Super*Zip (Linear Separation) Shock Characteristics." *The Shock and Vibration Bulletin* 56, part 1 (August 1986): 33–42.

Clow, David. "Flying in Deep Space: The Galileo Mission to Jupiter (Part II)."
 Quest 21, no. 1 (2014): 11–23.

"Clyde Rucker King, 1935–2016." Lietz-Fraze Funeral Home.com, accessed March
 25, 2021. https://lietz-frazefuneralhome.com/tribute/details/336/Clyde-King
 /obituary.html.

"A Conversation with Tom Gavin." *Jet Propulsion Laboratory Universe* 34, no. 14
 (July 16, 2004): 3.

Cowen, Robert C. "Manned Space Programs Get Message: Throttle Back." *Chris-
 tian Science Monitor*, December 19, 1990. https://www.csmonitor.com/1990
 /1219/aspace.html.

D'Amario, Louis A., Larry E. Bright, and Aron A. Wolf. "Galileo Trajectory
 Design." *Space Science Reviews* 60 (1992): 23–40.

"Deputy Assistant Laboratory Director of Flight Projects Appointed." Jet Propulsion Laboratory, February 1, 1988. https://www.jpl.nasa.gov/news/deputy
 -assistant-laboratory-director-of-flight-projects-appointed.

Dickson, David. "Europeans Decide on a Trip to Saturn." *Science* 242 (December 9,
 1988): 1375–76.

Dornheim, Michael A. "Magellan Begins Systems Checkouts after Entering Orbit
 around Venus." *Aviation Week & Space Technology*, August 20, 1990, 30.

Dye, Lee. "First Rockets Set to Go in 1992: Soviet Mars Mission Aims to Bring
 Samples Home." *Los Angeles Times*, May 21, 1987.

———. "Obituaries: Frederick L. Scarf; Space Scientist at TRW Had a Worldwide
 Impact." *Los Angeles Times*, July 20, 1988.

Easterbrook, Gregg. "Get Me Out of This Death Trap, Scotty." *Washington
 Monthly*, April 1980. https://web.archive.org/web/20040611101825/https://
 washingtonmonthly.com/features/2001/8004.easterbrook-fulltext.html.

"Eilene Theilig, Jupiter Explorer." Jet Propulsion Laboratory, January 1, 2001.
 https://www.jpl.nasa.gov/news/eilene-theilig-jupiter-explorer.

Evans, Ben. "Two Shuttles, Two Launches, One Planet . . . and a Five-Day Goal." *AmericaSpace*, May 17, 2012. http://www.americaspace.com/2012/05/17/two -shuttles-two-launches-one-planetand-a-five-day-goal/.

Fisher, Jack. "The Galileo Probe." *Our Space Heritage 1960–2000* (Hughes blog), May 30, 2012. https://hughesscgheritage.wordpress.com/2012/05/30/the -galileo-probe-jack-fisher/.

"A 4 Foot-Long Projectile That Looks Like a Golf Tee." UPI, October 23, 1986. https://www.upi.com/Archives/1986/10/23/A-4-foot-long-projectile-that -looks-like-a-golf-tee/3760530424000/.

Frazier, Ken. "Galileo's Epic Odyssey around Jupiter . . . and the Sandia Connection." *Sandia Lab News* 53, no. 3 (February 9, 2001). https://web.archive.org /web/20010727090913/https://www.sandia.gov/LabNews/LN02-09-01 /galileo/epic_story.htm.

———. "A Sandia/Galileo Timeline." *Sandia Lab News* 53, no. 3 (February 9, 2001). https://web.archive.org/web/20010727092016/https://www.sandia .gov/LabNews/LN02-09-01/galileo/timeline_story.htm.

Friedman, Lou. "World Watch." *Planetary Report* 15, no. 5 (September/October 1995): 6.

"Galileo's Telescope." Museo Galileo, accessed June 20, 2023. https://catalogue .museogalileo.it/object/GalileosTelescope_n01.html.

Gallet, R. M., and R. A. Helliwell. "Origin of 'Very-Low-Frequency Emissions.'" *Journal of Research at the National Bureau of Standards—D. Radio Propagation* 63D, no. 1 (July–August 1959): 21–25. https://nvlpubs.nist.gov/nistpubs /jres/63D/jresv63Dn1p21_A1b.pdf.

Givens, J. J., L. J. Nolte, and L. R. Pochettino. "Galileo Atmospheric Entry Probe System: Design, Development, and Test." AIAA-83-0098. AIAA 21st Aerospace Sciences Meeting, Reno, Nevada, January 10–13, 1983. https://arc.aiaa.org/doi /abs/10.2514/6.1983-98.

"Gorbachev's Visit to India." *Strategic Studies* 10, no. 1 (1986): 7–10. http://www .jstor.org/stable/45182348.

Gore, Rick. "What Voyager Saw: Jupiter's Dazzling Realm." *National Geographic Magazine* 157, no. 1 (1979): 11.

Gounley, Robert. "Design Lessons Learned from Temperature Management of Galileo's Retro-Propulsion Module." AIAA 2016-5285, September 9, 2016. https://arc.aiaa.org/doi/abs/10.2514/6.2016-5285.

———. "From Tragedy to Triumph: How Galileo Made It to Launch." Accessed October 3, 2024. https://d2pn8kiwq2w21t.cloudfront.net/documents /Universe_-_October_2024.pdf.

Greenstadt, E. W., S. L. Moses, C. F. Kennel, C. T. Russell, and F. V. Coroniti. "Frederick Scarf, 1930–1988." *American Geophysical Union EOS Magazine* 69, no. 48 (November 1988): 1601.

Gurnett, D. A. "The Origins of Space Radio and Plasma Wave Research at the University of Iowa." *Journal of Geophysical Research: Space Physics* 125, no. 2 (2020).

Hunter, Maxwell W., II. "Unmanned Scientific Exploration throughout the Solar System." *Space Science Reviews* 6, no. 5 (1967): 601–54. http://adsabs.harvard .edu/full/1967SSRv. . . . 6..601H.

"Huygens Probe Separation and Coast Phase." ESA. Last updated September 1, 2019. https://sci.esa.int/web/cassini-huygens/-/34956-huygens-probe-separation.

Jäkel, E., P. Rideau, P. R. Nugteren, J. Underwood, P. Faucon, and J.-P. Lebreton. "Drop Test of the Huygens Probe from a Stratospheric Balloon." *Advances in Space Research* 21, no. 7 (1998): 1033–39.

"JPL's John Casani Honored by Air and Space Museum." Jet Propulsion Laboratory, April 30, 2009. https://www.jpl.nasa.gov/news/jpls-john-casani-honored -by-air-and-space-museum.

Kacik, Alex. "ATK Aerospace Powers Space Exploration with Arrays of Technology." *Noozhawk (Santa Barbara)*, May 6, 2012. https://www.noozhawk.com /article/050612_atk_aerospace_solar_arrays_space_exploration.

"King, Clyde Rucker, 1935–2016." Legacy, accessed April 17, 2021. https://www .legacy.com/obituaries/pasadenastarnews/obituary.aspx?n=clyde-rucker&pid= 178095961&fhid=9193.

Krimigis, S. M. "Luncheon at the White House: On Comets and Uranus." *Johns Hopkins APL Technical Digest* 7, no. 4 (1986).

Kurth, Bill. "Personal Stories from the Mission." NASA Science, accessed August 16, 2020. https://voyager.jpl.nasa.gov/share/#group-5.

Launius, Roger D. "Planning the Post-Apollo Space Program: Are There Lessons for the Present?" *Space Policy* 28 (2012): 38–44. https://repository.si.edu /bitstream/handle/10088/24977/201214SH.pdf?sequence=1&isAllowed=y.

Lebreton, Jean-Pierre, Olivier Witasse, Claudio Sollazzo, Thierry Blancquaert, Patrice Couzin, Anne-Marie Schipper, Jeremy B. Jones, et al. "An Overview of the Descent and Landing of the Huygens Probe on Titan." *Nature* 438 (December 2005): 758–64.

Lenorovitz, Jeffrey. "ESA to Select New Space Science Mission for Mid-1990s Launch." *Aviation Week & Space Technology*, November 14, 1988.

Lerner, Eric J. "Galileo's Tortuous Journey to Jupiter." *Aerospace America*, August 1989, 21–28, 51.

Lorenz, Ralph D. "Planetary Penetrators: Their Origins, History and Future." *Advances in Space Research* 48 (2011): 403–31.

Lowes, Leslie L. "Ice, Water and Fire: The Galileo Europa Mission." *Planetary Report* 18, no. 1 (January/February 1998): 12–15.

"Magellan Fire Inquiry Board Urges New Test Procedures." *Aviation Week & Space Technology*, November 14, 1988, 35.

"Magellan's Radar Images of Venus to Unmask Cloud-Shrouded Planet." *Aviation Week & Space Technology*, October 9, 1989, 113–15.

Malina, Frank J. "Memoir on the GALCIT Rocket Research Project, 1936–1938." Prepared for the International Academy of Astronautics and International Union of the History and Philosophy of Science's First International Symposium on the History of Astronautics, Belgrade, September 25–26, 1967. http://archive.olats.org/pionniers/malina/aeronautique/memoir1.php.

Malloy, Brian. "Parker Says Shuttle O-Rings Met Standards." UPI, February 12, 1986. https://www.upi.com/Archives/1986/02/12/Firm-says-shuttle-seals-met -standards/1163508568400.

"Mars Climate Orbiter Mishap Investigation Board Phase I Report." November 10, 1999. https://pdf4pro.com/view/mars-climate-orbiter-mishap-investigation -board-phase-i-6c1631.html#google_vignette.

McKinnon, Mika. "The Cassini Team Reflects on How It Feels to Say Goodbye to Their Spacecraft." *Gizmodo*, September 17, 2017. https://www.gizmodo.com .au/2017/09/the-cassini-team-reflects-on-how-it-feels-to-say-goodbye-to-their -spacecraft/.

"Model Car Champs." *Boy's Life*, January 1952, 43.

Moreira, Alberto, Pau Prats-Iraola, Marwan Younis, Gerhard Krieger, Irena Hajnsek, and Konstantinos P. Papathanassiou. "A Tutorial on Synthetic Aperture Radar." *IEEE Geoscience and Remote Sensing Magazine*, March 2013, 6–43.

Morrison, David, Marcia Neugebauer, and Paul R. Weissman. "The Comet Rendezvous Asteroid Flyby Mission." *Abstracts from the Workshop on Analysis of Returned Comet Nucleus Samples*, January 1989, 54–55.

"NASA Board Nears End of Fire Review." *Aviation Week & Space Technology*, October 24, 1988, 24.

National Research Council. "On the Scientific Viability of a Restructured CRAF Science Payload: Letter Report." Washington DC: National Academies Press, 1990. https://doi.org/10.17226/12330.

"1997 AAE Distinguished Engineering Alumnus: William J. O'Neil, Manager, Project Galileo, Jet Propulsion Laboratory, BASE '61." School of Aeronautics and Astronautics, Purdue University, 1997. https://engineering.purdue.edu /AAE/people/alumni/distinguished/DEA/profiles/1997oneil.

Oberg, James. "The Spacecraft's Got Swing." *Astronomy Magazine* 27, no. 8 (August 1999): 50.

————. "Titan Calling." *IEEE Spectrum*, October 1, 2004, 28–33.

O'Donnell, Franklin. "JPL 101." JPL 4000-10048, Caltech, 2002, 2–11.

O'Toole, Thomas. "More Hurdles Rise in Galileo Project to Probe Jupiter." *Washington Post*, August 15, 1979.

————. "NASA Weighs Deferring 1982 Mission to Jupiter." *Washington Post*, September 4, 1979.

"People v. Acosta." District Court of Appeal, Second District, Division 1, California. Cr. 5306. Decided May 9, 1955. https://caselaw.findlaw.com/court/ca-court
-of-appeal/1803396.html.

"President Nixon's 1972 Announcement on the Space Shuttle." NASA Key Documents, January 5, 1972. https://history.nasa.gov/stsnixon.htm.

"Quest for Slice of Comet." *New York Times*, National Edition, November 3, 1987, C3.

Reeve, Ronald T. "Thermal Redesign of the Galileo Spacecraft for a VEEGA Trajectory." *Journal of Spacecraft and Rockets* 28, no. 2 (March 1991): 130–38. https://
arc.aiaa.org/doi/10.2514/3.26221.

"Report to the President by the Presidential Commission on the Space Shuttle Challenger Accident." June 6, 1986. https://history.nasa.gov/rogersrep
/genindex.htm.

"The Ride Report" advertisement. *Aviation Week & Space Technology*, February 29, 1988, 76.

Sandia National Laboratories. "Sandia Labs to Develop Custom, Radiation-Hardened Pentium Processor for Space and Defense Needs." December 8, 1998. https://www.sandia.gov/media/rhp.htm.

Sawyer, Kathy. "$1.4 Billion Galileo Mission Appears Crippled." *Washington Post*, December 18, 1991.

————. "Rewind Almost Unhinges Galileo Mission." *Washington Post*, October 29, 1995.

Schramm, Bill. "Penn Routed by Bears for First Loss, 40–0." *Daily Pennsylvanian* 67, no. 10 (October 12, 1953): 1.

Shapland, D. J., and W. Müller. "Study of Shuttle-Based Systems for High-Energy Planetary Missions." *Astronautical Research*, 1972, 103–11.

"Shearing Legend Colin Bosher Dies." *New Zealand Herald*, May 30, 2017. http://
www.nzherald.co.nz/nz/news/article.cfm?c_id=1&objectid=11866550.

"Slain Civil Rights Workers Found." History.com's *This Day in History*, August 4, 1964. https://www.history.com/this-day-in-history/slain-civil-rights-workers
-found.

Sollazzo, C., Jean-Pierre Lebreton, Kai Clausen, Thierry Blancquaert, Olivier Witasse, Anne-Marie Schipper, Joe Wheadon, et al. "The Huygens Probe Mission to Titan: Engineering the Operational Success." SpaceOps 2006 Conference, American Institute of Aeronautics and Astronautics, 2006. https://arc.aiaa.org/doi/abs/10.2514/6.2006-5503.

"Soviet Space Shuttle Facilities at Tyuratam Imaged by French Spot." *Aviation Week & Space Technology*, September 1, 1986, 42.

Sromovsky, L. A., F. A. Best, H. E. Revercomb, and J. Hayden. "Galileo Net Flux Radiometer Experiment." *Space Science Reviews* 60 (1992): 233–62.

Stacey, Kevin. "Head Honored for International Space Science Collaboration." News from Brown, October 23, 2014. https://news.brown.edu/articles/2014/10/vernadsky.

"Star Crossed: NASA's $1.5 Billion Blunder." *Newsweek*, July 9, 1990, cover.

Stephenson, R. Rhoads. "The Galileo Attitude and Articulation Control System: A Radiation-Hard, High Precision, State-of-the-Art Control System." Automatic Control in Space 1985: Proceedings of the Tenth IFAC Symposium, Toulouse, France, June 24–28, 1985. *IFAC Proceedings* 18, no. 4 (June 1985): 83–90.

"Steps to Mars II: A Conference Report." *Planetary Report* 15, no. 6 (November/December 1995): 10.

Stricker, R. E., A. G. F. Dingwall, S. Cohen, J. R. Adams, and W. C. Slemmer. "A Radiation-Hardened Bulk Si-Gate CMOS Microprocessor Family." SAND—79-1273C. 1979 IEEE Annual Conference on Nuclear and Space Radiation Effects, Santa Cruz, California, July 17–20, 1979. https://inis.iaea.org/collection/NCLCollectionStore/_Public/11/506/11506044.pdf?r=1&r=1.

Stultz, James W. "Thermal Design of the Galileo Bus and Retropropulsion Module." *Journal of Spacecraft and Rockets* 28, no. 2 (March 1991): 146–51. https://arc.aiaa.org/doi/abs/10.2514/3.26222.

Swift, Gary M., Gregory C. Levanas, J. Martin Ratliff, and Allan H. Johnston. "In-flight Annealing of Displacement Damage in GaAs LEDs: A Galileo Story." *IEEE Transactions on Nuclear Science* 50, no. 6 (December 2003): 1991–97.

Theilig, E. E., D. L. Bindschadler, K. A. Schimmels, and N. Vandermey. "Project Galileo: Farewell to the Major Moons of Jupiter." *Acta Astronautica* 53, no. 4-10 (August 2003): 329–52.

"Together in Space." *New Scientist* 111, no. 1516 (July 10, 1988): 18.

"University of Pennsylvania Two Hundred Thirty-Sixth Commencement for the Conferring of Degrees." May 19, 1992. https://archives.upenn.edu/wp-content/uploads/2018/04/commencement-program-1992.pdf.

Van Allen, James. "Is Human Spaceflight Obsolete?" *Issues in Science and Technology* (University of Texas at Dallas) 20, no. 4 (2004): 38–40.

———. "Space Science, Space Technology and the Space Station." *Scientific American* 254, no. 1 (January 1986): 32–39.

————. "Twenty-Five Milliamperes: A Tale of Two Spacecraft." *Journal of Geophysical Research* 110, no. A5 (May 1, 1996): 10479–95.

"Was 'Louie Louie' Banned in Indiana?" Indiana Local Government Information Website, accessed April 11, 2018. https://web.archive.org/web /20100613231530/https://www.agecon.purdue.edu/crd/localgov/Topics /Essays/Louie_Louie.htm.

Webster, Guy. "Galileo Gets One Last Frequent-Flyer Upgrade." Jet Propulsion Laboratory, March 15, 2001. https://www.jpl.nasa.gov/news/galileo-gets-one -last-frequent-flyer-upgrade.

————. "Spacecraft at Io Sees and Sniffs Tallest Volcanic Plume." Jet Propulsion Laboratory, October 4, 2001. https://www.jpl.nasa.gov/news/spacecraft-at-io -sees-and-sniffs-tallest-volcanic-plume.

Whalen, Mark, ed. "NASA Honor Awards Recognize Best of JPL." *JPL Universe* 28, no. 14 (July 10, 1998): 3.

————, ed. "Passings." *JPL Universe* 46, no. 4 (April 2016): 6.

Wilford, John Noble. "Bruce C. Murray, Who Helped Earth Learn of Mars, Dies at 81." *New York Times*, August 29, 2013, A20.

Witkin, Richard. "Agnew Proposes a Mars Landing." *New York Times*, July 17, 1969, 1, 22.

Interviews and Personal Communications

Adler, Mark. Interview by the author, January 28, 2016.

Allen, Lew, Jr. Interview by Heidi Aspaturian for California Institute of Technology Archives Oral History Project, Pasadena CA, June and July 1991 and March and April 1994. http://resolver.caltech.edu/CaltechOH:OH_Allen_L.

Boisjoly, Roger. Undated email exchanges with author, circa 1998.

Casani, John. Interview by the author, April 24, 2007; January 26–27, 2016; December 11, 2021; February 4, 2022; May 27, 2022; and November 23, 2024; and related correspondence.

Diehl, Denise. Interview by the author, October 1, 2018.

Diehl, Roger. Interview by the author, October 1 and 2, 2018; and related correspondence.

Esposito, Larry. Email exchanges with author, beginning May 19, 2022.

Evans, Michelle. Interview by the author, March 7, 2021; and related correspondence.

Fisher, Jack. Email exchanges with author, beginning July 22, 2020.

Fisk, Lennard A. Interview by Rebecca Wright for NASA Johnson Space Center Oral History Project, Ann Arbor MI, September 8, 2010. https://

historycollection.jsc.nasa.gov/JSCHistoryPortal/history/oral_histories
/NASA_HQ/Administrators/FiskLA/FiskLA_9-8-10.htm.

Friedman, Louis. Interview by the author, March 14, 2023; and related
correspondence.

Fuqua, Don. Interview by Catherine Harwood for NASA Johnson Space Center
Oral History Project, Houston, August 11, 1999.

Goldin, Dan. Email exchanges with author, beginning August 13, 2021.

Gounley, Robert. Telephone conversation with the author, February 18, 2025; and
related correspondence.

Gurnett, Don. Interview by the author, May 18 and May 25, 2020; and related
correspondence.

Head, James. Email exchanges with author, beginning October 18, 2021.

Horowitz, Norman H. Interview by Rachel Prud'homme for California Institute
of Technology Archives Oral History Project, Pasadena CA, July 9–10, 1984.
http://resolver.caltech.edu/CaltechOH:OH_Horowitz_N.

Hubbard, Scott. Email exchanges with author, beginning December 15, 2022.

———. Interview by Sandra Johnson for NASA Headquarters Science Mission
Directorate Oral History Project, Standford CA, August 22, 2017. https://
historycollection.jsc.nasa.gov/JSCHistoryPortal/history/oral_histories
/NASA_HQ/SMD/HubbardGS/HubbardGS_8-22-17.htm.

Ip, Wing-Huen. Email exchanges with author, beginning July 28, 2021.

Kennel, Charles. Email exchanges with author, beginning March 13, 2021.

King, James. Email exchanges with author, beginning April 12, 2021.

King, Seaton. Email exchanges with author, beginning April 4, 2021.

Kohlhase, Charley. Interview by the author, January 28, 2016; and related
correspondence.

Kurth, William "Bill." Interview by the author, January 12, 2021; and related
correspondence.

Lanzerotti, Louis. Email exchanges with author, beginning May 19, 2022.

Lebreton, Jean-Pierre. Email exchanges with author, beginning December 2, 2021.

Lind, Don. Interview by Rebecca Wright for NASA Johnson Space Center Oral
History Project, Houston, May 27, 2005. https://historycollection.jsc.nasa.gov
/JSCHistoryPortal/history/oral_histories/LindDL/LindDL_5-27-05.htm.

Ludwig, George. Letter to the author, June 18, 2007.

Manning, Rob. Interview by the author, January 28 and 29, 2016; and related
correspondence.

Marr, Jim. Email to the author, January 11, 2019.

McDonald, Allan. Email to the author, November 14, 2020.

Mitchell, Bob. Email exchanges with author, beginning August 12, 2021.

Mueller, George. Interview by Summer Chick Bergen, August 27, 1998.

Nieberding, Joe. Interview by the author, July 3, 2018; and related correspondence.

O'Neal, Mike. Interview by the author, May 29, 2021; and related correspondence.

O'Neil, Bill. Interview by the author, July 19 and 24, 2018; and related correspondence.

Reaves, Mary. Email exchanges with author, beginning April 14, 2022.

Roberts, Phil. Email exchanges with author, beginning January 21, 2022.

Ross, Larry. Interview by the author, July 3, 2018; and related correspondence.

Sagdeev, Roald. Interview by the author, May 24–26, 2010; and related correspondence.

Tatalias, Platon. Email exchanges with author, beginning January 3, 2022.

Van Allen, James. Interview by the author, May 3–5, 2005; and related correspondence.

Von Delden, Hugh. Email exchanges with author, beginning January 16, 2008.

———. Interview by the author, January 6, 2021; and related correspondence.

———. Telephone conversation with the author, March 1, 2018.

Von Kármán, Theodore. Interview by John Heilbron, Niels Bohr Library & Archives, American Institute of Physics, College Park MD, June 29, 1962. www .aip.org/history-programs/niels-bohr-library/oral-histories/4935.

Weiler, Edward J. Interview by Sandra Johnson for NASA Headquarters Science Mission Directorate Oral History Project, Vero Beach FL, April 4, 2017. https://historycollection.jsc.nasa.gov/JSCHistoryPortal/history/oral_histories /NASA_HQ/SMD/WeilerEJ/WeilerEJ_4-4-17.pdf.

Other Sources

DIEHL, ROGER

Diehl, Roger. Memo to Distribution. "Existence of Ballistic VEEGA Trajectories to Jupiter." August 12, 1986.

———. "S-TOUR Printouts." August 1, 1986.

GENERAL DYNAMICS CONVAIR DIVISION

General Dynamics Convair Division. "Centaur G Technical Description. A High-Performance Upper Stage for Use in the Space Transportation System." February 1982.

HUGHES AIRCRAFT COMPANY

Iorillo, A. J. "Analyses Related to the Hughes Gyrostat System." Hughes Aircraft Company, Space Systems Division, December 1967.

IP, WING-HUEN

Axford, W. I., W. Fillius, L. J. Gleeson, W-H. Ip, and A. Mogro-Campero. "Measurements of Cosmic Ray Anisotropies from Pioneers 10 & 11." University of California at San Diego, January 1, 1975.

Gautier, Daniel, and Wing-Huen Ip. "PROJECT CASSINI: A Proposal to the European Space Agency for a Saturn Orbiter/Titan Probe Mission in Response to the Call for Mission Proposals Issued on 6th July 1982."

JET PROPULSION LABORATORY

Aichele, Jean, and Jet Propulsion Laboratory. *Galileo Messenger*, no. 37 (September 1995).

Casani, John. "Galileo Report to the Administrator for the Month of October." November 7, 1985.

"Casani Appointed Chief Engineer." Press release, April 4, 1994.

Collins, Jeanne, ed., and Jet Propulsion Laboratory. *Galileo Messenger*, no. 16 (July 1986).

———. *Galileo Messenger*, no. 17 (December 1986).

———. *Galileo Messenger*, no. 18 (April 1987).

———. *Galileo Messenger*, no. 19 (December 1987).

———. *Galileo Messenger*, no. 21 (November 1989).

———. *Galileo Messenger*, no. 22 (February 1990).

———. *Galileo Messenger*, no. 23 (April 1990).

———. *Galileo Messenger*, no. 24 (July 1990).

———. *Galileo Messenger*, no. 26 (December 1990).

———. *Galileo Messenger*, no. 27 (April 1991).

———. *Galileo Messenger*, no. 28 (August 1991).

Fry, Lori, and Jeanne Collins, eds., and Jet Propulsion Laboratory. *Galileo Messenger*, no. 25 (September 1990).

Fulton, David, and Jean Aichele, eds., and Jet Propulsion Laboratory. *Galileo Messenger*, no. 38 (April 1996).

"Galileo End of Mission Press Kit." September 2003.

Harris, J. "Shuttle Crew Visits JPL." *Galileo Messenger*, no. 14 (September 1985): 2.

Kindt, D. H., G. G. Ball, and T. H. Bird. "Requirements and Capabilities for Planetary Missions: Venus Orbiter Imaging Radar 1983." Report no. 43-27, vol. 3 (August 1976).

Lockheed Missiles and Space Company. "Asteroid Belt and Jupiter Flyby Study: First Bimonthly Progress Report." Sunnyvale CA, September 15, 1964.

"Military Characteristics, Status of Corporal, and Corporal I Firings to 22 September 1952." Extracted from "Status Report on Corporal Guided Missiles" and

"Addenda Concerning Corporal Type I Firings to 30 June 1955 and Additional Military Characteristics." Pasadena: California Institute of Technology, Jet Propulsion Laboratory, September 22, 1952.

O'Donnell, Franklin. "NASA's Cassini Mission to Saturn Has Passed a Major Milestone." Press release, December 11, 1992.

"Orbiter Description Document for Jupiter Orbiter Probe 1981/1982 Mission." Publication 660-22. Pasadena: California Institute of Technology, Jet Propulsion Laboratory, September 1976.

Sohus, Anita, ed., and Jet Propulsion Laboratory. *Galileo Messenger* 1625-101, no. 1 (April 10, 1981).

———. *Galileo Messenger* 1625-101, no. 2 (July 10, 1981).

———. *Galileo Messenger* 1625-101, no. 3 (May 3, 1982).

———. *Galileo Messenger* 1625-101, no. 4 (August 1982).

———. *Galileo Messenger*, no. 5 (December 1982).

———. *Galileo Messenger* 1625-101, no. 6 (March 1983).

———. *Galileo Messenger* 1625-101, no. 7 (June 1983).

———. *Galileo Messenger* 1625-101, no. 8 (October 1983).

———. *Galileo Messenger* 1625-101, no. 9 (December 1983).

———. *Galileo Messenger* 1625-101, no. 10 (April 1984).

———. *Galileo Messenger*, no. 11 (September 1984).

———. *Galileo Messenger*, no. 12 (December 1984).

———. *Galileo Messenger*, no. 13 (July 1985).

———. *Galileo Messenger*, no. 14 (September 1985).

"Venus Radar Mapper (VRM) Project Plan." JPL Document D-814, November 1983.

MARSHALL SPACE FLIGHT CENTER

Miller, John Q. Memo, "Evaluation of SRM Clevis Joint Behavior." January 19, 1979.

———. Memo, "Restatement of Position on SRM Clevis Joint O-Ring Acceptance Criteria and Clevis Joint Shim Requirements." January 9, 1978.

Ray, Leon. Chart, "SRM Clevis Joint Leakage Study." October 21, 1977.

———. Memo, "Visit to Precision Rubber Products Corporation and Parker Seal Company." February 6, 1979.

MORTON THIOKOL—WASATCH DIVISION

Boisjoly, Roger, to R. K. Lund. Memo, "SRM O-Ring Erosion/Potential Failure Criticality." July 31, 1985.

Mulloy, Larry, to Larry Wear. Memo, "51C O-RING EROSION RE: 51-E FRR." January 31, 1985.

"Study of Solid Rocket Motors for a Space Shuttle Booster." Executive summary, vol. 1, Contract NAS 8-28430, March 15, 1972.

NATIONAL ACADEMY OF SCIENCES

Canizares, Claude, and John A. Wood. "Scientific Assessment of Options for the Disposal of the Galileo Spacecraft." With accompanying cover letter to John D. Rummel. June 28, 2000.

Goody, Richard M., E. Margaret Burbidge, A. G. W. Cameron, George R. Carruthers, Robert E. Danielson, Herbert Friedman, Robert A. Helliwell, et al. "Opportunities and Choices in Space Science, 1974." Washington DC: Space Science Board, National Research Council, 1975.

NATIONAL AERONAUTICS AND SPACE ADMINISTRATION

Bement, Laurence J., and Morry L. Schimmel. "Investigation of Super*Zip Separation Joint." NASA Technical Memorandum 4031, May 1988.

Clarke, Sir Arthur C. "Sir Arthur C. Clarke's 2007 Iapetus Flyby Greeting." September 10, 2007. https://science.nasa.gov/resource/sir-arthur-c-clarkes-2007-iapetus-flyby-greeting/.

Colin, Lawrence, et al. "Future Exploration of Venus (Post-Pioneer Venus 1978)." NASA Technical Memorandum No. TM X-62450, January 1976.

"Fifth Annual Summary Report." Report no. A-5. Astro Sciences Center of IIT Research Institute, Chicago, for Lunar and Planetary Programs, Office of Space Science and Applications, NASA Headquarters, Washington DC, July 1968.

"Final Report (NASW-2023)." Report no. A-9, Seventh Annual Summary Report. Astro Sciences Center of IIT Research Institute, Chicago, for Lunar and Planetary Programs, Office of Space Science and Applications, NASA Headquarters, Washington DC, November 1970.

McCullough, J. E., and J. F. Landis Jr. "Apollo Command Module Land-Impact Tests." NASA TN D-6979, October 1972.

Narin, F. "Digest Report: Missions to the Outer Planets." Report no. M-14. Astro Sciences Center of IIT Research Institute, Chicago, for Lunar and Planetary Programs, Office of Space Science and Applications, NASA Headquarters, Washington DC, May 1967.

NASA. "Galileo to Jupiter: Probing the Planet and Mapping Its Moons." NASA-CR-162142, January 1979.

———. "NASA Future Plans News Conference." January 13, 1970.

———. "NASA News, Press Kit: Release No. 77-136, Project: Voyagers 1 and 2." For release 12:00 Noon EDT, Thursday, August 4, 1977.

———. "Proceedings, Outer Planet Probe Technology Workshop, Summary Volume." May 21–23, 1974.

———. "Summary Report, Future Programs Task Group." January 1965.

Ride, Sally K. "Leadership and America's Future in Space: A Report to the Administrator." NASA, 1987.

Varghese, Philip L. "Investigation of Heat Transfer in Zirconium Potassium Perchlorate at Low Temperature: A Study of the Failure Mechanism of the NASA Standard Initiator." Grant no. NAG9-301. Period ending August 31, 1989. https://ntrs .nasa.gov/api/citations/19890020199/downloads/19890020199.pdf.

ROSS, LARRY

"Notes from Shuttle/Centaur Senior Manager's Panel." January 28, 1986.

Personal Day-Timer pages. January 27–29, 1986.

SCIENCE APPLICATIONS, INCORPORATED

"Advanced Planning Activity, February 1973–January 1974, SAI-120-M2." February 25, 1974.

Niehoff, John C. Letter to James Van Allen. "Pioneer Saturn and Uranus Entry Probe Mission Dates." May 21, 1973.

UNIVERSITY OF IOWA

"Fall 1962 Tuition Rate Schedule." University of Iowa Libraries, Special Collections. Accessed August 29, 2020. http://www.lib.uiowa.edu/sc/archives /tuition/tuitionrateschedule1962/.

Gurnett, Don. Letter to Mike Mitz. "A Plasma Wave Experiment for the MJS Mission." February 13, 1973.

Hinners, Noel. Letter to Dr. F. L. Scarf. "Selection of Your Plasma Wave Investigation." April 11, 1975.

———. Western Union mailgram to Dr. Donald A. Gurnett. "Selection as Principal Investigator on the Jupiter Orbiter Probe Mission." August 16, 1977.

Mitz, M. A. Letter to Don Gurnett. "Response to Plasma Wave Proposal." March 15, 1973.

———. Letter to Frederick Scarf. "Cover Letter for AFO Package." Including Appendix: Contents of AFO Package. June 1, 1972.

Office of Strategic Communication. "Kletzing Named Donald A. and Marie B. Gurnett Chair." June 25, 2019. https://now.uiowa.edu/2019/06/kletzing -named-donald-and-marie-b-gurnett-chair.

Scarf, Fred L. "Final Report of the OPGT/MJS Plasma Wave Science Team." Redondo Beach CA: Space Sciences Department, TRW Systems Group, September 1, 1972.

———. Letter to E. J. Smith. "Reponses to Five Questions." January 26, 1972.

———. Letter to OPGT Team Members and Associates. "Chance for MJS Mission in 1977." January 26, 1972.

Van Allen, James. Letter to Joseph Lepetich, NASA Ames. "A Note of Thanks for Your Fine Efforts." August 5, 1974. James Van Allen Papers, Special Collections Department, University of Iowa Libraries, Iowa City.

———. Open letter. "On the Targeting of Pioneers F and G at Jupiter." November 1, 1971. James Van Allen Papers, Special Collections Department, University of Iowa Libraries, Iowa City.

———. "Research in Space Physics at the University of Iowa: Annual Report 1974." University of Iowa Department of Physics and Astronomy, Iowa City, December 1974.

U.S. ARMY

Bragg, James W. "Development of the Corporal: The Embryo of the Army Missile Program." Vol. 1, "Narrative." Huntsville AL: Army Ballistic Missile Agency, Redstone Arsenal, April 1961.

"Development of the Corporal: The Embryo of the Army Missile Program." Vol. 2, "Supporting Data." Huntsville AL: Army Missile Command, April 1961.

U.S. CONGRESS

Congressional Record. Daily Digest, Monday, March 21, 1977, D183.

Congressional Record. Hearings before the House Subcommittee on Space Science and Applications. 95th Cong., February 16, 17, and 23, and March 4, 5, 6, and 7, 1977.

Congressional Record. Hearings before the House Subcommittee on Space Science and Applications. 95th Cong., no. 59, vol. 1, pt. 2, February 1, 2, and 7, 1978.

Congressional Record: Proceedings and Debates of the 95th Congress, First Session. Vol. 123, pt. 5. Senate, Thursday, February 24, 1977.

Congressional Record: Proceedings and Debates of the 95th Congress, First Session. Vol. 123, pt. 7. House, March 17, 1977, to March 23, 1977.

Congressional Record: Proceedings and Debates of the 95th Congress, First Session. Vol. 123, pt. 12. Senate, May 9, 1977, to May 17, 1977.

Congressional Record: Proceedings and Debates of the 95th Congress, First Session. Vol. 123, pt. 14. House, May 25, 1977, to June 6, 1977.

Congressional Record: Proceedings and Debates of the 95th Congress, First Session. Vol. 123, pt. 19. House, July 15, 1977, to July 21, 1977.

Public Law 100-685, 100th Congress. "National Aeronautics and Space Administration Authorization Act, Fiscal Year 1989." 102 Stat. 4083, November 17, 1988.

Public Law 101-611, 101st Congress. "National Aeronautics and Space Administration Authorization Act, Fiscal Year 1991." 104 Stat. 3188, November 16, 1990.

U.S. House. *1964 NASA Authorization: Hearings before the Committee on Science and Astronautics*. 88th Cong., 1st sess., no. 3, pt. 1, March 4 and 5, 1963.

U.S. Senate. Hearings before a Subcommittee of the Committee on Appropriations. 93rd Cong., 1st sess., April 9, 1973.

———. Hearings before the Subcommittee on Science, Technology, and Space of the Committee on Commerce, Science, and Transportation. 95th Cong., 1st sess., February 25, 1977.

U.S. GENERAL ACCOUNTING OFFICE

"Space Exploration: Cost, Schedule, and Performance of NASA's Galileo Mission to Jupiter." Washington DC: General Accounting Office, May 1988.

"Space Exploration: Cost, Schedule, and Performance of NASA's Magellan Mission to Venus." Washington DC: General Accounting Office, May 1988.

VIDEOS

Baggett, Blaine, writer, producer, and director. "Saving Galileo." *JPL and the Space Age*. Produced by Jet Propulsion Laboratory and California Institute of Technology. 2019. 59:35. https://www.youtube.com/watch?v=aezcXjKYZkM.

———. "Triumph at Saturn (Part I)." *JPL and the Space Age*. Produced by Jet Propulsion Laboratory and California Institute of Technology. 2021. 56:45. https://www.youtube.com/watch?v=SY-hQJ5pMd4.

———. "Triumph at Saturn (Part II)." *JPL and the Space Age*. Produced by Jet Propulsion Laboratory and California Institute of Technology. 2021. 1:29:52. https://www.youtube.com/watch?v=oGsajLIALJE.

Galileo at Jupiter, December 7, 1995. JPL TV program. Richard Terrile, host. With Wesley Huntress, Torrence Johnson, Robert Mitchell, Andrew Ingersoll, Edward Stone, Carl Sagan, and interviews with various personnel. Recorded at Jet Propulsion Laboratory, December 7, 1995. 89 min. https://www.youtube.com/watch?v=ucjg9TPDxSY.

Geoff Haines-Stiles, writer and producer. "Rocky Road to Jupiter." *NOVA*. WGBH Educational Foundation, Boston, April 7, 1987. 58 min.

Goodbye Galileo. Video recording of JPL live broadcast with Galileo mission team members. September 21, 2003. 70 min. https://www.jpl.nasa.gov/videos/galileo-end-of-mission-webcast.

O'Neil, Bill. *Galileo's Mission to Jupiter*. Public lecture in Theodore von Kármán Auditorium, JPL, Los Angeles, November 11, 1997. 60 min.

President Reagan's Remarks and Introduces Astronauts, Edwards AFB on July 4, *1982*. Ronald Reagan Presidential Library, Simi Valley CA. 20 min. https://www.youtube.com/watch?v=tx08mOtojgw.

WHITE HOUSE

Bush, George H. W. "Executive Order 12675—Establishing the National Space Council." April 20, 1989.

Huntsman, Jon M. Memo to George P. Shultz. "Future of NASA." September 13, 1971.

Nixon, Richard. Memo to the Vice President, Secretary of Defense, Acting and Space Administration, Science Adviser. "Recommendations for the Post-Apollo Period." February 13, 1969.

———. "President Richard Nixon's Daily Diary January 1, 1970–January 31, 1970." National Archives and Records Administration, Office of Presidential Papers and Archives, Box RC-4, 67–68. Richard Nixon Presidential Library and Museum, Yorba Linda CA.

———. "Statement about the Future of the United Stated Space Program." March 7, 1970. Available at the American Presidency Project. https://www.presidency.ucsb.edu/documents/statement-about-the-future-the-united-states-space-program.

Reagan, Ronald. "Remarks at Edwards Air Force Base, California, on Completion of the Fourth Mission of the Space Shuttle Columbia." July 4, 1982. Available at the American Presidency Project. http://www.presidency.ucsb.edu/ws/index.php?pid=42704.

Space Task Group. "The Post-Apollo Space Program: Directions for the Future." Report to President Richard Nixon, September 1969. NASA Historical Reference Collection, Washington DC. https://www.nasa.gov/history/the-post-apollo-space-program-directions-for-the-future/.

Weinberger, Caspar. Memo to the President. "Future of NASA." August 12, 1971.

Index

In the Outward Odyssey: A People's History of Spaceflight series

Into That Silent Sea: Trailblazers of the Space Era, 1961–1965
Francis French and Colin Burgess
Foreword by Paul Haney

In the Shadow of the Moon: A Challenging Journey to Tranquility, 1965–1969
Francis French and Colin Burgess
Foreword by Walter Cunningham

To a Distant Day: The Rocket Pioneers
Chris Gainor
Foreword by Alfred Worden

Homesteading Space: The Skylab Story
David Hitt, Owen Garriott, and Joe Kerwin
Foreword by Homer Hickam

Ambassadors from Earth: Pioneering Explorations with Unmanned Spacecraft
Jay Gallentine

Footprints in the Dust: The Epic Voyages of Apollo, 1969–1975
Edited by Colin Burgess
Foreword by Richard F. Gordon

Realizing Tomorrow: The Path to Private Spaceflight
Chris Dubbs and Emeline Paat-Dahlstrom
Foreword by Charles D. Walker

The X-15 Rocket Plane: Flying the First Wings into Space
Michelle Evans
Foreword by Joe H. Engle

Wheels Stop: The Tragedies and Triumphs of the Space Shuttle Program, 1986–2011
Rick Houston
Foreword by Jerry Ross

Bold They Rise: The Space Shuttle Early Years, 1972–1986
David Hitt and Heather R. Smith
Foreword by Bob Crippen

Go, Flight! The Unsung Heroes of Mission Control, 1965–1992
Rick Houston and Milt Heflin
Foreword by John Aaron

Infinity Beckoned: Adventuring Through the Inner Solar System, 1969–1989
Jay Gallentine
Foreword by Bobak Ferdowsi

Fallen Astronauts: Heroes Who Died Reaching for the Moon, Revised Edition
Colin Burgess and Kate Doolan with Bert Vis
Foreword by Eugene A. Cernan

To order or obtain more information on these or other University of Nebraska Press titles, visit nebraskapress.unl.edu.